RELATIVITY

RELATIVITY

Hilary. D. Brewster

Oxford Book Company

Jaipur ▪ New Delhi

Published By

Oxford Book Company

Regd. Off. "Bony Residency", Gate No. 2
Opp. Tilak Public School, Vishweswariya Nagar,
Gopalpura Road, Jaipur - 302018, India
Ph: 0141 - 2761280, Telefax: 2761381
email - oxfordbookjpr@gmail.com
website - oxfordbookcompany.com

Delhi Office: 102, 1st Floor, Satyam House,
4327/3, Ansari Road, Darya Ganj,
New Delhi - 110002, India
Ph: 011-45652440

Published in India, 2025

ISBN : 9789380179032

Price : ₹ 2295/-

Typeset by Shivangi Computers, Jaipur

Printed and bound in India by Thomson Press Ltd.,
New Delhi

Preface

The theory of relativity has become a cornerstone of modern physics. Over the course of time it has been scrutinized in a multitude of experiments and has always been verified with high accuracy. The correctness of this theory can no longer be called into question. Right after its discovery by Albert Einstein in 1905, special relativity was only gradually accepted because it made numerous predictions contradicting common sense, fervently castigated by Einstein, and also defied experiment for too long a time. It was only with the advent of particle or high energy physics that matter could be accelerated to very high velocities, close to the speed of light, which not only verified special relativity but also made it a requirement for machine construction.

The book opens with a description of the smooth transition from Newtonian to Einsteinian behaviour from electrons as their energy is progressively increased, and this leads directly to the relativistic expressions for mass, momentum and energy of a particle. The expansion of the physical research frontier toward astronomy and cosmology during the past ten to twenty years considerably increased the importance of special relativity and, above all, general relativity based thereupon.

Since astrophysics has in the same time become very popular among readers with a scientific background, the two theories of relativity have attained unprecedented publicity. The fascination with astronomy of children and youths shall only be mentioned incidentally, it is, however, one of the most impressive features of schools today. This book proceeds to do just that, offering a radically reoriented presentation of Einstein's Theory of Relativity that derives Relativity "from" Newtonian ideas, rather than "in opposition to" them.

Hilary. D. Brewster

Contents

	Preface	*v*
1.	Introduction to Relativity	1
2.	Relativity Made Simple	29
3.	Space, Time, and Newtonian Physics	70
4.	Minkowskian Geometry	113
5.	Accelerating Reference Frames	142
6.	Energy and Momentum in Relativity	156
7.	Relativity and Gravitational Field	177
8.	Relativity and Curved Spacetime	199
9.	Black Holes	226
10.	The Universe	272
	Index	297

Chapter 1

Introduction to Relativity

Relativity is a word that is used in a lot of different contexts to mean a lot of different things; however, in common usage, if you say "relativity," people think "Einstein."

Relativity is the idea that the laws of the universe are the same no matter what direction you are facing, no matter where you are standing, no matter how fast you are moving. Now, to say the laws of the universe are the same does not mean everything looks the same. A person standing in the Mojave desert sees very different things than an astronaut floating in space, or a diver 300 feet under water. But, they all see the same laws of physics and mathematics. Gravity always pulls you towards heavy objects. Objects in motion tend to stay in motion, unless something pushes on them. Electricity can give you a big shock.

Relativity is nearly always presented as a theory about the speed of light and black holes. The culmination of a 2400 year intellectual struggle to identify and understand mankind's place in the universe. In my opinion, by far the most important consequences of relativity are not the ability of physicists to calculate an extra decimal place or two, but rather the changes in philosophy, our understanding of religion and our relationship to Creation.

For most of western history, people have had their beliefs shaped by our experience on the Earth: the Earth seems to us to be enormously large and completely immovable. We jump up and down, throw large rocks, watch the tides come and go, but the Earth never seems to move. Historically, the stars were thought to be small objects, perhaps painted on a backdrop, but in any case visibly orbiting around the centre of the universe, the Earth. Similarly the Sun and the Moon also orbited around the Earth. This universe was thought to be fixed and unchanging, as is only appropriate for a perfect creation of a perfect God. It was only natural for mankind's original idea to be that the Earth was immobile and at the centre of this perfect universe.

Meanwhile, Euclid's geometry convinced everyone that mathematics was the most rigorous of all sciences, and therefore must be central to any description of the universe. Euclid, as it turned out, assumed that space was flat.

From about 1500 to 1916, these ideas were slowly examined and found wanting. Eventually, we found what we now consider to be a far more fundamental truth: the universe is a really quite large place, and the Earth is a very tiny little object in the universe. Everything in our universe is constantly in motion - nothing is fixed and unchanging. The Earth moves through the universe on a path determined by the laws of gravity, the very same laws that seem to determine the path of everything else. The Earth is in no way a special object, nor does it occupy a special position. This new understanding, that there seems to be no special place in the universe, no special direction, and no special speed which could be called "at rest," this understanding is called Relativity.

Just to put this into context, our current understanding is that the universe contains about 200 billion galaxies, and each galaxy contains about 200 billion stars. Our particular star, the Sun, seems to be a very average type of star, in no particular way distinguished from the other 40 thousand billion billion stars in the visible universe. We now know of many planets orbiting many stars, and seemingly find new ones on almost a weekly basis. Most of the planets we have found are enormous, imposing giants like Jupiter or even larger, with hundreds or even thousands of times the mass of the Earth - by comparison, the Earth is an all but invisible little rock.

Relativity is the story of mankind learning that we are not the centre of creation, at least as far as physics is concerned. This has been a rather upsetting and ego-deflating lesson.

Although this is the story of relativity, it must be emphasized that this story is not complete. At the time of this writing, there is no acceptable quantum theory of gravity. It is widely believed by many, including myself, that before we can build a quantum theory of gravity, first we need a quantum theory of relativity.

THE HISTORY OF RELATIVITY

Relativity is the culmination of 2400 years of human thought. Originally, we thought that the Earth was at rest at the centre of a flat universe that was constructed with a mathematical plan. 500 years ago, we figured out that the Earth was actually moving. 100 years ago we figured out that there is no centre to the universe, there is no place in the universe that is at rest, and the universe is not flat. We do continue to delude ourselves into thinking the universe was constructed with a purpose and a mathematical plan.

Today, we live in a world that is constantly changing - 100 years ago there were no airplanes and almost no cars. 50 year ago there were no lasers, no television, nothing we would recognize today a a computer. 25 years ago the Internet was a strange little network occupied by professors and military people. The computers from 5 years ago are doorstops today. We are very accustomed to change; in fact, today people almost don't believe in the idea that things might stay the same.

This is all new. Isaac Newton lived in a culture that believed the height of man's knowledge and understanding was reached between the time of Aristotle and Jesus. Newton himself believed that to really understand something, you had to go back to the original greek writings of Aristotle, Euclid, and the Bible. He believed that the modern knowledge of his day was currupted and watered- down versions of the original deeper understandings.

Western culture busied itself with the search for truth. This philosophical orientation led to the assumption that truth existed, and that is was perfect and unchanging.

Thus it was an easy step to believe that the Universe and indeed God himself were perfect and unchanging. Eastern philosophy was oriented on the search for what is real. Their presumption was that nothing was fixed and unchanging, that everything has a beginning and an ending. While this search led to many profound understandings, the search for what is real does not seem to lead to the development of science and mathematical logic. This is most curious, as we are now coming to understand that the universe is indeed a place of constant change, that nothing is fixed and unchanging, and that the universe itself has both a beginning and, presumably, an ending.

Our story begins with Aristotle, who compiled a set of ideas about how the world worked. Aristotle lived in Greece from 384 BC to 322 BC. Aristotle was primarily a teacher, and believed the highest calling for a man was to teach. He thought everything was made up of earth, water, air, and fire.

He thought things had a natural state - earth-like objects wanted to be at rest, air-like and fire-like objects wanted to rise up. This all apparently seemed intuitively obvious to him. Aristotle taught that heavier objects fall faster than light objects, as a rock falls faster than a feather. At the time, no one actually thought to do experiments and see if any of this was true - in fact, like his teacher Plato, Aristotle thought that the universe was guided by the rules of logic and mathematics, and therefore the laws of the universe could be deduced by logical thought and mathematical reasoning. Aristotle thought that the Earth was immobile at the centre of the universe.

Aristotle taught classes and basically invented the subjects of logic, physics, astronomy, meteorology, zoology, metaphysics, theology, psychology, politics, economics, and ethics. 250 years after Aristotle's death, his lecture notes were published by Andronicus of Rhodes. For the next 1500 years, the great thinkers of Europe and the Islamic world all traced their roots and primary influences back to Aristotle's teachings.

Euclid of Alexandria lived from 325 BC until about 265 BC. Euclid compiled many theorems and demonstrations that were already known, and carefully ordered them into what we now call an axiomatic system. An axiom is something you take as a given, take on faith as it were. An axiomatic system is a compilation of things you can conclude, demonstrate, or prove based on those axioms.

Euclid organized the theorems very carefully, and was able to reduce his assumptions to five axioms. Euclid's book, The Elements, was and is considered one of the greatest achievements of mankind. The beauty of his work is that once you accept the five axioms, you are led inexorably to his results. This was the first great work of logical reasoning. It has captivated many young thinkers over the ages, and motivated them to try to produce such a work of their own. It also convinced many people that this beauty was an element of the thoughts of God; that it must be the case that the entire universe is governed by a similar set of laws and their logical consequences. Physics is nothing more or less than the search for these laws.

Euclid's five axioms are:

- A straight line segment can be drawn joining any two points.
- Any straight line segment can be extended indefinitely in a straight line.
- Given any straight line segment, a circle can be drawn having the segment as radius and one endpoint as centre.
- All right angles are congruent.
- If two lines are drawn which intersect a third in such a way that the sum of the inner angles on one side is less than two right angles, then the two lines inevitably must intersect each other on that side if extended far enough.

The first three of Euclid's axioms are now called axioms of construction, that is, they tell you that you can build certain things. The fourth axiom is now recognized as being of a different nature - the axiom that all right angles are equivalent is the same as assuming that space is the same everywhere, and if you move a right angle around, it stays a right angle.

For 2,000 years it was recognized and agonized over that the first four axioms seem completely obvious and very simple, and the fifth axiom seems by contrast complicated and artificial. Many mathematicians spent substantially their entire careers trying to prove the fifth axiom was a logical consequence of the first four. All such attempts failed, for a simple reason which is now well understood. We now know that the fifth axiom is equivalent to the axiom that space is flat.

When Europe became interested in trying to climb out of the dark ages, the idea of learning and knowledge also became interesting. At this time, education meant learning Greek and Latin, and studying the Bible, Aristotle, and Euclid. It was considered that this was all of knowledge. It was also against this backdrop that the basic ideas of relativity had to evolve - according to Aristotle, the earth was in a special place, and material objects wanted to be at rest in this same special place. According to the interpretations of the Bible at the time, the Earth was a special part of God's creation, the very centre and purpose of existence.

So, the ideas that things were the same everywhere, that the stars were just like the Sun and the Sun was just like the stars, and that the Earth did not define and occupy the perfect centre of creation, but rather was just another part of creation, moving around and subject to the same forces as everything else, these ideas were very controversial and not very welcome. The people who promoted these ideas were similarly controversial and not very welcome.

The Greeks noticed that there were a few "stars" with rather peculiar habits - these stars moved forwards and backwards and then forwards again against the backdrop of the fixed stars. The Greeks called these moving stars "wanderers," or "planets" in Greek. Over the centuries, people continued to observe the planets and chart their courses - it became very fashionable to try to find a way to predict the positions of the planets. For most of this time, the prevailing religious view was that God's creation, the universe, was perfect, and the only perfect shape was the circle, therefore planets must move in circles. It proved impossible to describe the positions of the planets by assuming they moved in circles around the Earth.

As observations got better and better, people invented more and more complicated systems of planets moving in circles, which moved on greater circles, which moved on even greater and greater circles. However, although these systems of circles revolving in circles in circles in circles could be made to predict the planets positions reasonably well, the complexity of these artificial systems made a joke of the "perfection" and "simple beauty" of the circle.

Nicolaus Copernicus, a Polish economist and scientist who lived from 1473 to 1543, was the first person to openly challenge the Aristotelian view. In 1513, while in Italy, he published a short paper saying it was the Sun that was at rest at the centre of the universe, and everything else including the Earth moved around the Sun. Copernicus said the planets moved in circles around the Sun, as did the Earth. Using this simple system, he was able to predict the orbits of the planets with great precision, but at a great cost: the Earth had to be moved out of the centre of the universe.

This was a great leap forward, philosophically: the Earth was now considered a moving object, and not a special object at a special place in the universe. This idea was a threat to the established church, since it also effectively removed Earth from being the central object of Creation, and called into question the idea that mankind was God's greatest creation. Copernicus unintentionally started a conflict between religion and science, a conflict which unfortunately maintains to this day.

Tycho Brahe lived and worked in Denmark from his birth in 1546, and died in 1601 in the Czech Republic. On 11 November 1572, in the early evening, he saw a new star in the constellation of Cassiopeia, almost directly overhead. This was remarkable, because the fixed, unchanging stars had changed. This star is now called "Tycho's Supernova." With funding from the King of Denmark, Brahe set up an observatory and made the most accurate observations of the planets up to his time. Brahe was aware of Copernicus' theory, but did not like it. He invented a competing theory, where the Earth was at the centre of the universe, and the moon,

Sun, and stars revolved around the Earth. In his system, the other planets revolved around the Sun. Thus, in his mind, he melded the most important ideas of the day: the planets all moved in circles, and the Earth was the centre of creation again. But, the idea of an unchanging perfect universe was gone forever. And, Brahe set the stage for more and more precise measurements of our universe, and requiring that theories should agree with measurements.

Johannes Kepler was born in Germany in 1571 and lived until 1630. When Kepler was a young man, he was hired by Brahe as a mathematician. Thus Kepler had immediate access to the most precise astronomical observations of the day.

Kepler, like most scientists of his day, was convinced that God had made the Universe according to a mathematical plan, and that mathematics was therefore a strategy for understanding the universe. He spent most of his career calculating the orbits of the planets - for example, it took him nearly 1,000 sheets of paper to calculate the orbit of Mars. It was Kepler who first realized that the planets travel in ellipses, not in circles. Kepler also pointed out that Venus and Mercury are always seen near the Sun, which makes perfect sense if they orbit the Sun, but no sense if they orbit the Earth. Kepler also observed a supernova of his own in 1604, providing more evidence that the universe was a place of change.

Galileo Galilei lived in Italy from 1564 to 1642. Galileo studied the works of Copernicus, and was a great believer his views. In 1609, Galileo heard of a Dutch invention, the spyglass, and quickly made his own version, the first telescope. With his telescope, he discovered the moons of Jupiter, and he saw that Venus had phases like the moon. This proved that Venus orbited the sun, not the Earth.

Galileo also noticed that the moons of Jupiter orbited with a fixed period, just as our moon makes a complete revolution around the Earth every 28 days.

However, Galileo noticed that a few months later, his predictions of the orbits of Jupiter's moons were off by ten to fifteen minutes. Galileo correctly realized that this was because when the Earth is on the far side of the Sun from Jupiter, the light from the moons takes extra time to reach us over the extra distance. From this effect, Galileo was able to make a quite good estimate of the speed of light. The idea that light traveled at a finite speed, not instantaneously, was very new.

At the time, the views of Copernicus were quite controversial. Galileo had the habit of not only supporting these views, but trying to make his opponents look like fools, a habit which did not serve him particularly well. In 1616, Galileo was subjected to the Inquisition, and given a secret warning to recant his Copernican views.

In 1632, Galileo published his famous *Dialogue concerning the two greatest world systems*, which only got him into further trouble, first for continuing to support Copernicus' views, and second for continuing to try to make his opponents look like fools. Galileo was summoned to Rome, where he was found to be "vehemently suspected of heresy", and forced to recant his Copernican views and sentenced to house arrest for life. He is reputed to have muttered under his breath as he left the Inquisition, "Never the less, it still moves."

Galileo set the stage for modern physics by noticing that all things fall at the same rate without regard for their composition. Aristotle had taught that heavier things fall faster. Galileo realized that by rolling things down a ramp, he could slow the effects of gravity and make more careful measurements. In this fashion Galileo collected the data that Newton would use to create his great theory of gravity.

Sir Isaac Newton lived in England from 1643 to 1717. Newton took Galileo's work and put it into a mathematical form. Newton, as one of his axioms, said objects at rest tend to stay at rest, and objects in motion tend to stay in motion. This is in complete disagreement with Aristotle's view that moving objects tend towards their natural state of being at rest on the Earth. Newton codified Galileo's views on systems in mathematics - the idea that the laws of physics are the same no matter what direction you face, no matter where you stand, no matter how fast you move. In codifying these ideas, Newton made a distinction between moving at a constant velocity, and accelerating.

Newton claimed that the laws of physics were valid so long as you were moving at a constant velocity, but not if you were accelerating. These ideas are today called "Galilean Relativity." Although Newton undoubtedly knew of the speed of light, he did not think it was in any way special, and he thought that you could go as fast as you like, if you had the means to propel yourself.

Newton once said, "If I have seen far, it is because I have stood on the shoulders of giants." Here we have seen something of the giants whose shoulders he used. The mathematical abilities and physical intuition of Newton truly stand out in history and mark him as perhaps the most prominent physicist who ever lived. However, just as he indicated, much of the philosophical "heavy lifting" had already been done. The notion that to stand still on the Earth was to be perfectly at rest in the precise centre of God's perfect, unchanging creation was painful to give up. It took at least four noteworthy geniuses 150 years to set an appropriate stage for Newton's great work.

Johann Carl Friedrich Gauss lived in Germany from 1777 until 1855. He was one of the greatest mathematicians who ever lived, and made many important contributions to physics, also. Gauss considered non-Euclidean geometry, also called curved spaces, and worked out much of the math, but never published anything on the topic. He was afraid of the backlash from his church and the community. However, he though a lot about these issues, and at one point actually measured the angles in a triangle about 5 miles on a side in an attempt to determine if space was flat or curved. Today, we think Gauss and his student Riemann came very close to developing General Relativity.

Michael Faraday lived in England from 1791 until 1867. By today's standards, Faraday had little in the way of formal education or mathematical training or abilities, which makes his scientific accomplishment all the more impressive.

When Faraday was young, he was apprenticed to a book binder, and he used all his spare time educating himself by reading the scientific books laying around. Later, Faraday got himself a job as an assistant in a lab where he worked ceaselessly. Faraday almost single-handedly worked out the laws of the relationship between electricity and magnetism. Today, it's Faraday's work that tells us how to make electric motors and generators, the basic foundation of our entire technological civilization.

Georg Riemann lived in Germany from 1826 until 1866. He was Gauss' last and most famous student. To finish his doctoral degree, in 1854 Riemann was required to give a lecture. Gauss asked him to lecture on geometry. In this single lecture, Riemann laid almost all of the mathematical foundation for Einstein's General Relativity. Riemann went on to become an important mathematician, but did little work on curved spaces after this one lecture. Riemann was not a Catholic, and therefore was completely unconcerned about any backlash his lecture might generate.

James Clerk Maxwell lived in England from 1831 until 1879. Maxwell took the works of several other physicists including Coulomb, Ampere, Gauss, and especially Faraday, and brought them together into one set of equations that described all electric and magnetic phenomenon. Maxwell's equations are the first, and by far the most successful example of what we now call a Unified Field Theory. Prior to Maxwell it was thought that electricity, magnetism, and light were all completely unrelated. Maxwell showed that these phenomenon are all different aspects of the same thing.

Maxwell's equations made a very peculiar prediction: with his equations, he was able to calculate the speed of light. This is strange because up until that time it was thought that velocities simply added: if you were on a train moving at 100 miles per hour, and you turned on a flashlight pointing forwards, the light from that flashlight would go 100 miles per hour faster than light from a flashlight that was not moving. So, according to the physics of Newton, a theory should not predict the speed of light, since that speed should depend on how fast the observer and the source were moving. No one really knew quite what to make of this, but the longer people thought about it the more it bothered them.

At this time, people thought that since light is a wave, there must be something that is waving. Waves in the ocean are moving water. So light waves must be moving something. This something was given the name the Luminiferous Ether. From about 1865 until about 1920 people searched diligently for this Ether, but no one could ever find any. Today, we simply accept that light propagates, and we don't worry about what is moving.

We also know today that light travels in our universe at an absolute velocity - no matter how you produce it, no matter how you are moving when you see it, you always see light moving at the same speed. This is the single most important point of Einstein's relativity. This is also the reason why Maxwell's equations were able to predict a particular speed for light - Maxwell's equations turned out to be fully compatible with Einstein's relativity, even though Maxwell wrote his equations down 40 years before Einstein developed Special Relativity.

Maxwell knew of the rings of Saturn, and wondered what they were. He was able to show rather quickly that the rings could not be a solid disk, because the gravitational forces on a large solid disk would either tear the ring apart or make it crash into Saturn. Next he considered the idea that the rings were liquid, but again he was able to show that liquid rings could not orbit Saturn in a stable fashion. Lastly, he considered the possibility that the rings were made of dust, that is uncounted trillions of tiny individual particles.

This system turned out to be stable, and today we know that Saturn's rings are indeed made up of tiny little ice chips, grains of sand, and small rocks. This was one of the first times that it was shown that continuous systems can have stability problems, but quantized systems can work in the same circumstances. Today, all of our theories suffer from stability problems brought on by the continuum, and a need for a theory of quantized space and time is becoming more and more apparent.

Ernst Mach was an Austrian physicist and philosopher who lived from 1838 until 1916. Mach did important work on sound, so the speed of sound is called Mach 1. He also criticized the existing physical theories of his day on several grounds.

Mach said that since all we ever know of the universe comes to us through our senses, our theories should speak only of things which can be observed and measured. Mach also asserted that all phenomenon in our universe must have causes from within our universe. He proposed Mach's principle, which is that inertia, the tendency of a body in motion to stay in motion and to resist accelerations, must be a result of interactions with the other matter in the universe.

For example, if you whirl a bucket full of water about your head, the surface of the water will assume a curved shape. But, how does the water know to do that? How does the bucket know that it is accelerating in a circle, and you are standing still? From the bucket's point of view, it seems equally reasonable to assert that the bucket is just hanging out, and you are whirling around the bucket.

Mach's answer was that the key difference between you and the bucket is that you see the distant stars in the sky standing still, and the bucket sees the stars whirling about. Therefore, the surface of the water curves due to some interaction between the water and the distant stars. Mach therefore claimed that if there was nothing in the universe but you and the bucket, and you whirled the bucket about your head, the water would not curve. Mach further claimed that if you could spin the entire universe about the bucket, the water surface would curve. Unfortunately, we do not currently know how to test either of these very interesting ideas.

There is a logical problem with Newton's physical explanation of the water curving. We say the water surface curves because the bucket is accelerating due to an external force - you're pulling on the bucket handle. But, you may ask, how does the water know it's being pulled? According to Newton's theories, the answer is that the water knows its accelerating because its surface is curving, which is a sure sign of acceleration and forces. So, in the end, all we can really say is that the water surface is curving because it is curving. Not very satisfying. While physics can tell us with enormous accuracy precisely how much the water surface will curve, physics is an almost total failure at telling us why it will curve. Mach did not like this in the slightest.

The idea that things which happen in the universe must have causes from within the universe seems so obvious as to be tautological. However, in spite of this, many of our natural assumptions and theories fail this criteria. For example, we consider inertia and electric charge to be properties of material objects with names but no causes.

Perhaps more critically, we all know that time marches forwards inexorably, but we consider this most pervasive of effects to be without cause - it just is. Of course, the idea that everything has a cause is just an idea, and may be wrong. Or it may truly be that some things were simply chosen by God, so that their cause is not within our universe. However, it seems to me and most scientists that we should continue to try to explain what happens in our universe strictly in terms of things which are already in the universe until we're quite clear that this approach simply isn't working.

Einstein cited Mach as one of his primary inspirations. Hendrik Lorentz was a Dutch physicist who lived from 1853 until 1928. Lorentz worked out the basic mathematics of Special Relativity, which is now called the Lorentz transform. Einstein used the formulas invented by Lorentz to develop his theories.

Jules Henri Poincarı was a French physicist who lived from 1854 to 1912. His theories of coordinate transformations were also instrumental in the development of Special Relativity. Today, we consider that Special Relativity was simultaneously discovered by Lorentz, Poincarı, and Einstein.

Albert Einstein was born on March 14th, 1879 in Ulm, Germany and died on April 18th, 1955 in Princeton, USA. Einstein worked only in the field of theoretical physics because of "my disposition for abstract and mathematical thought, and my lack of imagination and practical ability."

Einstein created two theories of relativity, which he called the Special Theory of Relativity, and the General Theory of Relativity. Einstein said he started working on his theory of relativity when he was 16. He had just learned about Maxwell's equations and their predictions of the speed of light. Einstein liked to perform what he called "thought experiments", in his native German "Gedanken experiments." Here are two he thought up when he was 16.

First, he imagined himself holding a mirror in his hand at arms length, and looking at his own reflection. Then, he imagined starting to run faster and faster, until he was running at the speed of light. Would he be able to see his own reflection? Second, he imagined a light ray zooming past him, and he ran to catch up with it. What would the light ray look like when he was running right alongside of it?

Ten years later, in 1905, he figured out the answers to these two questions. You cannot run at the speed of light, and at any lesser speed, you simply see your own reflection. If you could run at the speed of light, the light ray would look like an electric field which changes in space but not in time. This is impossible according to Maxwell's equations, but you cannot run at the speed of light so you can never see such a thing.

We have seen that originally, people thought that the Earth was precisely at rest precisely in the centre of the universe. This belief, of course, would mean that we would have a very hard time figuring out what the laws of physics would look like if we were somewhere far from the centre of the universe traveling at some high rate of speed. From about 1500 until about 1680, several very smart people figured out what we now call Galilean Relativity, which is that the laws of physics are the same for anyone anywhere, traveling at any speed, so long as they were traveling at a constant speed and not accelerating.

Einstein's first contribution to relativity was to add to this, The speed of light is a constant, and does not depend on where you are or how fast you are moving. Once he understood this very counter-intuitive fact, he was able to quickly work out the laws of Special Relativity, which he published in 1905. Maxwell's equations were already compatible with Special Relativity, however Newton's equations were not. So, Einstein had to reformulate Newtonian Mechanics so as to be consistent with Special Relativity.

Special Relativity is the idea that the laws of physics are the same everywhere in the universe, no matter where you are, no matter how fast you are moving, so long as you are not accelerating, and one of the laws of physics is that light always travels at the same speed.

Very shortly after publishing Special Relativity, Einstein realized that the theory did not actually apply anywhere in our universe. The reason is very simple to understand: we live in a universe filled with thousands of billions of billions of planets and stars.

No matter where you go in the universe, you are being pulling in some direction by gravity. So, Einstein realized, there is actually no such thing as a place or a person who is not accelerating, because there is no such thing as a place or a person who is not being pulled on by gravity. Einstein mulled over this idea for another 11 years - one might almost think he was a bit slow. In 1907, two years after he had published Special Relativity, Einstein walked home from his job as a patent examiner in the Swiss Patent office, and said to his wife, "I had today the happiest thought of my entire life. I realized that a man freely falling in an elevator does not feel his own weight." Einstein had discovered a new axiom, that gravitational mass was equivalent to inertial mass.

This was already known as a measured fact, but was not understood. Einstein decided to elevate this fact from a curious coincidence to a principle, which he called the principle of equivalence. What Einstein had decided was that you could not tell the difference between the force of gravity and some other kind of force.

For example, if you were in an elevator with the doors closed and it was freely falling, you could not tell if you were close to the Earth or floating in space very far away from any other mass. Alternatively, if the elevator were sitting on the ground, you would feel your weight, but you could not tell if you were sitting on the Earth, or if the elevator were being pushed by a rocket motor with an acceleration of precisely one g, the acceleration of gravity on the Earth.

What Einstein had decided was that the closest thing there was to an inertial frame, that is moving at a constant velocity without acceleration, was to be freely falling. However, there are limits to this. Imagine you had an elevator which was 8,000 miles wide, and it was falling near the Earth. The Earth itself is only about 8,000 miles across, so clearly gravity will pull you straight down in the middle of the elevator, but at each end of the elevator you'll be pulled towards the middle. So, there's a limit to how big the elevator can be and still represent a freely falling frame.

Also, after a little while, the elevator will hit the Earth - this will be a big clue to the people inside that something has changed. So, there's a limit to how long this situation can last and still represent a freely falling frame.

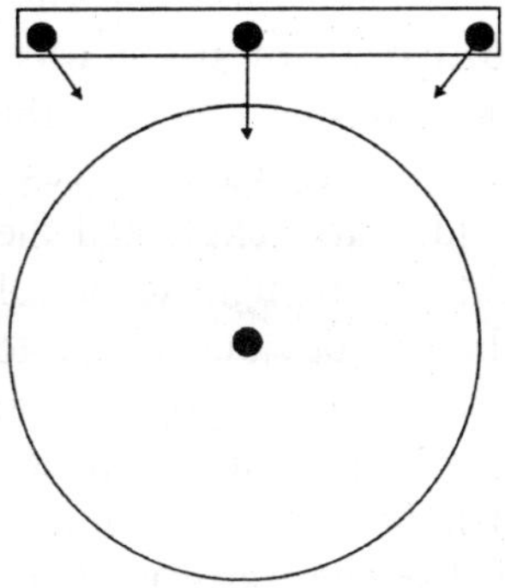

Fig. A Very Large Elevator Freely Falling Towards the Earth

Einstein realized that this was a lot like a curved space. If you're on a ship on the ocean, it looks like the Earth is flat. But, if there's another nearby ship and it sails away, after a time the ship seems to sink below the horizon. So, there's a limit to how big an area you can look at on the Earth and convince yourself that the Earth is flat.

Einstein spent much of the next 8 years learning the math invented by Riemann (invented in one week of work for a single afternoon lecture, which gives you an idea of their relative abilities at mathematics), and finally was able in 1916 to publish his General Theory of Relativity, which says that there is no centre to the universe and nothing is at rest. The only places in the universe that seem to be moving at a constant velocity are small areas that only last for a short time.

Is this the end of the story of Relativity? Well, yes and no. This brings us up to our best current understanding. However, many people are dissatisfied with the current situation, as was Einstein himself. There are a lot of pretty strong hints that we're still missing several important ideas. Einstein noted that gravity waves travel at the same speed as light waves. He could not believe this was a coincidence, and felt strongly that this indicated there was a link between gravity and electro-magnetism.

It is widely believed that we should somehow be able to make a quantum theory of gravity, but non-stop efforts from 1930 until 2004 have failed to produce a working theory. We have managed to prove that the techniques we currently use to build quantum field theories will not work for gravity, and we have no clue what techniques will work. All we're really certain of is that we have a lot left to learn.

THE SPEED OF LIGHT

In 1862, Maxwell calculated the speed of light. This seemed a very strange result at the time. In Galilean relativity, the speed of light should depend on the speed of the observer and the speed of the source.

Physicists thought about this for some time, and came up with an explanation: they decided that all of space was filled with a substance, which they called the Lumeniferous Ether. Light was then thought to be a disturbance of this ether, just as water waves are a disturbance of the surface of the water. This idea also explained the prediction of the speed of light - the speed of light was thought to be relative to this ether.

In 1887, Albert Michelson decided to try to prove the existence of this ether substance. He noticed that as the Earth revolves around the sun, the Earth travels through space at about 20 miles per second, about 70,000 mph. The speed of light was known to be about 186,000 miles per second, so the orbital speed of the Earth is about.01% the speed of light. Michelson decided he should be able to detect this - he would compare the travel time of two beams of light, one which traveled along the direction of the Earth's orbit, and a second beam which traveled sideways compared to Earth's orbit.

Today, we would also add in that the Sun is revolving around the centre of the Galaxy, with an orbital velocity of about 200 miles per second. This is about.1% the speed of light. So, actually, Michelson's experiment was about ten times more sensitive than he knew.

Michelson's daughter says he explained his experiment to her like this:

Suppose we have a river 100 feet wide flowing at 3 feet per second, and two swimmers who both swim at 5 feet per second. The swimmers have a race. One swims upstream 100 feet, then swims back to the start. The other swims directly across the river, then turns around and swims back. Who wins?

The swimmer going upstream is easiest to analyse. Going against the current, the swimmer makes only 5-3=2 feet per second, so the 100 feet takes 50 second. Coming back he's going 5+3=8 feet per second, so it takes him 12.5 seconds. His total time is 62.5 seconds for the 200 foot swim.

The swimmer going across the river has a different job. As he swims across the flow at 5 feet per second, the river is carrying him downstream at 3 feet per second. So, he has to swim at an angle in order to make it straight across the river. His net speed is the hypotenuse of a 3,4,5 triangle, so his net speed is 4 feet per second. He swims the 100 foot width in 25 seconds, then takes another 25 seconds to swim back, for a total time of 50 seconds for the 200 foot swim. So, this swimmer wins.

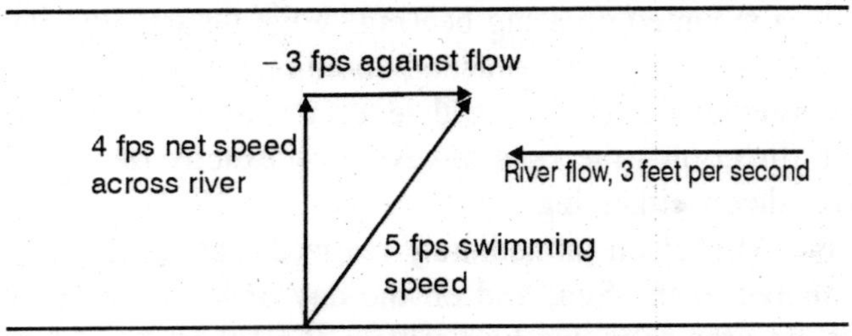

Fig. A Swimmer in a River.

Michelson realized the this exact same argument should apply to light moving along and against the flow of the Ether, and across the flow of the Ether. The math is pretty much the same as swimming across the stream. Let's suppose one light beam travels with and against the Earth's motion over a distance L each way. The other light beam travels normal (sideways) to the Earth's motion a distance L each way. The Earth is moving at a velocity v around the Sun.

The time required for the first beam is L, the length the light travels, divided by the speed of the light beam. Michelson figured the speed would be c + v going one way, and c - v going the other.

	L		L		2 c L		2 L
T1 =	-------	+	-------	=	---------	=	---------------
	c + v		c - v		$c^2 - v^2$		$c (1 - v^2 / c^2)$

The light beam going normal to the Earth's motion is following a path which is like the verticle leg of the triangle above. The three legs of the triangle now have length c (replaces 5), v (replaces 3), and $\Phi(c^2 - v^2)$ (replacing 4), so it's travel time is

	2 L
T2 =	----------------------
	$c \sqrt{(1 - v^2 / c^2)}$

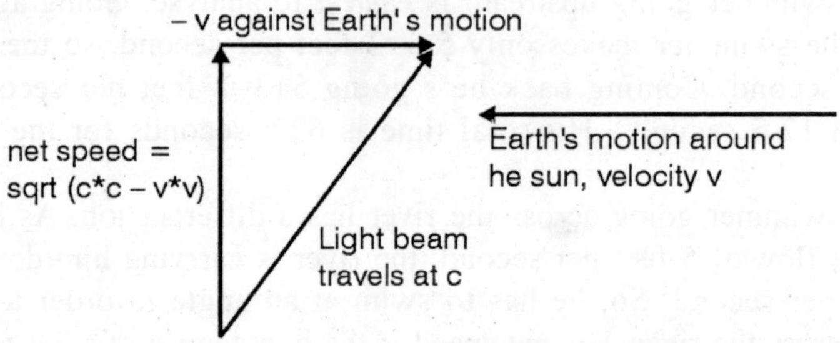

Fig. The Light that Travels Normal to the Earth's Velocity

Michelson set up an experiment and tried it. He found no effect - the two beams of light took exactly the same amount of time. Michelson's experiment was mounted on a big bearing. What he actually did was rotate the entire table as he was measuring, looking for a direction where the light took longer to travel in one direction than in the other. He never found such a direction - the light always took exactly the same amount of time to travel down either leg.

Of course, Michelson immediately realized that perhaps the Ether was moving compared to the Sun, and on the day of his experiment the Earth happened to be precisely standing still in the Ether. So, he repeated his experiments every couple of months for a year, and continued to find no effect. Michelson also wondered if perhaps the Earth was dragging the Ether around with it, so he repeated his experiment on top of a mountain. Again, no effect. Michelson also reasoned that if the Ether was being dragged along with the Earth, we should see the apparent position of stars move, depending on the angle their light entered the Earth's ether. No such effect has ever been observed.

Everyone found this result very confusing - as we have seen, the time taken by the light should be longer when the beam is aligned with the

Earth's motion than when the beam is at 90° to the Earth's motion. The understanding of the day was that the velocity of the Earth's motion should add and subtract from the speed of light, depending on the direction of the light. But, what was found was that the speed of light seemed to never change.

Hendrik Lorentz and George FitzGerald analyzed the Michelson-Morley experiment. They decided to postulate that when something is moving, it shrinks in the direction it is moving. This effect is called the Lorentz-FitzGerald contraction. This contraction can be calculated to exactly compensate for the velocity of the Earth. In other words, Lorentz and FitzGerald decided that the reason the beam aligned with the Earth's motion took the same time as the other beam was that it had a slightly shorter distance to traverse. So, they decided that Michelson's table shrunk in the direction of the Earth's motion, by exactly the right amount so that the two beams of light tied in their race.

We see immediately that if we were to multiply T_1 by $\Phi(1 - v^2/ c^2)$, then these equations give the same result. So, Lorentz and FitzGerald decided to assume that the table had contracted by this factor, Φ $(1 - v^2/ c^2)$, in the direction of Earth's motion. Then the math gave the right answer, which is that the travel time is the same in both directions.

What does this mean? To see this, we're going to learn how to make what are called Space-Time diagrams. This is an ordinary graph, but with time as the vertical axis. It's very inconvenient to label a graph with seconds running up the vertical axis, and units of 186,000 miles on the horizontal axis, so we're going to work in different units. We'll measure distance in feet, and time in nanoseconds. One nanosecond is one billionth of a second. One billion nanoseconds is one second. The speed of light is almost precisely one foot per nanosecond. That is, 186,000 miles times 5280 feet per mile is almost exactly one billion feet. On a space-time diagram, light is always drawn at a 45° angle, because light always moves at 1 foot per nanosecond.

By the way, you could ask "Why does light always move at one foot per nanosecond?" The answer is, we have not even the slightest clue. It just does. We've checked this a zillion times in as many ways as we can think of for over 125 years now, and it's always been true. That's why Einstein took this as an axiom. He could not prove it, he could not justify it, he could not even motivate it. He could only say that this seems to be true in our universe, so lets assume it's always true and see what the ramifications. Well, while this is a very important and interesting result, it's a trick from our current perspective. This light is traveling through a very peculiar medium in a very peculiar fashion. When we speak of the speed of light being a constant, we mean in a vacuum. If there are atoms nearby, the light can interact with the electrons and protons and start doing strange quantum things. It's these strange quantum things that cause rainbows and make lenses work and make the sky blue and make your eyes work. But, we're not studying quantum things in this book, so we're just going to think about the vacuum.

In figure space-time diagram, with a bunch of things drawn in it. Remember, the horizontal (X) axis is position, and the vertical (Y) axis is time. The units are feet and nanoseconds. We see three rays of light, one starting at x = 0, t = -2, and moving to the right. You can tell it's a ray of light because it's drawn at a 45° angle. Now, this is a space-time diagram, so there's something important to notice here: this particular ray of light comes into existence at -2 nanoseconds, and evaporates at +2 nanoseconds.

There's another ray of light which starts at x = 6, t = -3, and ends at x = 3, t = 0. Another ray of lights starts at x = 0, t = 2 and goes to at least x = -3, t = 5. There's a particle moving very quickly, at half the speed of light, from x = 1, t = 2 to x = 2, t = 4. This particle only exists for a short time. There's another particle at x = -5.

This particle is not moving at all, but it exists for at least as long as this graph exists. Finally, at about x = -2.5, t = 1.5, there's one of our new favourite characters, a rocket ship. This is not a book on art, so comments on the aesthetics of this particular rocket ship are not welcome.

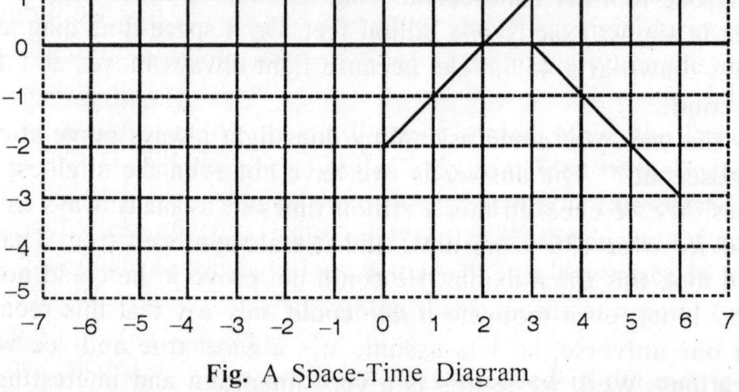

Fig. A Space-Time Diagram

So, we see that in a space-time diagram we know both where and when things are. This is a very different perspective than you are used to. For example, on a space-time diagram, you would look like a long pink tube, with one end attached to your mother and the other end just stopping somewhere about 75 years later. In between, the tube that is you twists and turns and wiggles to reflect where you went while you were alive. If you're female and have children, your children would start out as small pink tubes which branch off from you.

Space-time diagrams show "now" as the x-axis, and they show the past and the future below and above the x-axis. Things which are not moving are vertical lines. According to the rules, all lines which represent the motion of something with mass must be tipped from vertical less than 45°. Light is always at 45°. If a line is drawn which is tipped at more than 45°, it would represent something moving faster than light. We have a name for such objects: we call them tachyons. We also have names for things like "nice lawyers" and "honest politicians," but we've never actually seen any of these things, so don't get too excited.

Now, using our space-time diagram, we're going to try to understand what it means to say where and when something is. We're on unfamiliar ground here, so we're going to try to see how to do this without making any assumptions.

For example, you might look up in the sky and see an airplane fly by, maybe 3 miles up, going maybe 500 mph. You could look at your watch, and say "That airplane was right over my head at noon." However, this is an assumption that we're not going to make. What you can say is, "We saw the airplane at noon." We will not assume that we know how to tell time at places far away from us. In fact, what we would really like would be to have a clock hanging in the air 3 miles right above our head, and when the airplane flies past the clock, we can see the airplane and the clock next to each other and read the time off of that clock.

Now, we have the problem of trying to synchronize this clock 3 miles away with our wrist watch. How can we do this? Well, first we'll design a new clock. Our new clock has hands. The second hand ticks off nanoseconds, so to us mere humans it looks like a blur, but that's no big deal.

The light will flash, say, every micro-second, that is every 1,000 nanoseconds. Now, anyone anywhere can synchronize their clock with the flash from our clock. When you see the flash, you know the second hand is pointing straight up. Of course, we're trying to synchronize clocks here, so we have to account for the speed of light. If you're 100 feet away from my clock, and you see the flash, you know that my clock emitted the flash 100 nanoseconds ago. But, no problem, you just make sure your clock's second hand hits 100 nanoseconds exactly as you see the flash from my clock. The clock that's 3 miles up in the air is about 15,000 feet away, so that clock will be set 15,000 nanoseconds ahead of the flash.

Now, our clocks are synchronized. Remember, we're going to have a lot of clocks, strung out all over the place, all synchronized. We will know where each clock is. When we see something happen, we'll know where it happened and when it happened - the where is from knowing the position of the nearest clock, and the when is from reading the time off that clock, and no other. We have a special word for something happening, we call this an event. An event is something that happens at a particular time and place. Here's a space-time diagram of us synchronizing a couple of clocks. At x = 0 (that's where we're standing) and t = 0, we send out a flash.

The flash travels at the speed of light, which is a 45° line. Two feet away from us is another clock. Our trusty graduate student is standing there with another clock. He sees the flash at t = 2 nanoseconds, sets his clock, and sends a flash back. At t = 4 nanoseconds, we see our grad student's flash come back. Graduate students, by the way, are a very important part of physics: they're smart, educated, do what they're told, and work nearly for free. Without graduate students, all of science would come to a screeching halt.

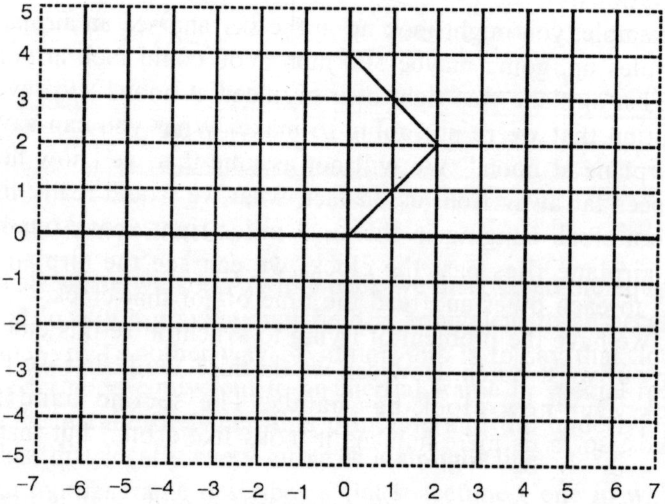

Fig. Synchronizing a Pair of Clocks with Light Flashes

Here's the situation we're going to imagine. Suppose there's some guy, George, who has a small lab. He has two clocks and he wants to synchronize them. So, we know how he's going to do this. He's going to have flashers on his clocks to help set the clocks. When the clocks are synchronized, he's going to just sit back and watch the clocks flash at each other - it's going to look like they're bouncing a little light ball back and forth between them, like they're playing ping-pong with light. Now, here's the trick.

George and his small lab are in a rocket ship, flying by us at half the speed of light. What do we see? Conveniently, his rocket ship is transparent, just like Wonder Woman's jet airplane, so we can see inside. We can turn on our clocks however we wish, so we'll agree that George will set the clock closest to him to read $t = 0$ exactly as he passes us. We'll also set our clock to read $t = 0$ just as George passes us. So, at the instant $t = 0$ our clock and one of George's clocks are right next to each other and read the same thing. Our job is to figure out what happens next. By the way, George's clocks happen to be bolted down to a large solid table which is 11.55 feet long.

Below is a space-time diagram of this situation. the scale to 5 feet and 5 nanoseconds per tick. George has two clocks. One of them flies right past us, so that's the line that goes through the origin. George is flying at half the speed of light, so in 10 nanoseconds he moves 5 feet - half as far as light would move. George has a second clock which is 11.55 feet away from him. But, we have to remember the Lorentz- FitzGerald contraction. Although George has carefully measured out 11.55 feet, we see his two clocks as being closer to each other. The distance is contracted by the factor $\Phi(1 - v^2/c^2)$.

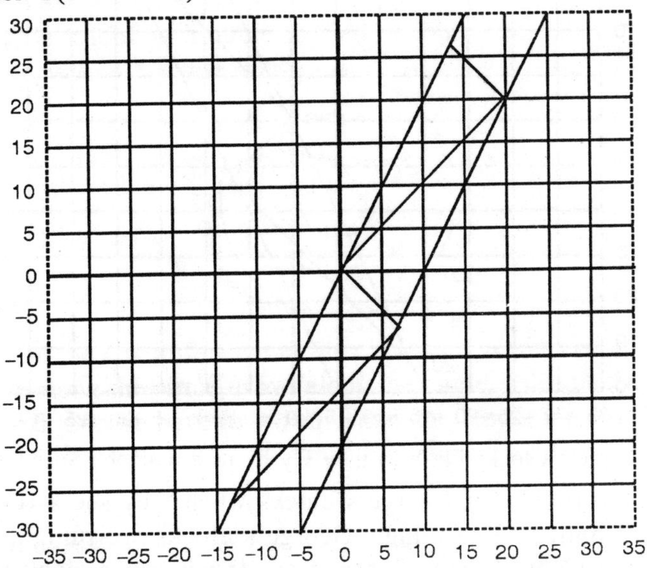

Fig. George and His Clocks Fly Past us at Half the Speed of Light

He's moving a c/2, so $v^2/c^2 = 1/4$. $\Phi(3/4) = .866..866 * 11.55$ feet = 10 feet. So we see George's two clocks as being 10 feet apart. In our space-time diagram, George's second clock is on a parallel line 10 feet away, just as I've drawn it. At our $t = 0$, we see one of George's clocks right on top of us, and one which is 10 feet away from us. I've also drawn a couple of light flashes emitted by George's clocks, as they look to us. Light always goes at 45°. When George's clock reads t = 0, it flashes. To us, it looks like it takes 20 nanoseconds for that flash to reach the clock at the other end of George's table. That clock then flashes, and it looks to us like it takes about 6.67 nanoseconds for that flash to reach George's first clock.

Now, remember, George has carefully set his clocks. He doesn't think there's anything strange about his lab, he's just setting his clocks to read the right thing. So, when his second clock sees the flash from the first clock, it's reading 11.55 nanoseconds.

After all, George has carefully synchronized his clocks. When the flash from that clock reaches George's first clock, the first clock is reading 23 nanoseconds. How can this be? To see how this works, we'll draw another space-time diagram where we show the light flashes that happened immediately before these flashes.

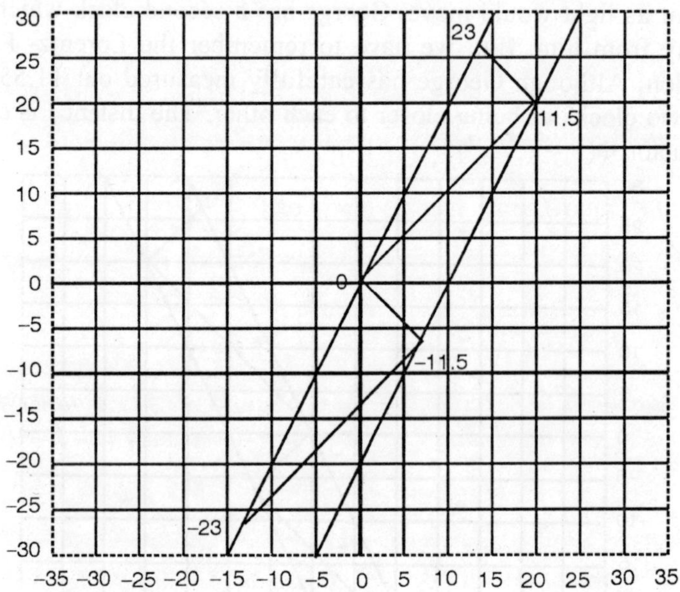

Fig. George's Clocks Flash as they Fly by us at half the Speed of Light

The time shown on George's clocks when they see a flash of light. We quickly notice some things. George's clocks are running slow. The clock nearest George read 23 nanoseconds when our clock reads 26.67 nanoseconds, and his reads -23 nanoseconds when ours reads -26.67 nanoseconds. The other thing we notice is that George's clock which is ten feet away from us at closest approach is not only running slow, but is also off.

Halfway between the points where George's second clock reads -11.55 and 11.55 nanoseconds happens when our clocks are reading 7.33

nanoseconds. So, George's second clock reads 0 when our clocks read 7.33 nanoseconds.

When our clocks read 0 and we see George's first clock as reading 0, we see George's second clock is reading -5.75 nanoseconds. So, this is big: George's clocks and our clocks are not synchronized. We see our clocks as synchronized, and George sees his clocks as synchronized, but we don't see George's clocks as synchronized, and he does not see our clocks as synchronized.

This special relativity idea has now cost us one of our most fundamental intuitions. It is not possible to synchronize clocks unambiguously. Or, we can say, there's no such thing as simultaneous. We see George's distant clock as being 5.75 nanoseconds behind his first clock, so events that George sees as simultaneous, like his two clocks both reading 0, we see as happening at very different times. We see our two clocks as reading zero at the same time - we worked very hard to arrange that - but George sees our two clocks as off by 5.75 nanoseconds.

The next thing we notice is that George's clocks are running slow by exactly the same factor as we think his table has contracted. That is, 26.67/ 23 = 11.5/ 10. This effect is called Lorentz-FitzGerald time dilation. We see George's clocks as running $\Phi\ (1 - v^2/\ c^2)$ as fast as our clocks.

This time dilation effect is the source of the "twin's paradox." If you stay on Earth, and your twin gets in a rocket ship and flies away at a very high speed for a year, then turns around and flies back to Earth, he has aged two years, and you've aged more than two years.

THE INVARIANT INTERVAL

We saw that the speed of light, c, is the same number for everyone, everywhere. The speed of light does not depend on how fast you are going, nor on how fast the source of the light is going, you always measure the same number.

We learned that two different observers moving at different speeds cannot synchronize their clocks with each other. We learned that moving clocks appear to be slow, and moving objects appear to contract in the direction of their motion.

The factor by which clocks slow down and objects contract is $\Phi(1 - v^2/\ c^2)$. We also learned that the space concept of position had to be replaced by the space-time concept of event, which is a particular position at a particular time.

We're used to using the Pythagorean theorem to calculate distances. So, in Figure below, $A^2 = B^2 + C^2$. We need to be able to calculate distances in special relativity, so we need to know how this formula should look in space-time.

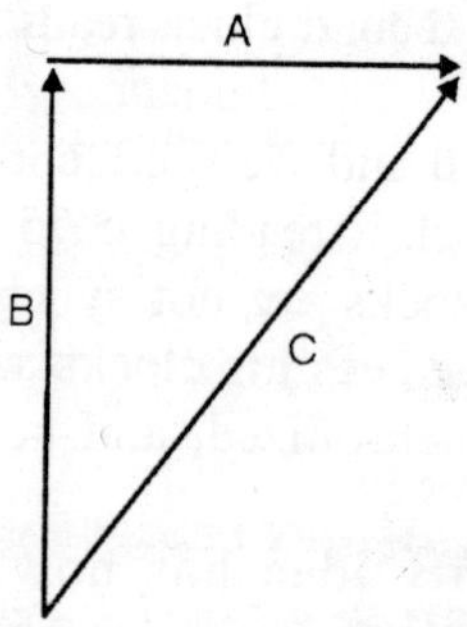

Fig. Caclulation of $A^2 = B^2 + C^2$

Here's what we know: the speed of light is always c. Speed is distance divided by time, so this means sqrt($X^2 + Y^2 + Z^2$)/ T = c. We can multiply both sides by T then square both sides to get $X^2 + Y^2 + Z^2 = c^2 T^2$. Or, $c^2T^2 - X^2 - Y^2 - Z^2 = 0$. We call this quantity $c^2T^2 - X^2 - Y^2 - Z^2$ the interval. We're going to see that in special relativity the interval takes the place of distance. But, there's a big difference - the interval is how far you went minus the time is took to get there. For a ray of light, the interval is always zero, because at light speed, 1 meter of distance takes 1 meter of time, and 1 - 1 = 0. This fact, that all people see the same speed of light, will be elevated from a curiosity to a fundamental axiom. The equation $c^2T^2 - X^2 - Y^2 - Z^2$ will similarly be elevated from a special equation about light to a fundamental equation about the distance between any two events.

We're used to $(\text{distance})^2 = X^2 + Y^2 + Z^2$. This formula is called a metric, and this particular type of metric is called positive-definite. Positive because the sum of three squares is always positive. Definite because for any X,Y,Z we can always calculate a unique number. So, in 3-space, we use the 3-distance sqrt($X^2 + Y^2 + Z^2$). But, we're working in 4-space now, so we need to figure out what the 4-distance is. In Special Relativity, the idea of distance will be replaced by the interval $c^2T^2 - X^2 - Y^2 - Z^2$, which is not positive definite.

We can see that if something moves a short distance in a long time, the interval is positive. If something moves at precisely the speed of light, the interval is zero. And if something moves faster than light, or more reasonably if we consider two points which are far apart in space but not in time, the interval is negative.

This is very different from flat Euclidean space. This is why we use a new word, interval: to help remind us that this is a very strange kind of distance that can be positive, zero, or negative. For example, the interval between the Earth and the Sun is about -7 minutes if we consider where the Earth and Sun are at the same time; the interval is zero if we consider the path that a ray of light would take from the Sun to the Earth; and the interval is about 6 months if we consider the path that a typical NASA satellite would take.

We're not used to a type of distance where how long you take to go somewhere counts as part of the distance. Similarly the interval from Los Angeles to New York is not the 3-distance of 3,000 miles. The interval from Los Angeles to New York is zero for a ray of light, it's about six hours for a traveller on a 747, and it's about five days for someone driving a car. If we consider where Los Angeles and New York are at exactly the same instant, so that $T^2 = 0$, then the interval is -3,000 miles, but now it's a negative number.

Right away we can see that these factors of c are going to be popping up all over the place. Also, we see that there's some confusion on whether the interval is measured in meters or seconds or hours or feet or light years or whatever. Actually, we're used to this for distance. If we asked someone, "How far is it from Los Angeles to New York," we would not be surprised to hear 3,000 miles, or 5,000 kilometers, or maybe even six million feet, or 50 million centimeters. But the idea that it could be two seconds or six hours or -3,000 miles from Los Angeles to New York seems very strange.

How can a distance be the same as a time? Why is the speed of light this strange number, 186,000 miles per second? Is there something special about this number, 186,000, that God particularly liked?

Let's think about horses for a minute. A horse's height is measured in hands, where a hand is four inches (don't ask me why, I've never owned a horse). So, the horse below stands about 15 hands high at the shoulders, and is about 7 feet long. When the horse rears up on its hind legs, if we were to measure the dumb way we might find that the horse is now 22 hands high but only 5 feet long.

The confusion here is because we're measuring height in hands, and length in feet. It would make a lot more sense if we were using the same units in both directions, like hands for both length and height. As it is, if we want to know the distance from the horse's rear hoof to his nose, we can't use Pythagoris' theorem, we can't say $(15 \text{ hands})^2 + (7 \text{ feet})^2 = \text{distance}^2$, because hands are not in the same units as feet. We could say something like $(15 \text{ hands} * 4 \text{ inches per hand}/ 12 \text{ inches per foot})^2 + (7 \text{ feet})^2 = \text{distance}^2$. You can see this is a real pain - it's really not very convenient to use different units for different dimensions.

Similarly, we have a built-in confusion about space and time: we measure time in seconds and distance in feet or meters. However, knowing that the speed of light is the same for everyone, we can use the speed of light to convert seconds into feet or meters. From now on, we'll agree that we're going to use the same units for time and space. For example, as we've already seen, one foot of distance equals one nanosecond at the speed of light, so if we say a foot of time we mean the same thing as if we say a nanosecond.

A meter of time is about 3 nanoseconds. Velocity, distance per time, is now meters per meter, so velocity has no dimensions. The speed of light is now just 1 with no units. The speed limit on most freeways is 65 miles per hour which equals about c/ 10,000,000. If physicists were running the highways, apparently highway signs would say "Speed limit 10^{-7}." A traffic ticket for going 85 would read "excessive speed: $1.3*10^{-7}$ in a 10^{-7} zone." That's it, no units.

APPLICATIONS OF DERIVATIVES

At the time of this writing, physics is rather annoyingly split into two completely different types of theories. Relativity is a theory about how space and time work, and very basic ideas about how particles must act in space-time. The other half of physics is quantum field theories, which are theories about what types of particles exist and how they interact with each other. One may say that relativity builds a stage, and quantum field theories fill the stage with players and a script.

In 3-space, we're used to putting X,Y,Z into a vector. Then the length of a vector squared is the dot product of the vector with itself, that is, $L^2 = V \bullet V$. But now we're working in 4-space, that is space-time, so at the very least we're going to need to have vectors which hold 4 things. Here's the rules we'll use for these 4-vectors:

- A vector will have a superscript index, as V^i. We always use superscripts for a vector. Subscripts will mean a different type of object.
- If the superscript is a latin letter like i,j,k, then the vector lives in 3-space and the index runs from 1 to 3. V^1, V^2, and V^3 are respectively X, Y, and Z.
- If the superscript is a greek letter like a,b,g, then the vector lives in 4-space and the index runs from 0 to 3. V^0 is time, and V^1, V^2, and V^3 are respectively X, Y, and Z. Notice that V^2 means the second entry in V, not V squared.
- V^0, the time coordinate, will be measured with the same units as the space coordinates, so the speed of light is 1.
- We can multiply a vector, that is something with a superscript, by something with a subscript. We never multiply two things that both have superscripts. We never multiply two things that both have subscripts.

In 3-space we have the dot product to help us calculate distance. The dot product makes no sense in space-time. A 4 vector dotted with itself gives us $X^2 + Y^2 + Z^2 + T^2$, which has no meaning. Why? Because we are looking for the interval, so we need $T^2 - X^2 - Y^2 - Z^2$. We have seen that $T^2 - X^2 - Y^2 - Z^2$ is zero for a ray of light for all observers. If we calculate $X^2 + Y^2 + Z^2 + T^2$ for a ray of light, we get a number which depends on the observer's velocity, as we'll see in a bit. We need a replacement for the dot product which gets us a minus sign in front of the space terms.

We could do this by remembering that the t term always gets subtracted instead of added, but this is just a recipe for trouble - we'll forget sometimes. We'll handle this with a matrix - a special matrix, called the Metric Tensor. The Metric Tensor, usually just called the Metric, is called h, which is read out loud as "eta." The Metric tensor will be the matrix:

$$(1, 0, 0, 0)$$
$$(0, -1, 0, 0)$$
$$(0, 0, -1, 0)$$
$$(0, 0, 0, -1)$$

If we multiply the vector $V = (V^0, V^1, V^2, V^3)$ by the matrix h, we get $(V^0, -V^1, -V^2, -V^3)$. Now, if we multiply our original V by this, we get $(V^0)^2 - (V^1)^2 - (V^2)^2 - (V^3)^2 = T^2 - X^2 - Y^2 - Z^2$, which is just what we're looking for. So, $V \cdot V$ gives us the wrong answer, but $V \cdot \eta \cdot V$ gives us the right answer. The purpose of h is to keep track of the minus signs in the space terms for us.

In Euclidean 3-space, the metric tensor is

$$(1, 0, 0)$$
$$(0, 1, 0)$$
$$(0, 0, 1)$$

which just turns a vector into itself, so the metric tensor is always ignored in 3-space. But it's still there, formally, and when we make the move to space-time we can't ignore it any longer. When we move to curved 4-space, we'll call the metric tensor g instead of h. This is because we'll find that the metric tensor g is not a constant, but depends on where we are. We'll find that in most of our universe, so long as we're not moving at close to the speed of light and we're not near any black holes, the metric tensor g is very nearly equal to η, so we can say that $g = \eta + h$, where h is a matrix containing only small numbers. We'll find out that h is the gravitational potential. So, this metric tensor stuff is very important, both formally and physically.

We've been cheating here for just a little bit - we've been ignoring our rules above, at least as far as notation goes.

The V we have been talking about is a vector in space-time, so by our rules it must have a superscript, like V^α.

The metric tensor h is an object with two subscripts, for example $\eta_{\alpha\beta}$. So V•η•V really means:

$$\sum_{\alpha=0}^{3}\sum_{\beta=0}^{3} V^{\alpha} * \eta_{\alpha\beta} * V^{\beta}$$

Now we can see explicitly that we followed our rules. There's two superscripts, and two subscripts, and we are always multiplying something with a superscript by something with a subscript. Things with superscripts are called vectors, or contravarients. Things with subscripts are called forms, or covarients. The metric tensor is called a 2-form, because it has two subscripts.

Einstein published a lot of papers, and one day the guy who did his typesetting said to Einstein, "Every time you have an index repeated, you have one of these sum symbols. Why bother?" So, Einstein invented the rule that whenever an index is repeated, it means multiply the terms containing that index, and sum from 0 to 3. Even though it was the typesetter who thought this up, Einstein gets the credit. We add to this the rule that if an index is repeated, one of them must be "up" (superscript) and one must be "down" (subscript). This saves us from writing a bunch of "*" and "S" characters. This is called the Einstein Convention.

$$\sum_{\alpha=0}^{3}\sum_{\beta=0}^{3} \eta_{\alpha\beta} * V^{\alpha} * V^{\beta} \equiv \eta_{\alpha\beta} * V^{\alpha}V^{\beta}$$

Next, we know that the speed of light is the same for all observers. That means that h_{ab} must be the same for all observers. So, if the transformation matrix which gets us from one person's frame of reference to another is L, then this must be true:

$$\eta_{\alpha\beta} = \Lambda_{\alpha}{}^{\gamma}\,\Lambda_{\beta}{}^{\delta}\,\eta_{\gamma\delta}$$

There are a couple different forms of notation for what we're learning here, so I'm going to take this chance to talk briefly about them. Some people use 4-vectors where the index runs from 1 to 4, and the 4th component is i times T, where i is the square root of -1. This notation was invented by Minkowski, one of Einstein's professors when Einstein was a college student.

Einstein always said he hated this notation, but in spite of that he sometimes used it. If you use Minkowski notation, you don't need the metric tensor, which somehow make us feel like space-time is more life space, but also makes the transition to General Relativity much harder. Some people use g for the metric tensor, that letter for the metric tensor in curved space-time. The g will remind us that there are gravity fields around. The notation I'm using is the modern notation, but not everyone has gotten with the programme yet.

Chapter 2

Relativity Made Simple

LORENTZ COORDINATE TRANSFORMATION

It is a common misconception that Einstein based his theory of special relativity on the Michaelson- Morley experiment. In the days of Michaelson and Morley it was thought that electromagnetic waves propagated by a medium which was called the Luminiferous Aether. The earth rotates on its axis and rotates around the sun with circles the galaxy which wanders about the local cluster group etc so it was expected that if a device could be made to detect the its motion with respect to the aether that it would yield a significantly nonzero result. Michaelson and Morley constructed just such a device and to their astonishment it yielded a null result.

This is enough to dispel the aether theory in most peoples minds, but at the time some people tried to explain why one would not be able to measure a speed based on the motion of light in the device even though they insisted the Earth was in motion with respect to the medium. Lorentz was one such person who empirically derived the Lorentz transformation equations by introducing length contractions and time dilations into a transformation which would leave the speed c invariant to frame so that one could not use it to determine a speed with respect to the medium.

Einstein is given so much credit because what he did different was to come up with two powerful postulates from a simple idea and from those postulates was able to derive the Lorentz transformations from first principles and developed special relativistic physics from there. The idea that led Einstein to his postulates is depicted in the figure.

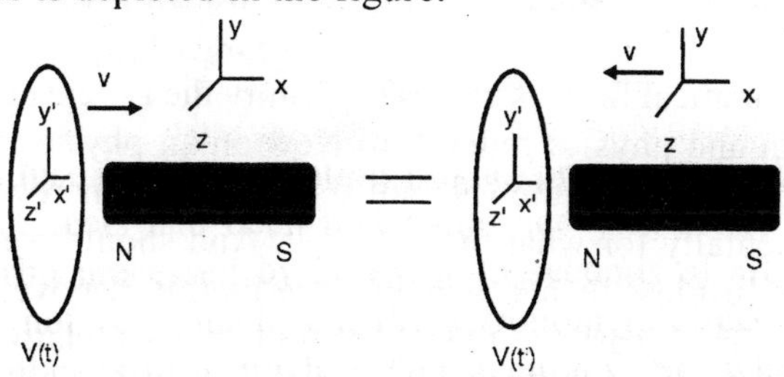

Fig. Einstein Postulates of Lorentz Transformations

Start out placing a magnet stationary on a table perpendicular to the plane of a loop of wire and pointed at the centre. Wire the loop in series with a resister and a current meter to calculate from the current and resitance the voltage induced in the loop. Move the loop at constant velocity and note the voltage function. Next fix the loop with respect to the table and move the magnet at the same velocity except for opposite in direction. Note the voltage function. Either way you do the experiment you get the same voltage function, the same physics.

Therefor it won't matter whether we say the magnet source is moving and the loop receiver is stationary or visa-versa as you get the same physics either way. This led Einstein to his first postulate. The electrons form a current in response to the changing magnetic flux, or the way the magnetic field changes across the loop over time. Either perspective yields the same physics so the information about the magnet as received by the electrons must travel at a speed independent of whether we say the source is stationary and the receiver is moving or visa-versa. This led Einstein to realise that there must be a speed that is invariant to frame at which the electromagnetic information transfers.

- The first postulate can be worded as: *The laws of physics are invariant to inertial frame transformation.*
- The second postulate can be worded as: *The invariant speed c, is finite and is the vacuum speed of light.*

These postulates are phrased a little different in virtually every text, but the core idea behind them as depicted above is the same.

The second in this manner for the reason that time dilation and such effects really have nothing directly to do with light itself. These are instead due to space time being structured such that the invariant speed c is finite. If we surprisingly discovered that light had some small amount of mass and thus really traveled at speeds just short of c, it would have no ramification on the physics of special relativity whatsoever.

It would merely mean that we are using bad terminology, for instance in calling c speed particles "light-like". What actually distinguishes Lorentz transformations and special relativistic physics from Galilean transformations and Newtonian physics is that in the Lorentz transformations the invariant speed c is finite.

In the mathematical limit as c goes to infinity the Lorentz transformations become Galilean and physics reduces to Newtonian physics. The statement that this invariant speed c is the vacuum speed of light merely tells us where to look experimentally for what that speed it. And should we find that light travels at speeds just less than c then one may merely remove the "*and is the vacuum speed of light*" part and the fact that physics is relativistic according to the remainder of the two postulates would be unaffected.

The first postulate tells us that it does not matter what inertial frames we take to be in motion, or what inertial frame we take to be stationary as the laws of physics do not depend on inertial frame. In a sense it is odd that relativity is called relativity at all because according to the first postulate physics is invariant to frame.

Thus relativity is really a theory of invariance. What is really relative in relativity are time and space coordinates and things defined in such a way that they depend on the coordinate frames. The equations used to model the "laws" of physics must be invariant equations if we are to be consistent with the first postulate. As we shall see tensor equations are invariant and so in relativity we write the laws of physics as tensor equations.

These two postulates imply that the coordinate transformation that correctly describes boosts between different inertial frames is the Lorentz coordinate transformation.

Lets say that one observer uses an inertial coordinate frame S with coordinates (ct,x,y,z). Another observer uses another inertial coordinate frame S' given by (ct',x',y',z'). They will be in motion with respect to each other so that the S frame observer observes the other to moving at speed v an the +x direction and the S' frame observer will observe the other to be moving at the same speed in the -x' direction. Lets say we know the location of an event according to one observer's coordinate frame and wish to determine the location according to the other coordinate frame. We transform the coordinates of the event from the one to the other by doing a *Lorentz coordinate transformation.* In this case the Lorentz coordinate transformation equations are

$$ct = \gamma(ct' + \beta x')$$
$$x = \gamma(x' + \beta ct')$$
$$y = y'$$
$$z = z'$$

where we make the definitions

$$\beta = v/c$$

and

$$\gamma = (1 - \beta^2)^{-1/2}$$

A more compact form of the transformation that allows the boost to be in any direction rather then restricting it to a coordinate axis is

$$ct = \gamma ct' + \gamma\beta\cdot r'$$
$$r = \gamma\beta ct' + r' + (\gamma - 1)(\beta\cdot r'/\beta^2)\beta$$

Inverted equation becomes

$$ct' = \gamma(ct - \beta x)$$
$$x' = \gamma(x - \beta ct)$$
$$y' = y$$
$$z' = z$$

and in differential form equation becomes

$$dct = \gamma(dct' + \beta dx')$$
$$dx = \gamma(dx' + \beta dct')$$
$$dy = dy'$$
$$dz = dz'$$

and the inverse differential form is

$$dct' = \gamma(dct - \beta dx)$$
$$dx' = \gamma(dx - \beta dct)$$
$$y' = y$$
$$z' = z$$

c is called the Lorentz invariant speed and according to Einstein's second postulate it is the finite vacuum speed of light.

(Exact by definition)

$$c = 299792458 m/s$$

If c were to be infinite then the Lorentz transformation equations would be the Galilean transformations

$$t = t'$$
$$x = x' + vt'$$
$$y = y'$$
$$z = z'$$

When speeds are large enough so that c can no longer be taken to be infinite then the Galilean transformations can no longer be used and the special relativistic phenomena such as time dilation and length contraction are observed

RELATIVE SPACE AND TIME

The differential form of one of the Lorentz coordinate transformation equations Eqn 1.1.5 is

$$dct = \gamma(dct' + \beta dx')$$

Thus extended to a finite interval this becomes.

$$\Delta ct = \gamma(\Delta ct' + \beta \Delta x')$$

Given the interval in time and space between two events according to the S' frame, this equation gives the interval in time between the events according to the S frame. Now consider the case that the events happen at the same location according to the S' frame, for instance the ticks on the S' frame observer's watch. In this case $\Delta x' = 0$ and we have

$$\Delta t = \gamma \Delta t'$$

To distinguish that the time interval is for events at constant location according to the proper frame we often write the proper time interval as $\Delta t' = \Delta\tau$ and call it proper time.

$$\Delta t = \gamma \Delta\tau$$

From this equation we see that the time interval between the events according to the S coordinate frame is longer than the time interval between the events according to the S' coordinate frame. This phenomenon is called time dilation. Time intervals between events are not absolute, but depend on whose coordinate frame you use.

We can extend this phenomenon to the case in which one of the observers is accelerated. Though special relativity is really only directly concerned with inertial frames, we can consider an accelerated state to be a state of transitions or boosts between different inertial frames. We will next let the S' frame observer enter a state of acceleration. *For small time intervals* we can say the S' frame observer is an inertial frame observer. Thus the equation

$$dt = \gamma d\tau$$

holds valid for describing how much time goes by according to the S frame observer given how much the S' observer has aged even if the S' frame observer is accelerating.

Now consider the case that the two events happen at the same time according to the S' frame. Then $\Delta ct' = 0$ and Eqn 1.2.1 becomes

$$\Delta ct = \gamma\beta\Delta x'$$

From this we see that if the events have a displacement in space along the x' direction, then they happen at *different* times according to the S frame. Thus the very notion of simultaneity is relative. Events simultaneous according to one coordinate frame are not all simultaneous according to another.

Next, consider the following differential Lorentz coordinate transformation equation from Eqn.1.1.6

$$dx' = \gamma(dx - \beta dct)$$

Extend over a finite interval to arrive at the corresponding equation for two displaced events

$$\Delta x' = \gamma(\Delta x - \beta\Delta ct)$$

Lets say that the S' observer puts a fire cracker on each end of a measuring stick of length L_0 oriented along the x' direction and sets them off timed so that the events occur at the same time according to the S frame($\Delta ct = 0$). Then the length of the stick according to the S frame L will be given by the spatial displacement between the events. Then we have $\Delta x' = L_0$ and $\Delta x = L$. Inserting these three inputs into the above equation results in

$$L_0 = \gamma L$$

or

$$L = (1/\gamma)L_0$$

This is called length contraction.

PARADOXES

Consider a ship that travels to another star at a constant velocity, then immediately whips around the star for the return trip at the same speed. We on Earth observe that the clocks on board the ship run slow due to time dilation throughout the entire trip both outgoing and incoming. On the way out, according to the ship frame it is the Earth that is moving away and so it is the

Earth clocks that run slow due to time dilation. Also on the way back, according to the ship frame it is the Earth that is approaching the ship and so again the Earth clocks run slow due to time dilation. This presents a problem called *the twin paradox*.

The problem is how to answer how the clocks read when the ship and earth arrive together again and what causes the difference.

The total time dilation can be calculated by 1.2.2 or 1.1.3, but only if the accelerated frame is taken as the primed frame. The question would be why this round trip case is not symmetrical. If the two remained in inertial states then each would observe the other as aging slower in a symmetric fashion. But in this round trip the end result couldn't work symmetrically because that would lead to a true paradox. One can explain this from various perspectives just as there are various frames that can be used to describe the situation.

The simplest explanation is that the accelerated frame of the ship is actually a piecewise construction of two different inertial frames for which there is a lack of simultaneity.

Because of the piecewise construction of the accelerated frame, the ship observer reckons that clocks in the direction of the acceleration undergo an advance during the acceleration by an amount that depends on how far away they are. Primes (′) will indicate the ship frame. T ‘will be the time it takes the ship to reach the star according to the ship frame and β as a function of proper time is.

$$\beta = \begin{vmatrix} \beta_0 & \text{for} & ct' < cT' \\ -\beta_0 & \text{for} & ct' > cT' \end{vmatrix}$$

The coordinate transformation from the accelerated ship frame coordinates to the inertial Earth frame coordinates is

$$ct = \gamma(ct' + \beta x')$$
$$x = \gamma[x' + \beta ct' + (\beta_0 - \beta)cT']$$
$$y = y'$$
$$z = z'$$

This set of transformation equations results in Lorentz transformation on the way out as well as on the way in and gives the solution that the ship clocks read less time upon their arrival back at Earth. Symmetric time dilation only occurs during the portions of the trip where both observers maintain inertial states. During acceleration the symmetry is broken and both observers will always agree on how much they should age differently in a round trip.

Consider a train whose proper length is greater than the proper length of a tunnel. The train moves at near c speeds so that it is extremely length contracted according to the frame of the tunnel. Let's say it is so length contracted that it fits inside the tunnel. Gates at the ends of the tunnel are set so that they each close once it's entirely inside.

According to the frame of the train observer the train is still while the tunnel moves the other direction. Therefor according to this frame it is the tunnel whose length is contracted. The proper length of the tunnel is itself shorter than the length of the train and so the tunnel is length contracted beyond this so that the train never fits to be enclosed by the gates.

This leads to a superficially apparent contradiction called the length contraction paradox. As with all the so-called paradoxes of relativity this only superficially seems to be a contradiction and is easily shown not to be a true contradiction. The solution is the realization that we did not account for relative simultaneity in this mind experiment. According to the tunnel frame the events that the two gates close are simultaneous. According to the train observer's frame the two events still occur, but they are not simultaneous. According to the train observer's frame the event that the gate closes behind the train happens after the front end of the train has smashed through the front gate which had already been closed.

REPRESENTING SIMPLE RELATIVITY

MINKOWSKI SPACETIME DIAGRAMS

To make a 2d Minkowski space-time diagram one typically first picks an inertial frame. The vertical axis of the graph is chosen to be the time ct axis. The horizontal axis is chosen to be a distance coordinate x. This means that velocity of a particle is obtained from the reciprocal of the slope of its path on the diagram instead of directly from the slope. Yes, this seems weird to do at first but its also the usual way to do it so you get used to it. So the path of an accelerating particle on a spacetime diagram can look like:

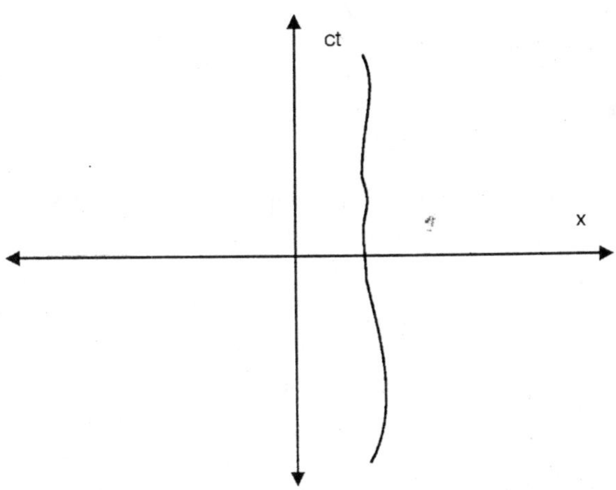

Now lets say the path represents the path of an accelerated observer. The events comprised of the ticks on the watch of this observer all lay along the path and so the observer's *world line* or the path the observer takes can be labeled as the time axis for this observer. So the graph looks like:

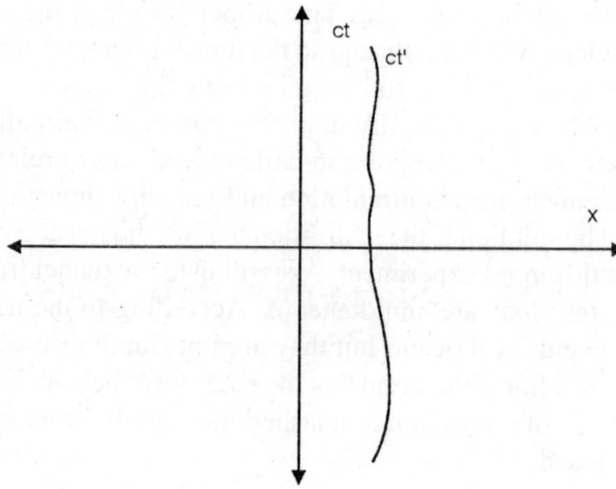

We might consider a case where the primed frame observer is in an inertial frame and the clocks are synchronized at the origin. In this case it becomes:

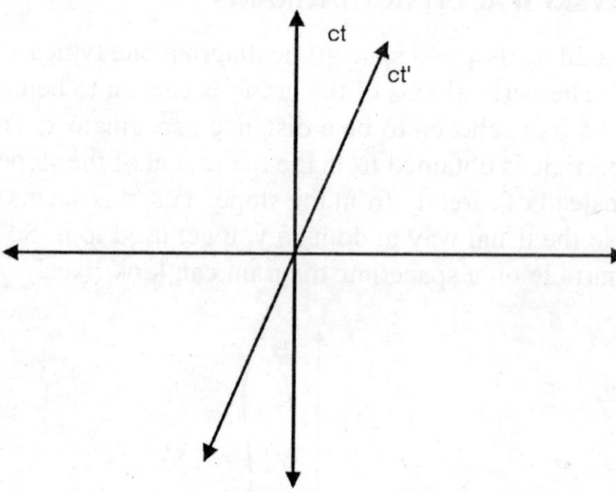

The time ct' axis consists of the events that are all at x' = 0. The Lorentz transformation equations describe this line parametrically.

$$ct = \gamma ct'$$
$$x = \gamma\beta ct'$$

The x' axis consists of events at ct' = 0. The Lorentz transformation equations also describe this parametrically

$$ct = \gamma\beta x'$$
$$x = \gamma x'$$

From these we verify that the reciprocal of the ct' axis slope results in the velocity β, but we also see that for the x' axis, the slope itself results in the velocity. So we can also now put the x' axis on the graph.

Lines representing events of constant position in x' are then drawn parallel to the ct' axis and lines representing events simultaneous in ct' are drawn parallel to the x' axis.

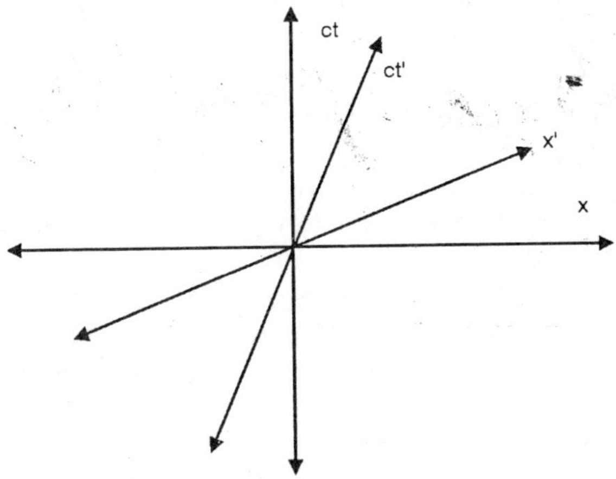

If we consider the paths of light moving along the x axis that cross the origin, they are drawn at 45°.

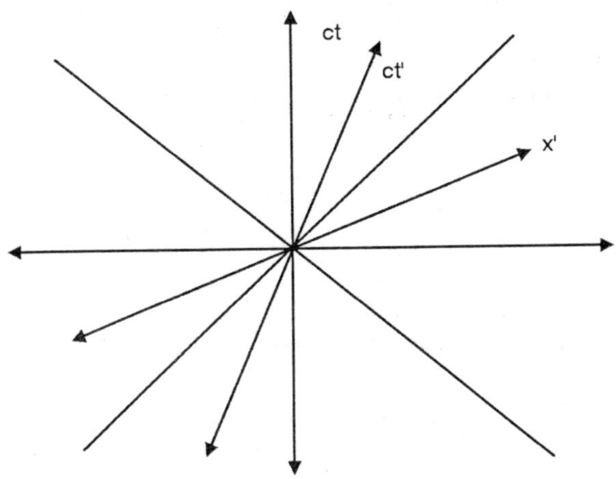

If a y axis is included coming out of the screen, the light paths sweep out a cone shape. This is known as a *light cone*.

TENSORS IN SR

The Pythagorean Theorem in 3 dimensions of space can be written

$$d\sigma^2 = dx^2 + dy^2 + dz^2$$

For the displacement between events in 4 dimensional space-time we should include the temporal displacement between the event in the interval. This can be done in such a way that the displacement calculated is the same according to any inertial frame. We define the *invariant interval* as

$$ds^2 = dct^2 - d\sigma^2$$

which can be written

$$ds^2 = dct^2 - dx^2 - dy^2 - dz^2$$

To verify that this interval has been constructed in an invariant way insert the expressions for the differentials from the differential form of the Lorentz coordinate transformation equations Eqn 1.1.5.

After simplification, the interval reduces to the same form

$$ds^2 = dct'^2 - dx'^2 - dy'^2 - dz'^2$$

When we observe an object in motion and describe the length of its path through space-time by the invariant interval we should realise that the object does not move according to its own frame

$$dx' = dy' = dz' = 0$$

therefor the invariant interval reduces to $ds = dc\tau$. Since the interval is an invariant and it is equal to the proper time of the object, we can look at proper time $d\tau$ as an invariant.

We can choose to define a displacement four-vector dx^μ with the following relations

$$dx^0 = dct$$
$$dx^1 = dx$$
$$dx^2 = dy$$
$$dx^3 = dz$$

Here we introduce the metric tensor for special relativity $\eta_{\mu\nu}$.

$$[\eta_{\mu\nu}] = \begin{bmatrix} 1 & 0 & 0 & 0 \\ 0 & -1 & 0 & 0 \\ 0 & 0 & -1 & 0 \\ 0 & 0 & 0 & -1 \end{bmatrix}$$

Using Einstein summation as discussed the invariant interval is written in more compact notation as

$$ds^2 = \eta_{\mu\nu} dx^\mu dx^\nu$$

Light paths are described by $ds = 0$. therefor any path given by $ds = 0$ is called a light-like path. A path where the overall sign of ds^2 is negative is called a space-like path. A path where the overall sign of ds^2 is positive is called a time-like path.

According to the first postulate of special relativity the laws of physics are the same for every inertial frame. Therefor when modeling the general laws of physics with equations we must use equations that do not change their basic form when transformed from one frame to another. For instance, if we use one coordinate system to write an equation like

$$F(ct,x,y,z) - G(ct,x,y,z) = 0,$$

Then in any other coordinate system it should also be

$$F'(ct',x',y',z') - G'(ct',x',y',z') = 0$$

And for example, it should not become

$$F'(ct',x',y',z') - G'(ct',x',y',z') = H(ct',x',y',z')$$

If such an equation does transforms like this then it is not one of the fundamental equations of physics.

We will define a tensor in terms of its transformation properties.

First recall the following differential form of the Lorentz transformation equations Eqn. 1.1.6

$$dct' = \gamma(dct - \beta dx)$$
$$dx' = \gamma(dx - \beta dct)$$
$$y' = y$$
$$z' = z$$

These can be written in matrix form as

$$\begin{pmatrix} dct' \\ dx' \\ dy' \\ dz' \end{pmatrix} = \begin{bmatrix} \gamma & -\gamma\beta & 0 & 0 \\ -\gamma\beta & \gamma & 0 & 0 \\ 0 & 0 & 1 & 0 \\ 0 & 0 & 0 & 1 \end{bmatrix} \begin{pmatrix} dct \\ dx \\ dy \\ dz \end{pmatrix}$$

From this we define the Lorentz transformation matrix

$$\Lambda = \begin{bmatrix} \gamma & -\gamma\beta & 0 & 0 \\ -\gamma\beta & \gamma & 0 & 0 \\ 0 & 0 & 1 & 0 \\ 0 & 0 & 0 & 1 \end{bmatrix}$$

And its inverse as

$$\Lambda^{-1} = \begin{bmatrix} \gamma & \gamma\beta & 0 & 0 \\ \gamma\beta & \gamma & 0 & 0 \\ 0 & 0 & 1 & 0 \\ 0 & 0 & 0 & 1 \end{bmatrix}$$

Now the Lorentz transformation equations can be written as matrix equations as

$$\overline{dx'} = \Lambda \overline{dx}$$

and

$$\overline{dx} = \Lambda^{-1} \overline{dx'}$$

In element notation, using the Einstein summation convention the transformation can be written

$$dx'^{\mu} = \Lambda^{\mu}{}_{\nu} dx^{\nu}$$

Note:

$$\Lambda^{\mu}{}_{\nu} = \partial x'^{\mu}/\partial x^{\nu}$$

and this transformation is just the ordinary chain rule of calculus.

For special relativity a tensor will be anything that Lorentz transforms. A contravariant tensor will be any quantity that transforms between frames according to

$$T' = \Lambda T$$

$$T'^{\mu} = \Lambda^{\mu}{}_{\nu} T^{\nu}$$

A covariant tensor will be any quantity that transforms between frames according to

$$T'_{\mu} = \Lambda_{\mu}{}^{\nu} T_{\nu}$$

There are also mixed tensors. For example

$$T'^{\mu}{}_{\nu} = \Lambda^{\mu}{}_{\sigma} \Lambda_{\nu}{}^{\rho} T^{\sigma}{}_{\rho}$$

From these transformation properties we can deduce that for an individual particle,

- A sum or difference of tensors is still a tensor.
- A product of tensors is still a tensor.
- A tensor multiplied or divided by an invariant is still a tensor.

Note: These rules apply only when the tensors involved describe that which is observed, not the state of the observer himself. So for example let $F_{\mu\nu}$ be a tensor describing something observed like say the electromagnetic field and U^{ν} is the four-vector velocity of the observer (c,0,0,0). It turns out that the electric field given by

$$E_{\mu} = F_{\mu 0} = F_{\mu\nu} U^{\nu}/c$$

is NOT a tensor. As U^{ν} is the four-vector velocity of whoever is the observer everyone uses (c,0,0,0) as a result and the expression does not transform as a four-vector. $E'_{\mu} = F'_{\mu 0} \neq (\partial x^{\lambda}/\partial x'^{\mu}) F_{\lambda 0}$. If U^{ν} were the four-vector velocity of one "particular" observer then the expression would transform as a tensor, but then it wouldn't represent the electric field to anyone except that observer and it would then only when $F_{\mu\nu}$ is the electromagnetic field already expressed according to his own frame. Likewise the magnetic field

$$B_{\mu} = -(1/2)\varepsilon_{\mu 0}{}^{\lambda\rho} F_{\lambda\rho}/c = -(1/2)\varepsilon_{\mu\nu}{}^{\lambda\rho} F_{\lambda\rho} U^{\nu}/c^2$$

where U^{ν} is the four-velocity of the observer (c,0,0,0) is also not a tensor.]

In relativity we write the fundamental equations of physics as tensor equations such as

$$T^{\mu\sigma\ldots}{}_{\nu\rho\ldots} = 0$$

because this doesn't change its form in a frame transformation. For instance, using the above transformation properties, it is easy to show that in any other frame this equation remains in the same form

$$T'^{\mu\sigma\ldots}{}_{\nu\rho\ldots} = 0$$

This is what satisfies the first postulate of special relativity, that the laws of physics are the same for all inertial frames.

Notice that since we model the laws of physics with tensor equations whose expressions are tensors defined by the transformation properties of the coordinates and since the Lorentz coordinate transformations are one to one invertable, there can be no true paradox's in special relativity.

SIMPLE RELATIVITY DYNAMIC IMPLICATIONS

In many texts mass has been defined in a circular manner. Some such texts have asserted a four-vector momentum in the form of $p^\mu = mU^\mu$ as a premise which doesn't work for massless particles and then defined mass as the contraction of that vector or visa-versa.

In order to avoid circularity and to include massless particles and also in order to facilitate a smoother transition to relativistic quantum mechanics this text will take a newer though not unique approach. For example, here there will be two different momentum four-vectors distinguished by capitalization p^μ, and P^μ. The lower case will be the momentum four-vector of the first kind and the upper case will be the momentum four-vector of the second kind. This is done in part because a particle's mass will be defined as the contraction of the momentum four-vector of the first kind which is the momentum four-vector referred to in classical relativistic (non-quantum relativistic) texts. The momentum four vector of the second kind is here defined mainly because its elements are what will correspond to quantum operators in relativistic quantum mechanics. (Some authors choose the capitals the other way around)

From experiments or due to quantum mechanics we know that the magnitude of the three component momentum of a particle can be related to a wavelength (whether or not the particle has mass).

$$p = \frac{h}{\lambda}$$

and the relation between the three element momentum and the three element k (whether or not the particle has mass) is

$$\vec{p} = \hbar\vec{k}$$

Also, a fourth element corresponding to a time coordinate can be related to a frequency (whether or not the particle has mass) and that element times c we will term relativistic energy E_R.

$$p^0 = \frac{\hbar\omega}{c}$$

$$E_R = p^0 c$$

where ω can be related to the wavelength by

$$\omega = 2\pi\frac{c}{\beta\lambda}$$

and from

$$\beta c = d\omega/dk:$$
$$\omega^2 = c^2k^2 + \text{constant}$$

The integration constant will turn out to be proportional to the square of the mass.

We will start with the premise that this four component definition of momentum constitutes a four-vector that will be called the momentum four-vector of the first kind p^μ.

Next consider the introduction of a four-vector potential ϕ^μ to which the test particle responds with a charge q. It does not matter at this point if this charge is electric, only that the vector potential to which it responds is a four-vector. We will define the momentum four-vector of the second kind by

$$P^\mu = p^\mu + (q/c)\phi^\mu$$

and we will call its time element P_0 total energy E

$$E = P_0c$$

As an artifact from physics texts that do not make this distinction, when the potential is not zero one might think of the total energy E as p^0/c even though it is really P_0/c which is $E_R + q\phi$ in special relativity. At the same time, one would think of the relativistic momentum as p^i. As a result one may think of energy E due to containing a potential as something that can have an arbitrary constant added to it, but would think of momentum as something that can not. This superficially seems to draw a distinction between time and space, as energy corresponds to a time element and momentum corresponds to spatial elements.

However here where the distinction between momentum four-vectors is made, one finds that there is no such distinction between time and space. This is because it is in P^μ that such an arbitrary constant can be added to the potential, and it is also in that four-vector that such arbitrary constants can be added to the spatial components of the vector potential. One can do this so long as one demands that they transform as the coordinates do, as a four-vector.

The special relativistic definition for the mass of a particle given those relations is

$$m^2c^2 = |\eta^{\mu\nu}[P_\mu - (q/c)\phi_\mu][P_\nu - (q/c)\phi_\nu]|$$

or

$$m = [(E_R/c^2)^2 - (p/c)^2]^{1/2} = E_0/c^2$$

This could just as well be expressed as

$$m^2c^2 = |\eta_{\mu\nu}p^\mu p^\nu|$$

In the second we define E_0 as the relativistic energy evaluated at zero velocity, $E_0 = E_R|_{v=0}$. Magnitude bars are included above merely so that the choice of sign convention for the metric's signature is arbitrary.

Though all equation are equivalent given the above relations, the definition in terms of the momentum four-vector of the second kind equation is preferable in quantum mechanics discussions because that is what yields relativistic quantum mechanics. For example, when one replaces the elements of the momentum four-vector of the second kind with the energy and momentum operators of quantum mechanics, and operates that on the wave function it yields the Klein-Gordon equation with the inclusion of a nonzero vector potential

$$\eta^{\mu\nu}[P_{op\mu} - (q/c)\phi_\mu]\,[P_{op\nu} - (q/c)\phi_\nu]\Psi = m^2c^2\Psi$$

which can also be written

$$[(H - \phi)^2 - (P_{op}c - q\phi)^2]\Psi = m^2c^4\Psi$$

where

$$H = i\hbar\frac{\partial}{\partial t}$$

and

$$\vec{P}_{op} = -i\hbar\vec{\nabla}$$

The above definition of mass, that a mass is *rest energy* $m = E_0/c^2$, $E_0 = E_R|_{v=0}$, or that its is *centre of momentum frame relativistic energy* $m = E_{cm}/c^2$, $E_{cm} = p^0{}_{cm}c|_{vcm=0}$, in the case of a system of particles, is the definition that we will use throughout the rest of the special relativity site where ever the letter m or the word mass is used unqualified. This is the m that goes into the relativistic version of Newton's second law in the form

$$F^\lambda = mA^\lambda$$

(four-force = mass times four-acceleration)

This mass is an invariant. It does not change with speed! Equations is called the mass-shell condition, because they are of isomorphic form to the equation of a spherical shell. Under the above definition of mass, a photon does not have mass. Due to quantum mechanical issues, virtual particles do not tend to have the expected value of energy for a given momentum. So sometimes it is said a particle lays off shell.

The coordinate velocity of a particle is simply given by

$$u^\mu = dx^\mu/dt$$

We write Four-Vector Velocity or Proper Velocity

$$U^\mu = dx^\mu/d\tau$$

where τ is called proper time, which is just time according to the frame of the particle at its location.

Through time dilation we can relate the two

$$U^\mu = \gamma u^\mu$$

Consider the following expression

$$\eta_{\mu\nu}m\ U^\mu mU^\nu$$

$$\eta_{\mu\nu}mU^\mu mU^\nu = [(mU^0)^2 - (mU^1)^2 - (mU^2)^2 - (mU^3)^2]$$

$$\eta_{\mu\nu}mU^\mu mU^\nu = m^2[(dct/d\tau)^2 - (dx/d\tau)^2 - (dy/d\tau)^2 - (dz/d\tau)^2]$$

$$\eta_{\mu\nu} mU^{\mu} mU^{\nu} = m^2(dt/d\tau)^2[c^2 - (dx/dt)^2 - (dy/dt)^2 - (dz/dt)^2]$$
$$\eta_{\mu\nu} mU^{\mu} mU^{\nu} = m^2\gamma^2c^2(1 - v^2/c^2)$$
$$\eta_{\mu\nu} mU^{\mu} mU^{\nu} = m^2c^2\gamma^2\gamma^{-2}$$
$$\eta_{\mu\nu} mU^{\mu} mU^{\nu} = m^2c^2$$

Next we refer to the definition of mass to arrive at

$$\eta_{\mu\nu} mU^{\mu} mU^{\nu} = \eta_{\mu\nu} p^{\mu} p^{\nu}$$

Final examination of this reveals the relation between four-vector velocity and the four-vector momentum of the first kind.

$$p^{\mu} = mU^{\mu}$$

We can then from equation discover the relation between four-momentum and coordinate velocity for massive particles

$$p^{\mu} = \gamma m u^{\mu}$$

or write the relation for particles that may or may not have mass

$$p^{\mu} = (E_R/c)(u^{\mu}/c)$$

The γ term is physically associated to the velocity term through time dilation. In the past a few physicists starting with Planck, Lewis, and Tolman, not Einstein, have miss-associated the γ term with the mass defining a new kind of mass

$$M = \gamma m \leftarrow \text{Bad}$$

This M is then inappropriately called "relativistic mass". In the absence of a potential, the zero^th element of the momentum four-vector is defined as the energy divided by c, resulting in

$$p^0 = Mu^0$$
$$E/c = Mc$$
$$M = E/c^2 \leftarrow \text{Bad}$$

Though much more complicated in the long run, the math is consistent and leads to consistent predictions concerning observation and so one might argue that the physics is therefor correct. But, in keeping with Occam's razor this definition and method must be done away. The m in this method is then inappropriately qualified and called the "rest mass".

It is wrong to do this for the following reason. Calling m the "rest mass" infers to the listener that m is not the mass according to other frames for which it is not at rest. We have already noted that m is an invariant as it is the same value as calculated according to any frame. It is not just the value for the rest frame. The relativistic mass method also leads to many erroneous conclusions. By that method light has zero "rest mass". For one of many examples, it has been argued that since light is not at rest in any frame, that the question of whether it has mass at rest or "rest mass" is unanswerable. No. m = 0 is observed as the contraction of a photon's four-momentum according to any frame, not just the "rest frame".

In short the terms "relativistic mass" and "rest mass" need to be done away and the real mass m which is actually observed is an invariant. It does not change with speed. Also, by this, the physically correct definition a photon, or anything that travels at the Lorentz invariant speed c, has zero mass.

We have

$$p^0 = E_R/c.$$

We have also demonstrated the relation between Four-Momentum and Four-Velocity equation resulting in

$$p^0 = mU^0.$$

Putting these together we have

$$E_R/c = mU^0$$
$$E_R/c = m(dct/d\tau)$$
$$E_R = (dt/d\tau)mc^2$$
$$E_R = \gamma mc^2 \leftarrow \text{Good}$$

This is the mass - relativistic energy relationship for a massive particle. Now this energy does not go to zero as v goes to zero so we see that a massive particle still has energy even when it is at rest. This tells us that mass is equivalent to rest energy meaning relativistic energy at zero velocity

$$E_0 = E_R|_{v=0} = mc^2 \leftarrow \text{Good}$$

The kinetic energy of a particle is the amount of energy that is associated with its motion only. Therefor

$$E_K = E_R - E_0$$

This results in

$$E_K = (\gamma - 1)mc^2$$

The stress energy tensor is a tensor that contains information about the density of energy, momentum, stresses, etc.. contained in the space. The energy tensor mass alone is Equation

$$T^{\mu\nu} = \rho_0 U^\mu U^\nu$$

The T^{00} component of this is

$$T^{00} = (dt/d\tau)^2 \rho_0 c^2$$

ρ_0 is the mass density according to a frame *moving with* that bit of mass, but because of special relativistic Lorenz length contraction on the local mass the coordinate frame mass density is then

$$\rho = (dt/d\tau)\rho_0$$

So this becomes

$$T^{00} = (dt/d\tau)\rho c^2$$

But this is just the coordinate frame energy density.

The simplest consistent general relativistic definition of coordinate frame energy density is then just

coordinate frame energy density $\equiv T^{00}$

For more general stress-energy tensors it is common to define ρ_0 as

$$\rho_0 = T^{\mu\nu} U_\mu U_{\nu/} c^4$$

If ρ_0 is to be positive then $T^{\mu\nu}U_\mu U_\nu$ must be greater than 0. For this not to be the case is called a violation of the weak energy condition. More generally speaking a the weak energy condition is

$$T^{\mu\nu}V_\mu V_\nu \geq 0$$

for any timelike vector V_μ. Matter may only violate this condition within limits set by the Pfenning inequality.

Other elements have other interpretations. For instance T^{ii} is a flow of momentum per area in the x^i direction *or* the pressure on a plane whose normal is in the x^i direction. T^{ij} is the x^i component of momentum per area in the x^j direction *or* describes a shearing from stresses. T^{0i} is the volume density of the ith component of momentum flow.

Next we will discuss the concept of system mass. We have seen that for a single particle mass is equivalent to rest energy. Equation

$$E_0 = mc^2.$$

For a system of particles the best concept for system mass m is defined as centre of momentum frame energy E_{cm}.

$$E_{cm} = mc^2.$$

The system mass does not turn out to be equal to the total or sum of masses m_{tot} of the constituent parts. Instead it is the total energy summed for all of the constituent parts according to the centre of momentum frame.

Consider for a moment a Lorenz invariant for the system consistent with the mass shell condition.

First define

$$p_{sys}' = [E_{cm}/c, 0, 0, 0]$$

as a four-element vector for the inertial centre of momentum frame. Then *define* P_{sys} as the Lorentz transform of this for any frame of interest.

$$p_{sys} = \Lambda p_{sys}'$$

Due to relative simultaneity p_{sys} as defined here is *not always* equal to the "simultaneous" sum of the four-momentum of the constituent parts when there are external forces acting at various locations on the system. The system mass is defined as the following invariant.

$$m^2c^2 = \eta_{\mu\nu} p_{sys}^{\ \mu} p_{sys}^{\ \nu}$$

or for the scenario described above,

$$m = [(E_{Rsys}/c^2)^2 - (p_{sys}/c)^2]^{1/2} = E_{cm}/c^2$$

Considering the time element of equation restores the relation

$$E_{Rsys} = \gamma_{cm} mc^2$$

Proof that this definition of system four momentum is the same as the sum of the four-momenta of the systems components goes as follows. Start with the sum of four-momentum for an arbitrary frame.

$$p_{sys} = \Sigma_i\, p_i$$

Lorentz boost to another frame

$$\Lambda p_{sys} = \Lambda \Sigma_i\, p_i$$

Interchange sum and transformation symbols

$$\Lambda p_{sys} = \Sigma_i \Lambda p_i$$

The Lorentz transform of each four-momentum is the four momentum according to the new frame

$$\Lambda p_{sys} = \Sigma_i p_i'$$

But the right side is the net four momentum according to the new frame

$$\Lambda p_{sys} = p_{sys}'$$

This proves that the system net four-momentum is indeed a four-vector itself and yields

$$E_{Rsys} = \gamma_{cm} mc^2$$

where m is the system's mass and is its centre of momentum frame energy as well as

$$p = \gamma_{cm} m u_{cm}$$

and

$$m^2c^2 = \eta_{\mu\nu} p_{sys}{}^{\mu} p_{sys}{}^{\nu} = E_{Rsys}{}^2/c^2 - p_{sys}{}^2 = E_{cm}{}^2/c^2$$

The reason that the mass of equation is not the same as the *"total"* of constituent masses, m_{tot}, is that the sum of masses of the constituent parts does not always equal the centre of momentum frame energy. For example, a system of massless particles have a zero mass shell condition when they all move the same direction while the system has a nonzero mass shell condition when they move in different directions.

One advantage the definition of centre of momentum frame energy for mass has over "total mass" is that by this definition, not only is mass an invariant, but this mass of a system is also conserved. Note that the concept of captive mass is equivalent to centre of momentum frame energy m and not the total of masses m_{tot}. In order to increase the system mass m one must increase the total centre of momentum frame energy E_{cm} equivalently. This demonstrates that mass defined by m is conserved in the same way that energy is. Transferring energy from some external matter to change the centre of momentum frame energy of an object will increase its individual system mass, but when you extend the system to include the matter from which the energy was transferred, it will always be found that centre of momentum frame energy or system mass m is ultimately conserved.

A sum of invariants is also an invariant and so one could just as well write the *total* of masses m_{tot} as a sum of the constituent parts. For a system of n particles this could be written·

$$m_{tot} = (\eta_{\mu\nu} p_1{}^{\mu} p_1{}^{\nu})^{1/2} + (\eta_{\mu\nu} p_2{}^{\mu} p_2{}^{\nu})^{1/2} + ... + (\eta_{\mu\nu} p_n{}^{\mu} p_n{}^{\nu})^{1/2}$$

or

$$m_{tot} = m_1 + m_2 + ... + m_n$$

The subscript indicates the particle number. Again, the major problem with thinking of a system mass as this is that this *total of masses* is not conserved. However, part of the reason this is brought up is that people do tend to think that mass is that kind of sum. This leads to another missunderstanding as to what is meant by "mass to kinetic energy conversion". Consider for example a massive particle that decays into two massless photons.

Because system energy is conserved and the system energy for the centre of momentum frame is the system mass, the system mass did not change in the decay. What did change was that the energy initially was associated with rest, the rest energy of the particle, but finally was associated with motion, the kinetic energies of the photons. In that light one should really not say there is mass-energy conversion.

The energy and system mass for the system is conserved. One should instead say that energy associated with the resting particles of the initial state becomes associated with the motion of the particles in the final state. Since this is cumbersome the term mass-energy conversion is used, but be wary that what it refers to is that a change in the sum of masses can be the cause of the change in kinetic energies of the remaining masses. Just remember that the system mass of a closed system doesn't change or "convert" into anything.

Sometimes it is more useful to define an invariant mass density instead. Just as there are two ways to describe the masses of a system above, m and m_{tot}, there are two important ways of describing an invariant mass density. The first is the ρ_0 in the following relation equation

$$T^{\mu\nu} = \rho_0 U^{\mu} U^{\nu}$$

This definition most closely corresponds to the mass m description of system mass above. It relates the stress energy tensor for matter composed of non-interacting constituents to the four-velocity of the unpressurized "fluid" at any given location.

The total energy for the system is conserved and could instead be defined by the following volume integral

$$E_{sys} = \int\int\int T^{00} dxdydz$$

For the example stress energy tensor this would become

$$E_{sys} = \int\int\int \gamma_{fluid}^2 \rho_0 c^2 \, dxdydz$$

One can also instead define the system momentum from the next integral

$$p_{sys} = \int\int\int (T^{0i} e_i/c) \, dxdydz$$

For the example stress energy tensor this would become

$$p_{sys} = \int\int\int (\gamma_{fluid}^2 \rho_0 u^i e_i) dxdydz$$

e_i is a unit vector in the direction of the ith momentum component.

One then still can define the system's mass as the centre of momentum frame energy. It is the energy for the frame according to which

$p_{sys} = 0$. So we still have

$$E_{cm} = mc^2.$$

The other invariant mass density concept corresponding to the total of masses m_{tot} would be

$$\rho_{Tot} = \eta_{\mu\nu} T^{\mu\nu}/c^2$$

This kind of mass density is an invariant, but its volume integral is not conserved. This kind of mass is what is meant when one refers to a field such as the electromagnetic field as a massless field, or when one refers to any system as massless. This is zero for any system of massless particles.

Of the two descriptions of system mass, the mass m concept is far more useful.

$$E_R = \gamma mc^2$$

where γ was given by

$$\gamma = (1 - v^2/c^2)^{-1/2}.$$

Notice that the energy becomes *divergent* at $v = c$ for nonzero mass. Thus no matter how hard or how long you push on a mass, you can never impart enough energy to it so that it reaches the speed c. The only way such a thing can travel at the speed c and still have finite energy is if it had zero mass. In that case, instead of a mass energy relation, there is a energy momentum relation from Equation resulting in,

$$E = E_R = pc,$$

where E and p are related to frequency and wavelength.

One might consider the case of a particle that instead of being pushed beyond the speed c, moves faster than c upon its creation. Such a hypothetical particle is called a tachyon. Notice that if v is greater than c then γ is imaginary. Since imaginary energy makes no physical sense, we would expect that the mass would also have to be imaginary so that the energy(and momentum) would be real.

The primary problem with the existence of such a particle is that it could be use to violate the *principle of causality*. The principle of causality is simply the statement that effect never precedes cause. Imagine setting up a tachyon emitter and a tachyon receiver at different points along an S frame x axis. Lets say that the signal travels arbitrarily fast so that the event of transmission and the event of reception are virtually simultaneous.

Next recall that events simultaneous in one frame are not all simultaneous in other frames. We could then easily pick a frame to look at the situation in which the event of reception *precedes* the event the event of transmission. This is a violation of causality.

Worse yet, it then leads to grandfather paradox's. The grandfather paradox is the idea that a time traveller goes back in time and kills his grandfather before his father was conceived. To set up a grandfather paradox with tachyons we simply set the receiver in motion *away* from the transmitter and give it a relay transmitter. We call the frame in which it is stationary the S' frame. We also connect a receiver to the S frame transmitter. We then programme the S frame transmitter so that in say 1hr it will send a signal *unless* it's receiver receives a signal.

To begin lets say that it receives no signal and so it sends one. The signal arrives at the relay, which is moving away and sets off the relay transmitter. Now the relay transmitter sends a return signal, but the return signal travels back to the S frame receiver/transmitter setup virtually instantaneously *according to the relay's S' frame*. Due to the Lack of simultaneity between the frames the return signal will be received back at the S frame transmitter at a time prior to the original transmission. But because we programmed it not to send a signal if it receives one it will now not send a signal. But then there is no signal to receive and so it sends one.

It is sometimes said that special relativity says that nothing can travel faster than the speed of light. As discussed, what it really implies is that long as we restrict our physics to special relativity, and we wish to preserve causality, information can not travel faster than c. Likewise, as long as we restrict our physics to special relativity, nothing with mass can travel at the speed c.

There have been experiments done in which the physicists involved say that they have indeed been able to get electromagnetic waves to propagate information faster than c through a dispersive medium.

In particular the controversy is over gain assisted faster than c group velocity transmission demonstrated in anomalous dispersive media.

If their claims that it was the "information" that has indeed been transferred at faster than c speeds are correct, then SR implies that we can find a frame according to which the reception of a signal at one end of the apparatus precedes the transmission of the signal at the other end. This would indeed be a causality violation and brings to question the validity of the principle of causality and causes us to reevaluate the (im)possibility of a physical grandfather paradox. However, it is conceivable that the universe may be structured in such a way that such a causality violation is attainable, but that grandfather paradoxes will still not be allowed. For example, if one of their dispersive media faster than c experiments were devised in attempt to simulate the relay-transmitter paradox discussed above, one could hypothetically set such a receiver in motion such a medium, but that medium is what determines the speed of the electromagnetic waves according to its rest frame. A relay transmitter sending the signal through the same medium would not end with the signal arriving at a time prior to transmission.

The following is an explanation why the experiments are not completely convincing of faster than c "information" transfer.

As an example, in a 6.0cm medium a laser pulse has been transmitted that transversed the distance at a speed of 310c. This is a group wave speed, not a phase wave speed. The below figures are a recreation representative of a receiver's data for two pulses. The blue dotted curve represents the intensity Vs time curve for the reception of the 310c speed pulse and the red curve represents the intensity Vs time curve for the arrival of a c speed pulse sent at the same time.

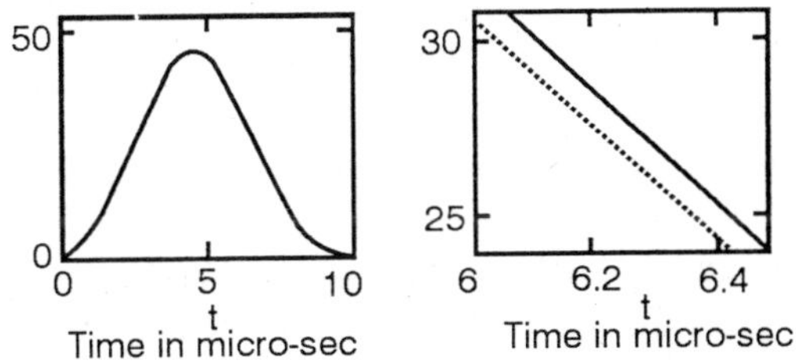

One can see on the close up second graph from the horizontal shift that the 310c curve arrived 62ns earlier than the c speed pulse. The question then arises whether this experiment is an example of a causality violation. In order for causality to be violated one must have faster than c "information" transfer. Under typical "long" transmissions one can consider the information transfer speed to be the group wave speed or the speed of the energy carried by the pulse. It has long been known that phase wave speeds often exceed c which is why it is often pointed out that the energy transmission in ordinary wave-guides occurs at the group speed which is less than c.

As such, the information transfer speed is less than c in ordinary wave guides. The reason that the group speed exceeding c for this experiment is not convincing of faster than c information transfer and the reason why the information transfer for this experiment can not be taken to be the group speed is because of the following.

Notice that the 62ns time shift between the two pulses is much less than the time it takes to receive the entirety of a pulse itself. Thus the time it took from the time the pulse began to enter the medium until the time the receiver read the entirety of the information was the sum of group transfer and read times. The full width at half-max FWHM of a pulse here is approximately 4.0μs. Take that to be read time. The group transfer time was 6.0cm/310c = 0.65ps. Taking the information transfer time to be the sum of these, approximately just 4.0μs, one finds that the information speed was 6.0cm/ 4.0μs = $5.0\text{x}10^{-5}$c, a mere measly fraction of the vacuum speed of light. There are two ways one might modify this experiment so that if successful it would clearly demonstrate faster than c information transfer.

First, one might made the medium of transfer much longer so that in the information transfer time it is the read time that is insignificant instead. The reason that this may be an impossible task is that due to the dispersive nature of the media itself, even with the gain assistance, there will be a trade off between the signal degradation and length. There may be a limiting trade off so that in a long enough transmission line so that the information transfer time yields a faster than c speed the signal would have been lost. Second, one might try to significantly narrow the pulse so that the information transfer time is approximately just the group transfer time.

The reason that this may be an impossible task two fold. There is a narrow frequency range at which the light must be sent through the medium in order for it to transfer with a faster than c group speed. This in itself puts a limit on how narrow the pulse may be. Also, the narrower the pulse is made the more rapidly it will tend to widen itself as it travels across the medium. By the time it gets to the other end the read time will always be longer than the send time so no matter how short the send time is made one will have to contend with a longer read time.

In conclusion, though such experiments do successfully demonstrate faster than c group transfer, they do not conclusively demonstrate the faster than c information transfer, which they would have to in order to show a causality violation.

THE SR DYNAMICAL EQUATIONS

In special relativity we define a four-vector force as

$$F^\lambda = dp^\lambda/d\tau$$

For a particle with mass we have

$$p^\lambda = mU^\lambda$$

The Acceleration Four-Vector for special relativity is given by

$$A^\lambda = dU^\lambda/d\tau$$

and so we can write the relativistic version of Newton's second law as

$$F^\lambda = mA^\lambda$$

Considering an inertial frame according to which the test mass is instantaneously at rest it is easy to show that

$$\eta_{\mu\nu}A^\mu U^\nu = 0$$

which yields

$$\eta_{\mu\nu}F^\mu U^\nu = 0$$

This serves as the work energy theorem for modern special relativity. Consider where $\eta_{\mu\nu}F^\mu U^\nu = 0$ leads:

$$\eta_{\mu\nu}F^\mu U^\nu = 0$$

$$\eta_{\mu\nu}(dp^\mu/d\tau)U^\nu = 0$$

$$\gamma^2\eta_{\mu\nu}(dp^\mu/dt)u^\nu = 0$$

$$\eta_{\mu\nu}(dp^\mu/dt)u^\nu = 0$$

$$(dp^0/dt)u^0 - (dp/dt)\cdot u = 0$$

$$u^0 = c \text{ and } p^0c = E_R$$

so

$$(dE_R/dt) - (dp/dt)\cdot u = 0$$

$$dE_R = (dp/dt)\cdot dx$$

$$dE_R = dE_k \text{ so}$$

$$\int dE_K = \int (dp/dt)\cdot dx$$
$$W = \int (dp/dt)\cdot dx$$

If we define another kind of force that is not a four-vector which we will call ordinary force as

$$f = dp/dt$$

then we arrive at

$$W = \int f\cdot dx$$

and there we see the work energy theorem of pre-modern relativistic physics.

So Newton's second law for special relativity in terms of ordinary force is

$$f^i = dp^i/dt$$

Using this force definition has its purposes, but in a lot of ways thinking of relativistic physics in terms of nontensor quantities very much complicates things. For example let us work out the relation between ordinary force and coordinate acceleration.
we can write as

$$f^i = m(d/dt)(dx^i/d\tau)$$

We then define α^λ by

$$\alpha^\lambda = (d/dt)(dx^\lambda/d\tau)$$

Which for acceleration in the direction of motion results in

$$\alpha = \gamma^3 a$$

and for that case of motion α will be equal to the proper acceleration A' which is the acceleration as observed from a frame according to which the particle is instantaneously at rest. The magnitude of this can be calculated from any frame as it is an invariant, $|A'| = (-\eta_{\mu\nu}A^\mu A^\nu)^{1/2}$. This is the amount of acceleration "felt" by the accelerated observer and according to the inertial frame in which the accelerated or "proper frame" observer is instantaneously at rest a = A'. We define α according equation as well because it is useful. α as calculated from any inertial frame according to which the force is in the direction of the motion turns out to be equal to the proper acceleration.)

This also restores a Newtonian form

$$f^i = m\alpha^i$$

(Note also - When the force is in the same direction as the motion, then the force felt by the object being pushed is equal to the ordinary force. In that case we have $F'^{\lambda}_{felt} = f^\lambda = m\alpha^\lambda$)
we can eliminate $d\tau$, in terms of dt from time dilation

$$f^i = m(d/dt)(\gamma dx^i/dt)$$

Use of the chain rule and simplification results in

$$f = \gamma m[a + \gamma^2 u(u\cdot a)/c^2]$$

where a^λ is the coordinate acceleration.

The four-vector force for special relativity is sometimes called the Minkowski force and is related to the electromagnetic field tensor $F^{\mu\nu}$

$$[F\mu\nu] = \begin{bmatrix} 0 & -E_x & -E_y & -E_z \\ E_x & 0 & cB_z & -cB_y \\ E_y & -cB_z & 0 & cB_x \\ E_z & cB_y & -cB_x & 0 \end{bmatrix}$$

by

$$F^\lambda = q\eta_{\mu\nu}(U^\mu/c)F^{\nu\lambda}$$

From this we can work out the relation between the components of the electromagnetic field, the coordinate velocity and the ordinary force, which yields

$$f = q(E + u\times B)$$

In the case that the force is in the direction of the motion yields,

$$f^i = \gamma^3 m a^i$$

Note that no matter what finite value the ordinary force is, as u approaches c, γ diverges and so the acceleration must vanish.

We expect this as nothing with mass can be pushed all the way up to the speed of light.

We have seen how c is a speed limit for the universe. Because of this, we must answer a question concerning velocity addition. Lets say an S' frame observer observes an object at speed u'. An S frame observer observes the S frame observer to be moving at speed v. u' can be any value less than c and v can be any value less than c. People tend to come to the wrong conclusion that the S frame observer observes that the object moves at a speed u = u' + v and that this speed should therefor be any speed less than 2c. They are using the wrong velocity addition formula. Consider the following Lorentz coordinate transformation equations in differential form.

$$dx = \gamma(dx' + \beta dct')$$

and

$$dct = \gamma(dct' + \beta dx')$$

To obtain the correct velocity addition equation divide equations and simplify.

$$dx/dct = [\gamma(dx' + \beta dct')]/[\gamma(dct' + \beta dx')]$$

simplified

$$dx/dt = [dx'/dt' + v]/[1 + (dx'/dt')v/c^2]$$

Now making the replacements u = dx/dt and u' = dx'dt' we arrive at

$$u = (u' + v)/(1 + u'v/c^2)$$

This is the correct equation to use for that velocity addition. u' and v can be any values less than c but the result will always be that the speed of the object according to the S frame, u, will always be less than c. One can also use the same method to find the velocity addition equations for the case that the object moves in the y or z directions.

ROTATIONS, ROCKETS, AND FREQUENCY SHIFTS

We have shown that velocities do not add linearly in special relativity. For motion along one direction velocities were adding nonlinearly according to Equation

$$u = (u' + v)/(1 + u'v/c^2)$$

Rapidity θ as a function of v is given by

$$\tanh\theta = v/c$$

This definition is useful as it simplifies much of dynamics equations. It does this because, unlike velocity, rapidity does add linearly.

$$\theta_u = \theta_{u'} + \theta_v$$

The Lorentz transformation matrix

$$\Lambda = \begin{bmatrix} \gamma & -\gamma\beta & 0 & 0 \\ -\gamma\beta & \gamma & 0 & 0 \\ 0 & 0 & 1 & 0 \\ 0 & 0 & 0 & 1 \end{bmatrix}$$

Rapidity also has the following relations to γ and $\gamma\beta$

$$\gamma = \cosh\theta$$

$$\gamma\beta = \sinh\theta$$

From these, the Lorentz transformation matrix becomes

$$\Lambda = \begin{bmatrix} \cosh\theta & -\sinh\theta & 0 & 0 \\ -\sinh\theta & \cosh & 0 & 0 \\ 0 & 0 & 1 & 0 \\ 0 & 0 & 0 & 1 \end{bmatrix}$$

Comparing this to an ordinary rotation matrix makes it clear why Lorentz transformations can be thought of as a rotation in space-time. At this point the relation between Lorentz transformation and rotation may still seem to be a superficial one, but once one becomes familiar with spinor calculus a much more intimate relation is revealed..

Here we will derive and discuss the implications of single stage relativistic rocket equations. The non-relativistic rocket equation is

$$\Delta v = v_{ex}\ln(m_i/m)$$

This gives the change in velocity Δv a rocket undergoes accelerating in one direction given a measure of exhaust speed v_{ex} which is a constant and the initial mass of the rocket m_i and the final mass of the rocket m after some of the ships mass in fuel has been burnt off.

The relativistic version of this equation in terms of rapidity θ is similar

$$\Delta\theta = (v_{ex}/c)\ln(m_i/m)$$

The speed of the rocket is then calculated from the rapidity Equation.

$$v = c\tanh\theta$$

Notice that since $\tanh\theta < 1$ for any θ, v is always less than c no matter how much of the ships mass is burnt off as fuel and no matter how fast the exhaust speed is. We can even consider tachyon exhaust where $v_{ex} > c$ and yet the rocket still never reaches the speed of light.

Start with conservation of momentum and energy relating the initial and final states of the rocket and exhaust for a small element m_{fex} burned off.

$$\gamma mv = (m + dm)[\gamma v + d(\gamma v)] + m_{fex}\gamma_{fex}v_{fex}$$
$$\gamma mc^2 = (m + dm)(\gamma + d\gamma)c^2 + m_{fex}\gamma_{fex}c^2$$

Simplified

$$0 = \gamma vdm + md(\gamma v) + m_{fex}\gamma_{fex}v_{fex}$$
$$0 = \gamma dm + md\gamma + m_{fex}\gamma_{fex}$$

Eliminate $m_{fex}\gamma_{fex}$

$$0 = \gamma vdm + md(\gamma v) - (\gamma dm + md\gamma)v_{fex}$$

Insert relativistic velocity addition

$$0 = \gamma vdm + md(\gamma v) - (\gamma dm + md\gamma)[(v - v_{ex})/(1 - vv_{ex}/c^2)]$$

Simplify

$$0 = [\gamma vdm + md(\gamma v)]\,(1 - vv_{ex}/c^2) - (\gamma dm + md\gamma)(v - v_{ex})$$

Switch variables to rapidity

$$0 = [\sinh\theta dm + md(\sinh\theta)]\,[1 - \tanh\theta(v_{ex}/c)]$$
$$c - [\cosh\theta(dm) + md\,(\cosh\theta)]\,(\tanh\theta - v_{ex}/c)c$$

Simplify

$$0 = [\sinh\theta dm + m\cosh\theta d\theta]\,[1 - \tanh\theta(v_{ex}/c)]$$
$$- [\cosh\theta dm + m\sinh\theta d\theta](\tanh\theta - v_{ex}/c)$$
$$0 = \sinh\theta dm + m\cosh\theta d\theta - (\sinh\theta dm + m\cosh\theta d\theta)\tanh\theta\,(v_{ex}/c)$$
$$- (\cosh\theta dm + m\sinh\theta d\theta)\tanh\theta + (\cosh\theta dm + m\sinh\theta d\theta)(v_{ex}/c)$$
$$0 = \sinh\theta dm + m\cosh\theta d\theta - (v_{ex}/c)\sinh\theta\tanh\theta dm - (v_{ex}/c)m\sinh\theta d\theta$$
$$- \sinh\theta dm - m\sinh\theta\tanh\theta d\theta + (v_{ex}/c)\cosh\theta dm + (v_{ex}/c)m\sinh\theta d\theta$$
$$0 = m\cosh\theta d\theta - (v_{ex}/c)\sinh\theta\tanh\theta dm - m\sinh\theta\tanh\theta d\theta + (v_{ex}/c)\cosh\theta\, dm$$
$$0 = (m\cosh\theta - m\sinh\theta\tanh\theta)d\theta + (v_{ex}/c)(\cosh\theta - \sinh\theta\tanh\theta)dm$$
$$0 = m(\cosh^2\theta - \sinh^2\theta)d\theta + (v_{ex}/c)(\cosh^2\theta - \sinh^2\theta)dm$$
$$0 = md\theta + (v_{ex}/c)dm$$
$$d\theta = -(v_{ex}/c)dm/m$$

After integration equation is obtained

$$\Delta\theta = (v_{ex}/c)\ln(m_i/m)$$

Now consider the ships proper acceleration for motion in one direction refer to equation:

$$\alpha = \gamma^3 a = \cosh^3\theta dv/dt = \cosh^3\theta(dv/d\theta)(d\theta/dm)(dm/dt')(dt'/dt)$$

$$\alpha = \cosh^3\theta(c\,\mathrm{sech}^2\theta)\ (-(v_{ex}/mc))(dm/dt')\mathrm{sech}\theta$$

$$\alpha = (v_{ex}/m)(dm/dt')$$

If the proper acceleration is kept constant then integration results in

$$\alpha\Delta t'/c = (v_{ex}/c)\ln(m_i/m) = \Delta\theta$$

Consider initial conditions of $v = 0$ at $t = t' = 0$.

$$\alpha t'/c = \theta$$

If the rocket starts at rest and is run at a constant proper acceleration α, then the equation can be written

$$v = c\tanh(\alpha t'/c)$$

These initial conditions also result in

$$v = c\tanh[(v_{ex}/c)\ln(m_i/m)]$$

equivalently

$$v = c\frac{\left(\frac{m_i}{m}\right)^{2\frac{v_{xy}}{c}} - 1}{\left(\frac{m_i}{m}\right)^{2\frac{v_{xy}}{c}} + 1}$$

Inverting these results in

$$m_i/m = \exp[(c/v_{ex})\tanh^{-1}(v/c)]$$

equivalently

$$m_i/m = [\gamma(1 + \beta)]^{c/vex}$$

Running it at a constant proper acceleration also results in

$$\beta = \tanh(\alpha t'/c)$$

$$\gamma = \cosh(\alpha ct'/c^2)$$

$$\gamma\beta = \sinh(\alpha ct'/c^2)$$

Using the Lorentz like transformation equations

$$ct = \int^{ct'}\gamma dct' + \gamma\beta x'$$

$$x = \gamma x' + \int^{ct'}\gamma\beta dct'$$

$$y = y'$$

$$z = z'$$

Results in a good global coordinate transformation from the accelerated ship frame to an inertial frame. Take the inertial frame to be the one in which it starts instantaneously at rest at $t' = 0$ and these become

$$ct = (c^2/\alpha + x')\sinh(\alpha ct'/c^2)$$

$$x = (c^2/\alpha + x')\cosh(\alpha ct'/c^2) - c^2/\alpha$$
$$y = y'$$
$$z = z'$$

There is a difference between what frequency one *observes* as being emitted from a source and what frequency an observer actually *sees* as coming from the source. This is true even in nonrelativistic physics. For instance, as a car drives past you will hear a shift in the tone of the engine as it goes from coming toward you to going away from you. This is the frequency you *hear*. You may use the ordinary Doppler shift formula with the speed it was traveling to then extrapolate what frequency it really emits according to your coordinate frame. This is the frequency you *observe*.

The relativistic Doppler shift formula is really the same thing as the ordinary Doppler shift formula except that it is usually written in terms of *the source frame's* emitted frequency instead of the observed emitted frequency. Its just that in the nonrelativistic case these are the same. In relativistic Doppler shift, you accounts for the fact that due to time dilation the frequency you observe to be emitted is different then the frequency according to the frame of the object.

If you are at rest with respect to the medium of propagation for a wave, then the ordinary Doppler shift formula is

$$f = f_0/[1 + (v/c)\cos\theta]$$

In terms of sound we make the following relations.

f is the frequency you *hear* (for instance if this was sound).

f_0 is the frequency you *observe* have been emitted according to your frame at the time of the emission. It is the transverse or

$\theta = \pi/2$ frequency for f.

v is the speed the emitter travels with respect to the medium of the waves at the time the wave was actually emitted.

c is the speed of the waves of the medium with respect to the medium.

θ is the angle it was traveling off of strait away from you at the time the heard frequency was actually emitted.

Relativistic Doppler shift IS ordinary Doppler shift. This formula happens to stand correct for the relativistic Doppler shift of light with the following adjustments to the relations.

f is the frequency you *see*.

f_0 is the frequency you *observe* to have been emitted according to your frame at the time of the emission. It is the transverse or

$\theta = \pi/2$ frequency for f.

v is the speed the emitter travels with respect your frame at the time the light was actually emitted.

c is the Lorentz invariant vacuum speed of light.

θ is the angle it was traveling off of strait away from you at the time the heard frequency was actually emitted according to your frame. The angle is different according to the other frame and so use of the other frames angle changes the form of the equation.

Now to write it in terms of the frequency emitted according to the frame of the object we start by writing the periods according to the two frames in te rms of time dilation.

$$T_0 = T_0'(1 - v^2/c^2)^{-1/2}$$

We then relate frequency to period.

$$f_0 = 1/T$$
$$f_0' = 1/T_0'$$

Putting these together results in

$$f_0 = f_0'\,(1 - v^2/c^2)^{1/2}$$

Inserting this into the Doppler shift formula results in

$$f = f_0'\,(1 - v^2/c^2)^{1/2}\,/[1 + (v/c)\cos\theta]$$

The wavelength of the light will be $\lambda = c/f$, or...

$$\lambda = \lambda_0'\,[1 + (v/c)\cos\theta]\,/(1 - v^2/c^2)^{1/2}$$

Next consider the case that the object travels strait toward the observer. $\theta = \pi$. Then after algebraic simplification these becomes

$$f = f_0'\,[(c + v)/(c - v)]^{1/2}$$
$$\lambda = \lambda_0'\,[(c - v)/(c + v)]^{1/2}$$

If the object traveled strait away from the observer $\theta = 0$, it would have become

$$f = f_0'\,[(c - v)/(c + v)]^{1/2}$$
$$\lambda = \lambda_0'\,[(c + v)/(c - v)]^{1/2}$$

STARTING GENERAL RELATIVITY

THE CONCEPTUAL PREMISES FOR GENERAL RELATIVITY

Lets say that there is a space-lab out in the depths of space sealed up so that there is no way for its crew to see anything outside of the lab. There are two experimentalists, Terrance and Stella, inside the space-lab. In this environment they are weightless and Terrance is still with respect to the ship walls. Stella is also initially still with respect to the lab walls, but she can maneuver around without touching the walls because she wears a rocket pack. They both also carry with them cesium watches that keep time accurate to within a millionth of a second and a computer that can read off such small time differences in their displays.

They then do the following experiment. They synchronize their watches to start and they start at the same location within the space-lab. Terrance stays there and Stella travels away and back to him along any number of paths so long as she arrives back when his watch says an hour has gone by to within its millionth of a second accuracy. The watch's times are then compared and a path is sought for which as much time as possible goes by on Stella's watch. Finally they experimentally discover what we knew from special relativity which is that the path that maximized her watches time was simply where she stayed put weightless next to Terrance and didn't go anywhere else. Every other path she took she underwent special relativistic time dilation while in motion with respect to Terrance.

Next we shift perspectives to a third party, Lois, who is for the moment moving in a state of constant velocity through the ship. According to Lois the path that Stella followed that maximized Stella's time between the events of the experiment's start and stop next to Terrance was a path of constant velocity. So we see that in special relativity the paths things tend to take which are paths of constant velocity are also the paths that maximize proper time intervals between events along the path.

Next they do another experiment. Lois releases two balls of different mass. They are both unacted on by forces in the ship so they just keep their same motion of constant velocity right along with Lois without deviating away from each-other.

Next we go to a fourth observer, Clark Kent, who is far out in the depths of space, but can see through the walls of the space-lab into the experiments. He also sees that their space-lab is falling toward a planet which they didn't realise because they were in free fall and couldn't see outside their lab. According to Clark the path of maximal proper time that Stella took between the events of the beginning and the ending of her experiment was not a path of a constant velocity state at all, but was the path of a body accelerating in the presence of a gravitational field. So we note that the path that things tend to follow in gravitational fields are still paths of maximal proper time even though they are not paths for a constant velocity state.

He also notices that the balls of the experiment though they have different masses, accelerate at the same rate.

Through this mind experiment we have discovered the core essence of general relativity.

The equivalence principle comes in different strengths.

The weak version of the equivalence principle boils down to the equivalence of gravitational and inertial mass. "Gravitational mass" and "inertial mass" are Newtonian concepts refering to variables that enter into equations for Newtonian physics. In Newtonian the gravitational force f from a point active gravitational mass M_1 acting on a point passive gravitational mass M_2 at a distance r comes from

$$f_r = -GM_1M_2/r^2$$

In Newtonian physics we also write the relation between the f_r acting on an inertial mass M_{2i} and a_r as

$$f_r = M_{2i}a_r$$

Putting these together we have

$$a_r = (-GM_1/r^2)(M_2/M_{2i})$$

We noted the balls of different masses fell at the same rate of acceleration according to Clark. In order for this acceleration to be independent of the ball mass as Clark saw that it was, with the correct choice for the value of G it becomes clear that the gravitational mass M_2 must be equivalent to inertial mass M_{2i}. Then we have

$$a_r = -GM_1/r^2$$

In general relativity we will have an invariant definition of mass. There will also be a four-vector force equation for general relativity in the form

$$F^\lambda = mA^\lambda$$

where m is the mass as invariant for general relativity.

Gravitation acting alone corresponds to $F^\lambda = 0$. This yields:

$$mA^\lambda = 0$$

The Acceleration four-vector for general relativity is a combonation of two parts, resulting in

$$mdU^\lambda/d\tau + m\Gamma^\lambda_{\mu\nu}U^\mu U^\nu = 0$$

The m in the term on the left corresponds to the "inertial mass" in Newtonian physics. The m in the term on the right corresponds to the passive "gravitational mass" in Newtonian physics. As these are really the same thing that was just multiplied through it is obvious that indeed the inertial and gravitational masses are identically equivalent.

The semi-strong level of the equivalence principle comes from the realization that the crew never knew that they were actually falling in a gravitational field. The experiments of a local free fall frame have results indistinguishable from the same experiments done in inertial frames. This is an equivalence of inertial and local free fall frames.

We could also extend this to the realization that if the lab had rocket engines burning, keeping them at a constant proper acceleration, they wouldn't have known the difference between being accelerated by the rocket engines or sitting on the surface of a planet in the presence of a gravitational field.

The strong level of the equivalence principle comes from the realization that any local free fall frames are equivalent for doing the physics. The laws of physics were the same for Lois as they were for Terrance. When the equivalence principle is mention unqualified it is usually this level of equivalence that is being referred to.

Above this strength we find the level of equivalence that is really required to result in the form of general relativity that we have today. This is sometimes called the general principle of relativity and sometimes the general principle of covariance. That is simply the statement that the general laws of physics are frame covariant. In other words the equation form that the laws of physics take are the same, invariant, according to every frame whether accelerated or not, whether in the presence of a gravitational field or not, whether rotating or not. To ensure this we must model the general laws of physics with tensor equations. The equations for the general laws of physics are then unchanged by transformations.

TENSORS IN GENERAL RELATIVITY

What defines a vector in any physics is its vector transformation properties. Not everything that merely has a magnitude and a direction is a vector, even in non-relativistic physics. For instance angular *displacement* is not really a vector because it doesn't always obey the vector property

$$A + B = B + A.$$

The vectors of relativity obey tensor transformation properties. In general, a four-vector is a rank one tensor. In element notation is has only one index, so it is a tensor with only four elements.

Some of the things we like to think of as individual properties of nature are incomplete as physical properties being only a component of a tensor. For instance, the electric field by itself does not obey tensor transformation properties. The magnetic field by itself also does not obey tensor transformation properties. In the context of this text a pseudovector will be anything that has multiple elements like a vector, but lacks any of the tensor transformation properties. These two pseudo-vectors can be combined into a *unified field* called the electromagnetic field tensor. Thus we see that the electric and magnetic fields are actually incomplete parts of the actual unified field called the electromagnetic field. This is a rank two tensor.

In the same sense, momentum by itself is not a complete physical quantity as it does not obey tensor transformation properties and so it is not really a vector in the relativistic sense. But, when we combine it with a fourth element, energy, we get a tensor called the *momentum four-vector*.

Likewise there are displacement four-vectors, velocity four-vectors, acceleration four-vectors, force four-vectors, etc...

According to a general principle of relativity the laws of physics are frame covariant. Therefor when modeling the general laws of physics with equations we must use expressions that are also frame covariant. For instance, if we use one coordinate system to write an equation like

$$F(ct,x,y,z) - G(ct,x,y,z) = 0,$$

Then in any other coordinate system it should also be

$$F'(ct', x',y',z') - G'(ct', x', y', z') = 0$$

It should not change its basic form. For example, it should not become

$$F'(ct', x', y', z') - G'(ct', x', y', z') = H'(ct', x', y', z')$$

If such an equation does transforms like this then it is not one of the fundamental equations of physics.

Here we will define a tensor in terms of its transformation properties. A contravariant tensor will be any quantity that transforms between frames according to

$$T'^{\mu} = (\partial x'^{\mu}/\partial x^{\nu})T^{\nu}$$

A covariant tensor will be any quantity that transforms between frames according to

$$T'_{\mu} = (\partial x^{\nu}/\partial x'^{\mu})T_{\nu}$$

There are also mixed tensors. For example

$$T'^{\mu}{}_{\nu} = (\partial x'^{\mu}/\partial x^{\sigma})(\partial x^{\rho}/\partial x'^{\nu})T^{\sigma}{}_{\rho}$$

From these transformation properties we can deduce that for an individual particle,

- A sum or difference of tensors is still a tensor.
- A product of tensors is still a tensor.
- A tensor multiplied or divided by an invariant is still a tensor.

Note: These rules apply only when the tensors involved describe that which is observed, not the state of the observer himself. So for example let $F_{\mu\nu}$ be a tensor describing something observed like say the electromagnetic field and U^{ν} is the four-vector velocity of the observer (c,0,0,0). It turns out that the electric field given by

$$E_{\mu} = F_{\mu 0} = F_{\mu\nu}U^{\nu}/c$$

is NOT a tensor. As U^{ν} is the four-vector velocity of whoever is the observer everyone uses (c,0,0,0) as a result and the expression does not transform as a four-vector. $E'_{\mu} = F'_{\mu 0} \neq (\partial x^{\lambda}/\partial x'^{\mu})F_{\lambda 0}$. If U^{ν} were the four-vector velocity of one "particular" observer then the expression would transform as a tensor, but then it wouldn't represent the electric field to anyone except that observer and it would then only when $F_{\mu\nu}$ is the electromagnetic field already expressed according to his own frame. Likewise the magnetic field

$$B_{\mu} = -(1/2)\varepsilon_{\mu 0}{}^{\lambda\rho}F_{\lambda\rho}/c = -(1/2)\varepsilon_{\mu\nu}{}^{\lambda\rho}F_{\lambda\rho}U^{\nu}/c^2$$

where U^{ν} is the four-velocity of the observer (c,0,0,0) is also not a tensor.]

In relativity we must write the fundamental equations of physics as tensor equations such as

$$T^{\mu\sigma\ldots}{}_{\nu\rho\ldots} = 0$$

because this remains frame covariant. For instance, using the above transformation properties, it is easy to show that in any other frame this equation remains in the same form

$$T'^{\mu\sigma\ldots}{}_{\nu\rho\ldots} = 0$$

THE METRIC AND INVARIANTS OF GENERAL RELATIVITY

The invariant interval can be expressed in the form equation

$$ds^2 = dct^2 - dx^2 - dy^2 - dz^2$$

Or in a more compact notation it can be written equation

$$ds^2 = \eta_{\mu\nu}dx^{\mu}dx^{\nu}$$

If we were to express this in a curvilinear coordinate system it will take on a form different from the top equation. For example do the following transformation to cylindrical coordinates

$$x = r\cos\theta$$

$$y = r\sin\theta$$

The invariant interval will then take the form

$$ds^2 = dct^2 - dr^2 - r^2 d\,\theta^2 - dz^2$$

Notice that in curvilinear coordinate systems functions of the coordinates may appear as coefficients of the differential quantities within the interval such as the $-r^2$ appears front of the $d\theta^2$ term above. Another possibility is the appearance of cross terms such as a dctdz term. To write this as a more compact and general form it is expressed

$$ds^2 = g_{\mu\nu}dx^{\mu}dx^{\nu}$$

When there is matter or fields of any type in the space it effects the form that $g_{\mu\nu}$ can take globally. So the popular interpretation for gravitation is simply that matter gives space-time an intrinsic curvature. In a situation where matter curves the space-time one can not globally transform $g_{\mu\nu}$ to $\eta_{\mu\nu}$. However one can always do the transformation locally.

We again express the invariant interval in the form

$$ds^2 = g_{\mu\nu}dx^{\mu}dx^{\nu}$$

Given that the interval is invariant we know that

$$g_{\mu\nu}dx^{\mu}dx^{\nu} = g'_{\lambda\rho}dx'^{\lambda}dx'^{\rho}$$

We also know that dx^{μ} transforms according to the calculus chain rule

$$dx'^{\mu} = (\partial x'^{\mu}/\partial x^{\nu})dx^{\nu}$$

This results in

$$g_{\mu\nu}dx^{\mu}dx^{\nu} = (\partial x'^{\lambda}/\partial x^{\mu})(\partial x'^{\rho}/\partial x^{\nu})g'_{\lambda\rho}dx^{\mu}dx^{\nu}$$

And therefor

$$g_{\mu\nu} = (\partial x'^{\lambda}/\partial x^{\mu})(\partial x'^{\rho}/\partial x^{\nu})g'_{\lambda\rho}$$

Now this is how a rank 2 covariant tensor transforms. Therefor if ds^2 is to be invariant then $g_{\mu\nu}$ is a rank 2 covariant tensor. This has been given the name "the metric tensor"

As we shall cover in the sections on gravitational pseudo forces the metric tensor is analogous to the gravitational potential for non-relativistic physics. In non-relativistic physics the gravitational force or other fields are often describable as the gradient of a potential. The gravitational pseudo forces will be related to affine connections which contain the metric tensor and its first order derivatives.

For special relativity we have

$$\eta_{\mu\nu'\nu} = 0$$

We can always transform to a local frame according to which the metric is $\eta_{\mu\nu}$ so we know so far that for a local frame also

$$\eta_{\mu\nu'\nu} = 0$$

Now consider the transformation to be to a local free fall frame so that the affine connections vanish. In that case we also have

$$\eta_{\mu\nu'\nu} = 0$$

Now transform this result to an arbitrary frame and we also find

$$g_{\mu\nu'\nu} = 0$$

(Summation still implied on all four above)

Next consider the quantity

$$g_{\mu\rho}g^{\rho\nu}$$

as arrived at for any point in spacetime by a transformation to an arbitrary set of Coordinates from a local Cartesian coordinate frame:

$$g_{\mu\rho}g^{\rho\nu} = (\partial x^{\alpha}/\partial x'^{\mu})(\partial x^{\beta'}/\partial x'^{\rho})\eta_{\alpha\beta}(\partial x'^{\rho}/\partial x^{\lambda})(\partial x'^{\nu}/\partial x^{\sigma})\eta^{\lambda\sigma}$$

Rearrange terms

$$g_{\mu\rho}g^{\rho\nu} = (\partial x^{\alpha}/\partial x'^{\mu})(\partial x^{\beta}/\partial x'^{\rho})(\partial x'^{\rho}/\partial x^{\lambda})(\partial x'^{\nu}/\partial x^{\sigma})\eta_{\alpha\beta}\eta^{\lambda\sigma}$$

Yielding

$$g_{\mu\rho}g^{\rho\nu} = (\partial x^{\alpha}/\partial x'^{\mu})\delta^{\beta}{}_{\lambda}(\partial x'^{\nu}/\partial x^{\sigma})\eta_{\alpha\beta}\eta^{\lambda\sigma}$$

Simplify

$$g_{\mu\rho}g^{\rho\nu} = (\partial x^{\alpha}/\partial x'^{\mu})(\partial x'^{\nu}/\partial x^{\sigma})\eta_{\alpha\beta}\eta^{\beta\sigma}$$

From the matrix equation for $\eta_{\mu\nu}$ it is easy to verify the next step

$$g_{\mu\rho}g^{\rho\nu} = (\partial x^{\alpha}/\partial x'^{\mu})(\partial x'^{\nu}/\partial x^{\sigma})\delta_{\alpha}{}^{\sigma}$$

Simplify

$$g_{\mu\rho}g^{\rho\nu} = (\partial x^{\alpha}/\partial x'^{\mu})(\partial x'^{\nu}/\partial x^{\alpha})$$

This yields

$$g_{\mu\rho}g^{\rho\nu} = \delta_{\mu}{}^{\nu}$$

Contract this and we have

$$g_{\mu\rho}g^{\rho\mu} = \delta_{\mu}{}^{\mu}$$

Which results in

$$g_{\mu\nu}g^{\mu\nu} = 4$$

The covariant metric tensor also acts as a lowering index operator and the contravariant metric tensor acts as a raising index operator. For example,

$$T_{\mu} = g_{\mu\nu}T^{\nu} \text{ and}$$
$$T^{\mu} = g^{\mu\nu}\,T_{\nu}$$

It is easy to verify this property based how contravariant and covariant tensors are defined by how they transform. For example consider the following expression,

$$(\partial x^{\lambda}/\partial x'^{\mu})\,(g_{\lambda\nu}T^{\nu})$$

based on how tensors transform this becomes

$$(\partial x^{\lambda}/\partial x'^{\mu})(\partial x'^{\alpha}/\partial x^{\lambda})(\partial x'^{\beta}/\partial x^{\nu})g'_{\alpha\beta}(\partial x^{\nu}/\partial x'^{\rho})T'^{\rho} = (\partial x^{\lambda}/\partial x'^{\mu})(g_{\lambda\nu}T^{\nu})$$

Rearranging:

$$(\partial x^{\lambda}/\partial x'^{\mu})(\partial x'^{\alpha}/\partial x^{\lambda})(\partial x'^{\beta}/\partial x^{\nu})(\partial x^{\nu}/\partial x'^{\rho})g'_{\alpha\beta}T'^{\rho} = (\partial x^{\lambda}/\partial x'^{\mu})(g_{\lambda\nu}T^{\nu})$$

Recognizing these result in delta Kroneckers and collecting the priming it becomes,

$$\delta_{\mu}{}^{\alpha}\delta^{\beta}{}_{\rho}(g_{\alpha\beta}T^{\rho})' = (\partial x^{\lambda}/\partial x'^{\mu})(g_{\lambda\nu}T^{\nu})$$

This simplifies to

$$(g_{\mu\rho}T^{\rho})' = (\partial x^{\lambda}/\partial x'^{\mu})(g_{\lambda\nu}T^{\nu})$$

But then we recognize that this is how a covariant tensor transforms and so we name T_{μ} by calling it,

$$T_{\mu} = g_{\mu\nu}T^{\nu}$$

Thus we've verified the lowering index property of the covariant metric tensor. Verifying the raising index property of the contravariant metric tensor is easier at this point. Start with the expression,

$$g^{\mu\nu}T_{\nu}$$

We've named our previous expression T_{ν} and so we insert it.

$$g^{\mu\nu}g_{\nu\rho}T^{\rho} = g^{\mu\nu}T_{\nu}$$

But we've already verified that $g^{\mu\nu}g_{\nu\rho} = \delta^{\mu}{}_{\rho}$ so we have

$$\delta^{\mu}{}_{\rho}T^{\rho} = g^{\mu\nu}T_{\nu}$$

Which results in

$$T^{\mu} = g^{\mu\nu}T_{\nu}$$

This verifies the raising of index property of the contravariant metric tensor.

With the exception of the locations of physical singularities, the space-time for the universe in which we live is an everywhere locally Lorentzian spacetime.

A locally Lorentzian spacetime is a spacetime for which we can locally transform $g_{\mu\nu}$ to $\eta_{\mu\nu}$ where $\eta_{\mu\nu}$ is given by Equation

$$[\eta_{\mu\nu}] = \begin{bmatrix} 1 & 0 & 0 & 0 \\ 0 & -1 & 0 & 0 \\ 0 & 0 & -1 & 0 \\ 0 & 0 & 0 & -1 \end{bmatrix}$$

A locally Euclidean Space-time is a spacetime for which we can locally transform $g_{\mu\nu}$ to $w_{\mu\nu}$ where $w_{\mu\nu}$ is given by

$$[w_{\mu\nu}] = \begin{bmatrix} -1 & 0 & 0 & 0 \\ 0 & -1 & 0 & 0 \\ 0 & 0 & -1 & 0 \\ 0 & 0 & 0 & -1 \end{bmatrix}$$

In other words all the dimensions of a Euclidean "spacetime" are spacelike.

Either type of spacetime can have Riemannian Curvature as these are only *locally* Euclidean, or Lorentzian.

Note: Sometimes it is said that our Universe is everywhere locally Euclidean. This basically means that we can do local transformations to arrive at .

$$[h_{ij}] = \begin{bmatrix} -1 & 0 & 0 \\ 0 & -1 & 0 \\ 0 & 0 & -1 \end{bmatrix}$$

This is correct, but to prevent confusion it is really more appropriate to say that our universe is everywhere locally Lorentzian.

Our universe is also described as being a globally Riemannian spacetime. This means that it globally takes the quadradic form of Equation.

$$ds^2 = g_{\mu\nu} dx^{\mu} dx^{\nu}$$

and is the same thing as saying it is everywhere locally Lorentzian.

An invariant as defined for this text is a quantity whose value does not depend on speed, location with respect to gravitational sources etc... nor upon whose frame it was calculated from. Invariants are said to be invariant to frame transformations, or frame invariant.

This does not imply that the value of an invariant must be the same everywhere (for example invariant "densities") nor that it must be conserved. In this context an invariant can be thought of as short for invariant scalar though there are tensor expressions such as the delta kronecker tensor whose elements are all frame invariant.

Some people also think of tensors in general as invariants as they represent physical entities and physical entities will not depend in any intrinsic way on our choice of frame. From this perspective the "elements of a tensor" are thought of as "projections of the tensor" onto a coordinate dependent template. The paradigm for this text will instead be that the tensor is the template onto which the projections have been made.

It is not invariant, but transforms according to the transformation properties of an infinitesimal displacement vector. Some relativity authors use the word scalar to be short for invariant scalars or what are just called invariants in this text. This is popular, but extremely inappropriate. The reason that it is inappropriate is that if people continue to redefine things without good reason so that they have a different meaning for whatever theory comes along then when they are used in general, eventually a student will practically have to learn a different dialect of the spoken language for every theory encountered. This is complication beyond reason. Here are a few examples of invariants

- c The local vacuum speed of light
- m Mass
- p The pressure scalar [$p = (1/3)(T^{\mu\nu}U_\mu U_\nu c^{-2} - g_{\mu\nu}T^{\mu\nu})$, for example the pressure of a gass]
- τ

The proper time between events along a world line.

- q Charge

An example of how one of these invariants might not be conserved would be to consider the pressure of the gas after a balloon is popped in space. As it expands the pressure decreases and so it is not conserved.

An example from special relativity of a quantity that is conserved, *but not invariant* would be the total energy of a particle E.

An example of a quality that is both invariant and conserved would be total charge q.

Consider the transformation of the full contraction of a tensor T^μ.

$$g'_{\mu\nu}T'^\mu T'^\nu = [(\partial x^\lambda/\partial x'^\mu)(\partial x^\rho/\partial x'^\nu)g_{\lambda\rho}][(\partial x'^\mu/\partial x^\alpha)T^\alpha][(\partial x'^\nu/\partial x^\beta)T^\beta]$$

$$g'_{\mu\nu}T'^\mu T'^\nu = (\partial x^\lambda/\partial x'^\mu)(\partial x'^\mu/\partial x^\alpha)(\partial x^\rho/\partial x'^\nu)(\partial x'^\nu/\partial x^\beta)g_{\lambda\rho}T^\alpha T^\beta$$

$$g'_{\mu\nu}T'^\mu T'^\nu = \delta^\lambda{}_\alpha \delta^\rho{}_\beta g_{\lambda\rho}T^\alpha T^\beta$$

$$g'_{\mu\nu}T'^\mu T'^\nu = g_{\lambda\rho}T^\lambda T^\rho$$

So we note that the full contraction of a tensor is an invariant.

THE AFFINE CONNECTIONS AND THE COVARIANT DERIVATIVE

We want to make equations for the general laws of physics out of tensor equations. So in developing a differentiation operator for general relativity we must assure that when it is operated on a tensor it results in something that is still a tensor.

We find that many of the special relativistic laws of physics are described by equations involving ordinary differentiation and so this operator must also reduce to the ordinary differentiation operator in local free fall frames. Consider the chain rule for the ordinary differentiation of a tensor.

$$dT^\lambda = (\partial T^\lambda/\partial x^\rho)dx^\rho$$

Using the transformation property of a contravariant tensor we find

$$dT^\lambda = \{(\partial/\partial x^\rho)[(\partial x^\lambda/\partial x'^\sigma)T'^\sigma]\}dx^\rho$$

Using the product rule we come to

$$dT^\lambda = (\partial^2 x^\lambda/\partial x^\rho \partial x'^\sigma)T'^\sigma dx^\rho + (\partial x^\lambda/\partial x'^\sigma)(\partial T'^\sigma/\partial x^\rho)dx^\rho$$

And again from the chain rule we finally have

$$dT^{\lambda} = (\partial^2 x^{\lambda}/\partial x^{\rho}\partial x'^{\sigma})T'^{\sigma}dx^{\rho} + (\partial x^{\lambda}/\partial x'^{\sigma})dT'^{\sigma}$$

Now if on the right hand side we only had the second term then the differentiation of a tensor would still transform as a tensor, but we have the extra first term so we know it does not. Thus to find a differentiation operator which maps tensors to tensors we introduce a second term in the operation. The new differential operator is called the covariant derivative opperator.

$$DT^{\lambda} = dT^{\lambda} + \delta T^{\lambda}$$

For a contravariant vector the second term necessary to keep DT^{λ} a tensor is

$$\delta T^{\lambda} = \Gamma^{\lambda}{}_{\mu\nu}T^{\mu}dx^{\nu}$$

where the affine connection(sometimes called the Christophel symbol of the second kind) $\Gamma^{\lambda}{}_{\mu\nu}$ is given by

$$\Gamma^{\lambda}{}_{\mu\nu} = (1/2)g^{\lambda\rho}(g_{\mu\rho'\nu} + g_{\nu\rho'\mu} - g_{\mu\nu'\rho})$$

For covariant four vectors we can write it in the same form

$$DT_{\lambda} = dT_{\lambda} + \delta T_{\lambda}$$

But here we have

$$\delta T_{\lambda} = -\Gamma^{\mu}{}_{\lambda\nu}T_{\mu}dx^{\nu}$$

In the case of the differentiation of a multiple mixed rank tensor we find

$$DT^{\lambda\ldots}{}_{\kappa\ldots} = dT^{\lambda\ldots}{}_{\kappa\ldots} + \Gamma^{\lambda}{}_{\mu\nu}T^{\mu\ldots}{}_{\kappa\ldots}dx^{\nu} + \ldots - \Gamma^{\rho}{}_{\kappa\nu}T^{\lambda\ldots}{}_{\rho\ldots}dx^{\nu} - \ldots$$

Also it is important to make note that though the affine connection is a part of a covariant derivative operator, it is not a tensor itself.

So, for example, the covariant derivative of a tensor T^{λ} with respect to some invariant parameter such as $d\tau$ is

$$DT^{\lambda}/d\tau = dT^{\lambda}/d\tau + \Gamma^{\lambda}{}_{\mu\nu}T^{\mu}(dx^{\nu}/d\tau)$$

As mentioned, a comma will represent a partial derivative and a semicolon will represent a partial covariant derivative. So for example

$$T^{\lambda}{}_{'\rho} = T^{\lambda}{}_{'\rho} + \Gamma^{\lambda}{}_{\mu\nu}T^{\mu}(\partial x^{\nu}/\partial x^{\rho})$$

$$T^{\lambda}{}_{'\rho} = T^{\lambda}{}_{'\rho} + \Gamma^{\lambda}{}_{\mu\rho}T^{\mu}$$

Chapter 3

Space, Time, and Newtonian Physics

The fundamental principle of relativity is the constancy of a quantity called c, which is the speed of light in a vacuum:

$$c = 2.998 \times 10^8 \ m/s, \text{ or roughly } 3 \times 10^8 \ m/s.$$

This is fast enough to go around the earth along the equator 7 times each second. This speed is the same as measured by "everybody." We'll talk much more about just who "everybody" is. But, yes, this principle does mean that, if your friend is flying by at 99% of the speed of light, then when you turn on a flashlight both of the following are true:

- The beam advances away from you at 3×10^8 m/s.
- Your friend finds that the light beam catches up to her, at 3×10^8 m/s.

'Newtonian' physics is the stuff embodied in the work of Isaac Newton. Now, there were a lot of developments in the 200 years between Newton and Einstein, but an important conceptual framework remained unchanged. It is this framework that we will refer to as Newtonian Physics and, in this sense, the term can be applied to all physics up until the development of Relativity by Einstein. Reviewing this framework will also give us an opportunity to discuss how people came to believe in such a strange thing as the constancy of the speed of light and why you should believe it too.

Many people feel that Newtonian Physics is just a precise formulation of their intuitive understanding of physics based on their life experiences. But still, the basic rules of Newtonian physics ought to 'make sense' in the sense of meshing.

COORDINATE SYSTEMS

We're going to be concerned with things like speed (e.g., speed of light), distance, and time. As a result, coordinate systems will be very important.

How many of you have worked with coordinate systems?

Let me remind you that a coordinate system is a way of labeling points; say, on a line. You need:

A zero

A positive direction

A scale of distance

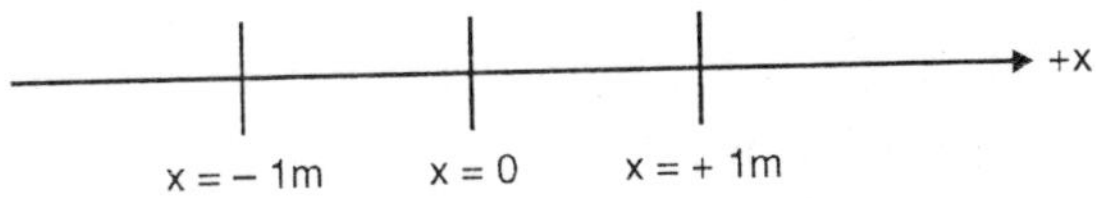

We're going to stick with one-dimensional motion most of the time. Of course, space is 3-dimensional, but 1 dimension is easier to draw and captures some of the most important properties.

In this course, we're interested in space and time:

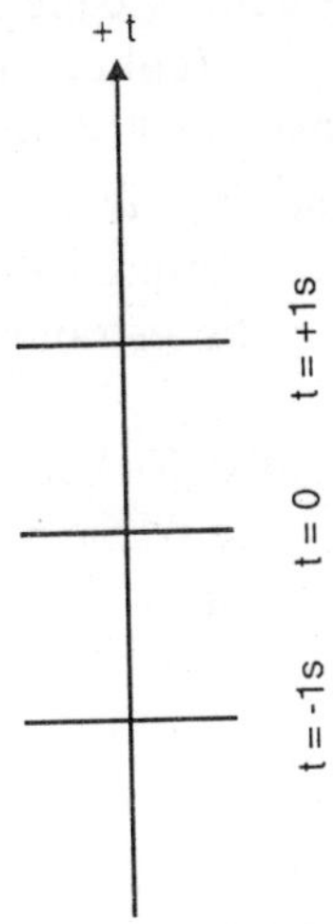

Put these together to get a 'spacetime diagram'

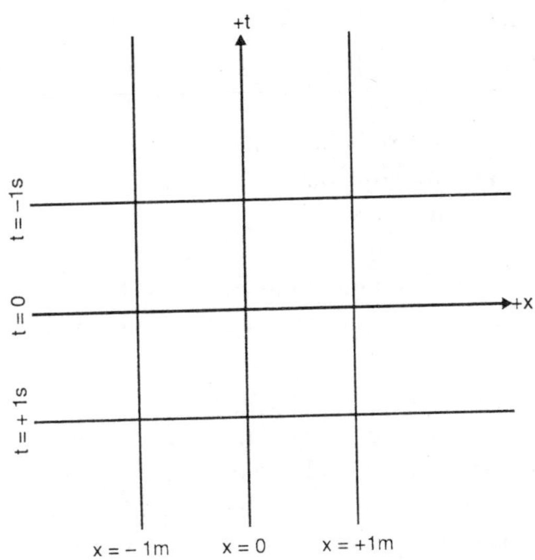

Note: For this class, t increases upward and x increases to the right. This is the standard convention in relativity and we adopt it so that this course is compatible with all books.

Also note:

- The x -axis is the line $t = 0$.
- The t-axis is the line $x = 0$.

REFERENCE FRAMES

A particular case of interest is when we choose the line $x = 0$ to be the position of some object: *e.g.* let $x = 0$ be the position of a piece of chalk.

In this case, the coord system is called a 'Reference Frame'; i.e., the reference frame of the chalk is the (collection of) coordinate systems where the chalk lies at $x = 0$ (All measurements are 'relative to' the chalk.)

We can talk about the chalk's reference frame whether it is "at rest," moving at constant velocity, or wiggling back and forth in a chaotic way. In both cases we draw the $x = 0$ line as a straight line in the object's own frame of reference.

Also: The reference frame of a clock has $t = 0$ whenever the clock reads zero. (If we talk about the reference frame of an object like a piece of chalk, which is not a clock, we will be sloppy about when $t = 0$.)

Note: A physicist's clock is really a sort of stopwatch. It reads $t = 0$ at some time and afterwards the reading increases all the time so that it moves toward $+ \infty$.

Before $t = 0$ it reads some negative time, and the distant past is $- \infty$. A physicist's clock does not cycle from 1 to 12.

Unfortunately, we're going to need a bit more terminology. Here are a couple of key definitions:

- *Your Worldline:* The line representing you on the spacetime diagram. In your reference frame, this is the line $x = 0$.
- *Event:* A point of spacetime; i.e., something with a definite position and time. Something drawn as a dot on a spacetime diagram. Examples: a firecracker going o, a door slamming, you leaving a house.

That definition was a simple thing, now let's think deeply about it. Given an event (say, the opening of a door), how do we know where to draw it on a spacetime diagram?

Suppose it happens in our 1- D world.

- How can we find out what time it really happens?: One way is to give someone a clock and somehow arrange for them be present at the event. They can tell you at what time it happened.
- How can we find out where (at what position) it really happens?: We could hold out a meter stick (or imagine holding one out). Our friend at the event in question can then read offhow far away she is.

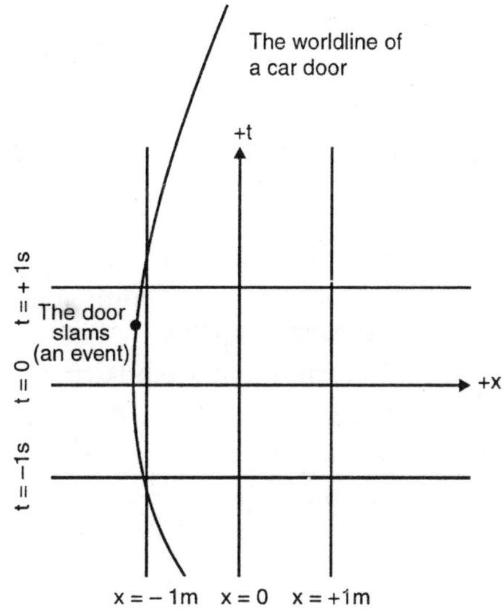

Note that what we have done here is to really define what we mean by the position and time of an event.

This type of definition, where we define something by telling how to measure it (or by stating what a thing does) is called an operational definition. They are very common in physics.

Now, the speed of light thing is really weird. So, we want to be very careful in our thinking. You see, something is going to go terribly wrong, and we want to be able to see exactly where it is.

Let's take a moment to think deeply about this and to act like mathematicians. When mathematicians define a quantity they always stop and ask two questions:

- Does this quantity actually exist? (Can we perform the above operations
  and find the position and time of an event?)
- Is this quantity unique or, equivalently, is the quantity "well-defined?" (Might there be some ambiguity in our definition? Is there a possibility that two people applying the above definitions could come up with two different positions or two different times?)

Well, it seems pretty clear that we can in fact perform these measurements, so the quantities exist. This is one reason why physicists like operational definitions so much.

Now, how well-defined are our definitions are for position and time? Z" One thing you might worry about is that clocks and measuring rods are not completely accurate.

Maybe there was some error that caused it to give the wrong reading. We will not concern ourselves with this problem. We will assume that there is some real notion of the time experienced by a clock and some real notion of the length of a rod. Furthermore, we will assume that we have at hand 'ideal' clocks and measuring rods which measure these accurately without mistakes. Our real clocks and rods are to be viewed as approximations to ideal clocks and rods.

Let's take the question of measuring the time. Can we give our friend just any old ideal clock? No. it is very important that her clock be synchronized with our clock so that the two clocks agree.

And what about the measurement of position? Well, let's take an example. Suppose that our friend waits five minutes after the event and then reads the position offof the meter stick. What if, for example, she is moving relative to us so that the distance between us is changing?

So, perhaps a better definition would be:

Time: If our friend has a clock synchronized with ours and is present at an event, then the time of that event in our reference frame is the reading of her clock at that event.

position: Suppose that we have a measuring rod and that, at the time that some event occurs, we are located at zero. Then if our friend is present at that event, the value she reads from the measuring rod at the time the event occurs is the location of the event in our reference frame.

But, how can we be sure that they are well-defined? There are no certain statements without rigorous mathematical proof. So, since we have agreed to think deeply about simple things (and to check all of the subtleties), let us try to prove these statements.

NEWTONIAN ASSUMPTIONS ABOUT SPACE AND TIME

Of course, there is also no such thing as a proof from nothing. This is the usual vicious cycle. Certainty requires a rigorous proof, but proofs proceed only from axioms (a.k.a. postulates or assumptions). So, where do we begin?

We could simply assume that the above definitions are well-defined, taking these as our axioms. However, it is useful to take even more basic statements as the fundamental assumptions and then prove that position and time in the above sense are well-defined. We take the fundamental Newtonian Assumptions about space and time to be:

T All (ideal) clocks measure the same time interval between any two events through which they pass.

S Given any two events at the same time, all (ideal) measuring rods measure the same distance between those events.

What do we mean by the phrase 'at the same time' used in *(S)* This, after all requires another definition, and we must also check that this concept is welldefined. The point is that the same clock will not be present at two different events which occur at the same time. So, we must allow ourselves to define two events as occurring at the same time if any two synchronized clocks pass through these events and, when they do so, the two clocks read the same value. To show that this is well-defined, we must prove that the definition of whether event A occurs 'at the same time' as event *B* does not depend on exactly which clocks (or which of our friends) pass through events.

Corollary to T: The time of an event (in some reference frame) is well-defined. Proof: A reference frame is defined by some one clock ±. The time of event A in that reference frame is defined as the reading at A on any clock β which passes through A and which has been synchronized with ±. Let us assume that these clocks were synchronized by bringing β together with ± at event *B* and setting β to agree with ± there. We now want to suppose that we have some other clock (γ) which was synchronized with ± at some other event *C*. We also want to suppose that γ is present at A. The question is, do β and γ read the same time at event *A*?

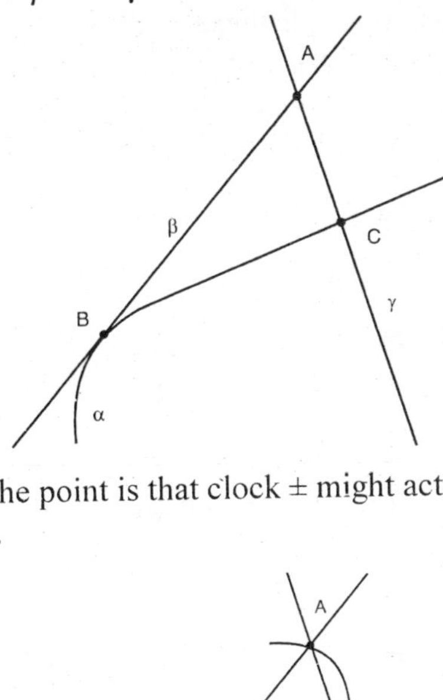

Yes, they will. The point is that clock ± might actually pass through ± as well as shown below.

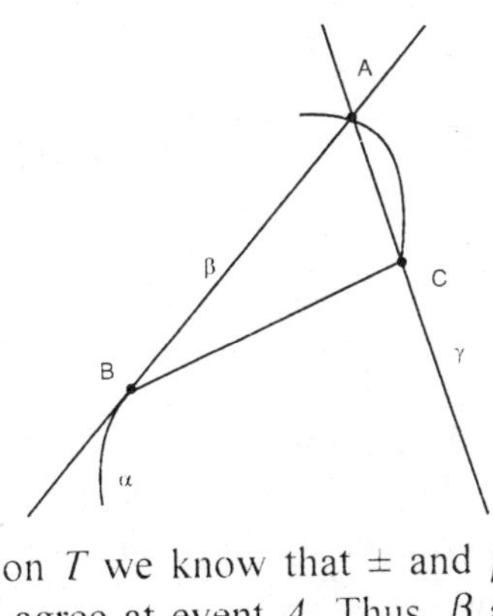

Now, by assumption *T* we know that ± and β will agree at event *A*. Similarly, ± and γ will agree at event *A*. Thus, β and γ must also agree at event *A*. Finally, a proof! We are beginning to make progress! Since the time of any event is well defined, the difference between the times of any two events is well defined. Thus, the statement that two events are 'at the same time in a given reference frame' is well-defined.

But, might two events be at the same time in one reference frame but not in other frames ?

Second Corollary to T: Any two reference frames measure the same time interval between a given pair of events.

Proof: A reference frame is defined by a set of synchronized clocks. From the first corollary, the time of an event defined with respect to a synchronized set of clocks is well-defined no matter how many clocks are in that synchronized set.

Thus, we are free to add more clocks to a synchronized set as we like. This will not change the times measured by that synchronized set in any way, but will help us to construct our proof.

So, consider any two events E_1 and E_2. Let us pick two clocks β_X and γ_X from set *X* that pass through these two events. Let us now pick two clocks β_Y and γ_Y from set *Y* that follow the same worldlines as β_X and γ_X. If such clocks are not already in set Y then we can add them in. Now, β_X and β_Y were synchronized with some original clock α_X from set *X* at some events *B* and C.

Let us also consider some clock α_Y from set Y having the same worldline as α_X. We have the following spacetime diagram:

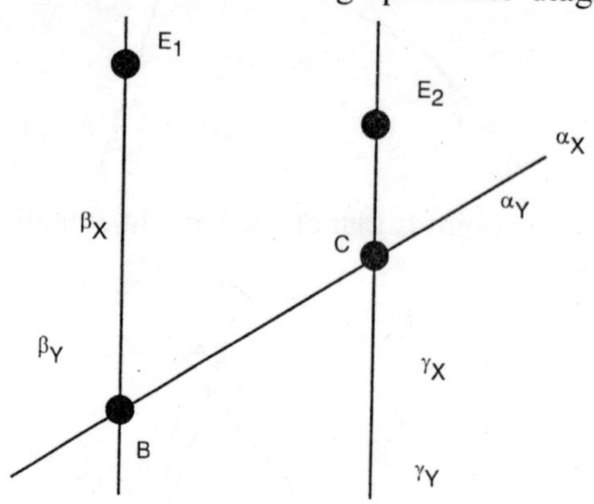

Note that, by assumption *T*, clocks $\pm X$ and $\pm_Y$ measure the same time interval between *B* and *C*. Thus, sets *X* and *Y* measure the same time interval between *B* and *C*.

Similarly, sets *X* and *Y* measure the same time intervals between *B* and E_1 and between *C* and E_2. Let $T_X(A, B)$ be the time difference between any two events A and *B* as determined by set *X*, and similarly for *TY* (*A*, *B*). Now since we have both $T_X(E_1, E_2) = T_X(E_1, B) + T_X(B, C) + T_X(C, E_2)$ and $T_Y(E_1, E_2) = T_Y(E_1, B) + T_Y(B, C) + T_Y(C, E_2)$, and since we have just said that all of the entries on the right hand side are the same for both X and Y, it follows that $T_X(E_1, E_2) = T_Y(E_1, E_2)$. In contrast, note that (*S)* basically states directly that position is well-defined.

NEWTONIAN ADDITION OF VELOCITIES?

Let's go back and look at this speed of light business. Remember the 99 %c example? Why was it confusing?

Let V_{BA} be the velocity of *B* as measured by A (i.e., "in A's frame of reference").

Similary V_{CB} is the velocity of C as measured by *B* and V_{CA} is the velocity of C as measured by A. What relationship would you guess between V_{BA}, V_{CB} and V_{CA}?

Most likely, your guess was:

$$V_{CA} = V_{CB} + V_{BA},$$

and this was the reason that the 99%c example didn't make sense to you. But do you know that this is the correct relationship?

The answer (still leaving the speed of light example clear as coal tar) is follows from assumptions *S* and T. Proof: Let A, *B,* C be clocks. For simplicity, suppose that all velocities are constant and that all three clocks pass through some one event and that they are synchronized there. The more general case where this does not occur will be one of your homework problems, so watch carefully!! Without Loss of Generality (WLOG) we can take this event to occur at $t = 0$.

The diagram below is drawn in the reference frame of A:

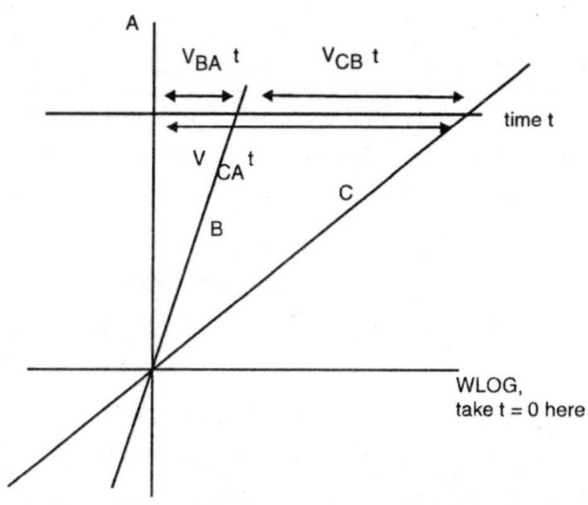

At time t, the separation between A and C is $V_{CA}t$, but we see from the diagram that it is is also $V_{CB}t + V_{BA}t$. Canceling the t's, we have

$$V_{CA} = V_{CB} + V_{BA}.$$

Now, our instructions about how to draw the diagram (from the facts that our ideas about time and position are well-defined) came from assumptions *T* and *S,* so the Newtonian formula for the addition of velocities is a logical consequence of *T* and *S*. If this formula does not hold, then at least one of *T* and *S* must be false. It is a good idea to start thinking now, based on the observations we have just made, about how completely any such evidence will make us restructure our notions of reality.

- *Q:* Where have we used T?
- *A:* In considering events at the same time (i.e., at time *t* on the diagram above).
- *Q:* Where have we used *S*?
- *A:* In implicitly assuming that d_{BC} is same as measured by anyone (A, *B*, or C).

NEWTON'S LAWS: ARE ALL REFERENCE FRAMES EQUAL?

The above analysis was true for all reference frames. It made no difference how the clock that defines the reference frame was moving.

However, one of the discoveries of Newtonian Physics was that not all reference frames are in fact equivalent. There is a special set of reference frames that are called Inertial Frames. This concept will be extremely important for us throughout the course.

Here's the idea: Before Einstein, physicists believed that the behaviour of almost everything (baseballs, ice skaters, rockets, planets, gyroscopes, bridges, arms, legs, cells,...) was governed by three rules called 'Newton's Laws of Motion.' The basic point was to relate the motion of objects to the 'forces' that act on that object. These laws picked out certain reference frames as special.

The first law has to do with what happens when there are no forces. Consider someone in the middle of a perfectly smooth, slippery ice rink. An isolated object in the middle of a slippery ice rink experiences zero force in the horizontal direction. Now, what will happen to such a person? What if they are moving?

Newton's first law of motion: There exists a class of reference frames (called inertial frames) in which an object moves in a straight line at constant speed (at time t) if and only if zero (net) force acts on that object at time t.

Note: When physicists speak about velocity this includes both the speed and the direction of motion. So, we can restate this as: There exists a class of reference frames (called inertial frames) in which the velocity of any object is constant (at time t) if and only if zero net force acts on that object at time t.

This is really an operational definition for an inertial frame. Any frame in which the above is true is called inertial.

The qualifier 'net' (in 'net force' above) means that there might be two or more forces acting on the object, but that they all counteract each other and cancel out. An object experiencing zero net force behaves identically to one experiencing no forces at all.

We can restate Newton's first law as: Object A moves at constant velocity in an inertial frame T! Object A experiences zero net force.

Here the symbol (T!) means 'is equivalent to the statement that.' Trust me, it is good to encapsulate this awkward statement in a single symbol.

AN OBJECT IS IN AN INERTIAL FRAME

Newton's first Law: There exists a class of reference frames (called inertial frames). If object A's frame is inertial, then object A will measure object B to have constant velocity (at time t) if and only if zero force acts on object B at time t.

To tell if you are in an inertial frame, think about watching a distant (very distant) rock floating in empty space. It seems like a safe bet that such a rock has zero force acting on it.

Examples: Which of these reference frames are inertial? An accelerating car? The earth? The moon? The sun? Note that some of these are 'more inertial' than others. Probably the most inertial object we can think of is a rock drifting somewhere far away in empty space.

It will be useful to have a few more results about inertial frames. To begin, note that an object never moves in its own frame of reference. Therefore, it moves in a straight line at constant (zero) speed in an inertial frame of reference (its own). Thus it follows from Newton's first law that, if an object's own frame of reference is inertial, zero net force acts on that object. Is the converse true? To find out, consider some inertial reference frame. Any object A experiencing zero net force has constant velocity v_A in that frame. Let us ask if the reference frame of A is also inertial.

To answer this question, consider another object C experiencing zero net force (say, our favourite pet rock). In our inertial frame, the velocity v_C of C is constant. Note that the velocity of C in the reference frame of A is just v_C "v_A, which is constant. Thus, C moves with constant velocity in the reference frame of A!!!! Since this is true for any object C experiencing zero force, A's reference frame is in fact inertial.

We now have: Object A is in an inertial frame T! Object A experiences zero force T! Object A moves at constant velocity in any other inertial frame. Note that therefore any two inertial frames differ by a constant velocity.

NEWTON'S OTHER LAWS

We will now complete our review of Newtonian physics by briefly discussing Newton's other laws. We'll start with the second and third laws. The second law deals with what happens to an object that does experience a new force. Definition of acceleration, a, (of some object in some reference frame): a = dv/dt, the rate of change of velocity with respect to time. Note that this includes any change in velocity, such as a change in direction. Z" In particular, an object that moves in a circle at a constant speed is in fact accelerating in the language of physics.

Newton's Second Law: In any inertial frame, (net force on an object) = (mass of object)(acceleration of object) F = ma.

The phrase "in any inertial frame" above means that the acceleration must be measured relative to an inertial frame of reference. By the way, part of your homework will be to show that calculating the acceleration of one object in any two inertial frames always yields identical results. Thus, we may speak about acceleration 'relative to the class of inertial frames.'

Note: We assume that force and mass are independent of the reference frame. On the other hand, Newton's third law addresses the relationship between two forces.

Newton's Third Law: Given two objects (A and *B*), we have (force from A on *B* at some time t) = – (force from *B* on A at some time t)

This means that the forces have the same size but act in opposite directions. Now, this is not yet the end of the story. There are also laws that tell us what the forces actually are. For example, Newton's Law of Universal Gravitation says: Given any two objects *A* and *B,* there is a gravitational force between them (pulling each toward the other) of magnitude

$$F_{AB} = \frac{m_A m_B}{d_{AB}^2}$$

with $G = 6.673 \times 10^{-11} Nm^2/kg^2$.

Important Observation: These laws hold in any inertial frame. There is no special inertial frame that is any different from the others. It makes no sense to talk about one inertial frame being more 'at rest' than any other. You could never find such a frame, so you could never construct an operational definition of 'most at rest.' Why then, would anyone bother to assume that a special 'most at rest frame' exists? As you will see in the reading, Newton discussed something called 'Absolute space.' However, he didn't need to and no one really believed in it. We will therefore skip this concept completely and deal with all inertial frames on an equal footing. The above observation leads to the following idea, which turns out to be much more fundamental than Newton's laws.

Principle of Relativity: The Laws of Physics are the same in all inertial frames. This understanding was an important development. It ended questions like 'why don't we fall offthe earth as it moves around the sun at 67,000 mph?'

Since the acceleration of the earth around the sun is only 0.006 m/s^2, the motion is close to inertial. This fact was realized by Galileo, quite awhile before Newton did his work (actually, Newton consciously built on Galileo's observations. As a result, applications of this idea to Newtonian physics are called 'Galilean Relativity').

MAXWELL, ELECTROMAGNETISM, AND ETHER

Newtonian physics (essentially, the physics of the 1700's) worked just fine. And so, all was well and good until a scientist named Maxwell came along. The hot topics in physics in the 1800's were electricity and magnetism. Everyone wanted to understand batteries, magnets, lightning, circuits, sparks, motors, and so forth (eventually to make power plants).

THE BASICS OF E & M

Let me boil all of this down to some simple basics. People had discovered that there were two particular kinds of forces (Electric and Magnetic) that acted only on special objects. (This was as opposed to say, gravity, which acted on all objects.) The special objects were said to be charged and each kind of charge (Electric or Magnetic) came in two 'flavors':

Electric: + and –

Magnetic: N and *S* (north and south)

Like charges repel and opposite charges attract.

There were many interesting discoveries during this period, such as the fact that 'magnetic charge' is really just electric charge in motion. As they grew to understand more and more, physicists found it useful to describe these phenomena not in terms of the forces themselves, but in terms of things called "fields". Here's the basic idea:

Instead of just saying that *X* and Y 'repel' or that there is a force between them, we break this down into steps:

- We say that *X* 'fills the space around it with an electric field *E*'
- Then, it is this electric field E that produces a force on *Y*.

(Electric force on *Y*) = (charge of *Y*)(Electric field at location of *Y*)

$$F_{on}\ Y = q_Y E$$

Note that changing the sign (±) of the charge changes the sign of the force. The result is that a positive charge experiences a force in the direction of the field, while the force on a negative charge is opposite to the direction of the field.

The arrows indicate the field. Red (positive charge) moves left with the field. Blue (negative charge) moves right against the field.

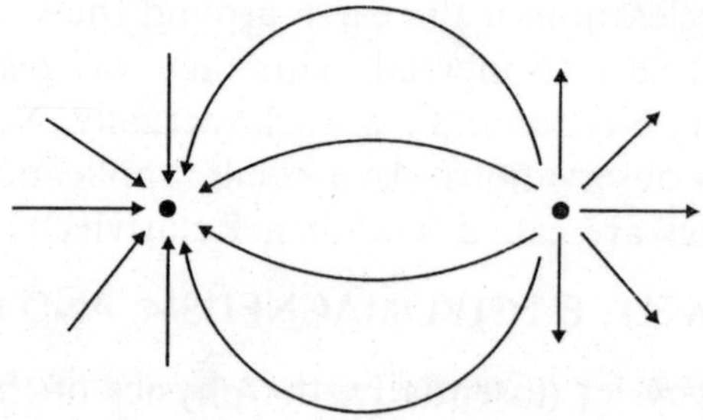

Fig. Blue = Negative Charge red = Positive Charge

Similarly, a magnetic charge fills the space around it with a magnetic field B that then exerts a force on other magnetic charges.

Now, you may think that fields have only made things more complicated, but in fact they are a very important concept as they allowed people to describe phenomena which are not directly related to charges and forces.

For example, the major discovery behind the creation of electric generators was Faraday's Law. This Law says that a magnetic field that changes in time produces an electric field. In a generator, rotating a magnet causes the magnetic field to be continually changing, generating an electric field. The electric field then pulls electrons and makes an electric current.

By the way: Consider a magnet in your (inertial) frame of reference. You, of course, find zero electric field. But, if a friend (also in an inertial frame) moves by at a constant speed, they see a magnetic field which 'moves' and therefore changes with time.

Thus, Maxwell says that they must see an electric field as well! We see that a field which is purely magnetic in one inertial frame can have an electric part in another. But recall: all inertial frames are supposed to yield equally valid descriptions of the physics.

Conversely, Maxwell discovered that an electric field which changes in time produces a magnetic field. Maxwell codified both this observation and Faraday's law in a set of equations known as, well, Maxwell's equations. Thus, a field that is purely electric in one reference frame will have a magnetic part in another frame of reference.

It is best not to think of electricity and magnetism as separate phenomena. Instead, we should think of them as forming a single "electromagnetic" field which is independent of the reference frame. It is the process of breaking this field into electric and magnetic parts which depends on the reference frame.

There is a strong analogy with the following example: The spatial relationship between the physics building and the Hall of Languages is fixed and independent of any coordinate system. The relationship is fixed, but the description differs. For the moment this is just a taste of an idea, but we will be talking much more about this in the weeks to come. In the case of electromagnetism, note that this is consistent with the discovery that magnetic charge is really moving electric charge.

Not only do we find a conceptual unity between electricity and magnetism, but we also find a dynamical loop. If we make the electric field change with time in the right way, it produces a magnetic field which changes with time. This magnetic field then produces an electric field which changes with time, which produces a magnetic field which changes with time..... and so on. Moreover, it turns out that a changing field (electric or magnetic) produces a field (magnetic or electric) not just where it started, but also in the neighboring regions of space.

This means that the disturbance spreads out as time passes! This phenomenon is called an electromagnetic wave. For the moment, we merely state an important property of electro-magnetic waves: they travel with a precise (finite) speed.

Maxwell's Equations and Electromagnetic Waves

Maxwell's equations lead to electromagnetic waves, the important point here is just to get the general picture of how Maxwell's equations determine that electromagnetic waves travel at a constant speed.

Maxwell's equations (Faraday's Law) says that a magnetic field (*B*) that changes is time produces an electric field (E). I'd like to discuss some of the mathematical form of this equation. To do so, we have to turn the ideas of the electric and magnetic fields into some kind of mathematical objects. Let's suppose that we are interested in a wave that travels in, say, the x direction.

Then we will be interested in the values of the electric and magnetic fields at different locations (different values of x) and a different times t. We will want to describe the electric field as a function of two variables $E(x, t)$ and similarly for the magnetic field $B(x, t)$.

Now, Faraday's law refers to magnetic fields that change with time. How fast a magnetic field changes with time is described by the derivative of the mag- netic field with respect to time. For those of you who have not worked with 'multivariable calculus,' taking a derivative of a function of two variables like $B(x, t)$ is no harder than taking a derivative of a function of one variable like $y(t)$. To take a derivative of $B(x, t)$ with respect to t, all you have to do is to momentarily forget that x is a variable and treat it like a constant. For example, suppose $B(x, t) = x^2 t + xt^2$. Then the derivative with respect to t would be just $x^2 + 2xt$. When B is a function of two variables, the derivative of B with respect to t is written $\frac{\partial B}{\partial t}$.

It turns out that Faraday's law does not relate $\frac{\partial B}{\partial t}$ directly to the electric field. Instead, it relates this quantity to the derivative of the electric field with respect to x. That is, it relates the time rate of change of the magnetic field to the way in which the electric field varies from one position to another. In symbols,

$$\frac{\partial B}{\partial t} = \frac{\partial E}{\partial x}.$$

It turns out that another of Maxwell's equations has a similar form, which relates the time rate of change of the electric field to the way that the magnetic field changes across space. Figuring this out was Maxwell's main contribution to science. This other equation has pretty much the same form as the one above, but it contains two 'constants of nature' – numbers that had been measured in various experiments. They are called μ_0 and μ_0 ('epsilon zero and mu zero'). The first one, $\in_0$ is related to the amount of electric field produced by a charge of a given strength when that charge is in a vacuum.

Similarly, μ_0 is related to the amount of magnetic field produced by a certain amount of electric current (moving charges) when that current is in a vacuum. The key point here is that both of the numbers are things that had been measured in the laboratory long before Maxwell or anybody else had ever thought of 'electromagnetic waves.

Their values were $\varepsilon_0 = 8.854 \times 10^{-12}\frac{C^2}{Nm^2}$ $\mu_0 = 4\pi \times 10^{-7}\frac{Ns^2}{C^2}$.

Anyway, this other Equation of Maxwell's looks like:

$$\frac{\partial E}{\partial t} = \in_0 \mu_0 \frac{\partial B}{\partial x}.$$

Now, to understand how the waves come out of all this, it is useful to take the derivative (on both sides) of equation with respect to time. This yields some second derivatives:

$$\frac{\partial^2 B}{\partial t^2} = \frac{\partial^2 E}{\partial x \partial t}$$

Note that on the right hand side we have taken one derivative with respect to t and one derivative with respect to x.

Similarly, we can take a derivative of equation on both sides with respect to x and get:

$$\frac{\partial^2 B}{\partial x^2} = \in_0 \mu_0 \frac{\partial^2 E}{\partial x \partial t}.$$

The interesting fact that it does not matter whether we first differentiate with respect to x or with respect to t: $\frac{\partial}{\partial t}\frac{\partial}{\partial x}E = \frac{\partial}{\partial x}\frac{\partial}{\partial t}E$.

Note that the right hand sides of equations and differ only by a factor of $\varepsilon_0\ \mu_0$. So,divide equation by this factor and then subtract it from to get $\frac{\partial^2 B}{\partial t^2} - \frac{1}{\epsilon_0\mu_0}\frac{\partial^2 B}{\partial x^2} = 0$

This is the standard form for a so-called 'wave equation.' To understand why, let's see what happens if we assume that the magnetic field takes the form

$$B = B_0 \sin(x - vt)$$

for some speed v. Note that equation has the shape of a sine wave at any time t. However, this sine wave moves as time passes. For example, at $t = 0$ the wave vanishes at $x = 0$. On the other hand, at time $t = \pi/2v$, at $x = 0$ we and other materials are associated with somewhat different values of electric and magnetic fields, and that depend on the materials. This is due to what are called 'polarization effects' within the material, where the presence of the charge (say, in water) distorts the equilibrium between the positive and negative charges that are already present in the water molecules. This is a fascinating topic (leading to levitating frogs and such) but is too much of a digression to discuss in detail here.

The subscript 0 on ?0 and 0 indicates that they are the vacuum values or, as physicists of the time put it, the values for 'free space.' have $B = -B_0$. A 'trough' that used to be at $x = -\pi/2$ has moved to $x = 0$. We can see that this wave travels to the right at constant speed v.

Taking a few derivatives shows that for B of this form we have

$$\frac{\partial^2 B}{\partial t^2} - \frac{1}{\epsilon_0\mu_0}\frac{\partial^2 B}{\partial x^2} = \left(v^2 - \frac{1}{\epsilon_0\ \mu_0}\right)B_0 \sin(x - vt)\ .$$

This will vanish (and therefore solve equation if (and only if) $v = \pm 1/\sqrt{\epsilon_0\ \mu_0}$. Thus, we see that Maxwell's equations do lead to waves, and that those waves travel at a certain speed4 given by $1/\sqrt{\epsilon_0\ \mu_0}$. Maxwell realized this, and was curious how fast this speed actually is. Plugging in the numbers that had been found by measuring electric and magnetic fields in the laboratory, he found (as you can check yourself using the numbers above!) $1/\sqrt{\epsilon_0\ \mu_0} = 2.99... \times 10^8$m/s. Now, the kicker is that, not too long before Maxwell, people had measured the speed at which light travels, and found that (in a vacuum) this speed was also 2.99... × 108m/s.

Maxwell didn't think so. Instead, he jumped to the quite reasonable conclusion that light actually was a a kind of electromagnetic wave, and that it consists of a magnetic field of the kind we have just been describing (together with the accompanying electric field). We can therefore replace the speed v above with the famous symbol c that we reserve for the speed of light in a vacuum.

THE ELUSIVE ETHER

The Laws of Physics are the same in all inertial frames. So, the laws of electromagnetism (Maxwell's equations) ought to hold in any inertial reference frame, right?

But then light would move at speed c in all reference frames, violating the law of addition of velocities... And this would imply that T and S are wrong! How did physicists react to this observation?

They said "Obviously, Maxwell's equations can only hold in a certain frame of reference."

Consider, for example, Maxwell's equations in water. There, they also predict a certain speed for the waves as determined by ε and μ in water (which are different from the ε_0 and μ_0 of the vacuum).

However, here there is an obvious candidate for a particular reference frame with respect to which this speed should be measured: the reference frame of the water itself. Moreover, experiments with moving water did in fact show that $1/\sqrt{\epsilon\,\mu}$ gave the speed of light through water only when the water was at rest.

The same thing, by the way, happens with regular surface waves on water (e.g., ocean waves, ripples on a pond, etc.). There is a wave equation not unlike which controls the speed of the waves with respect to the water.

So, clearly, c should be just the speed of light 'as measured in the reference frame of the vacuum.' Note that there is some tension here with the idea we discussed before that all inertial frames are fundamentally equivalent. If this is so, one would not expect empty space itself to pick out one as special. To reconcile this in their minds, physicists decided that 'empty space' should not really be completely empty.

After all, if it were completely empty, how could it support electromagnetic waves? So, they imagined that all of space was filled with a fluid-like substance called the "Luminiferous Ether." Furthermore, they supposed that electromagnetic waves were nothing other than wiggles of this fluid itself.

So, the thing to do was to next was to go out and look for the ether. In particular, they wanted to determine what was the ether's frame of reference. Was the earth moving through the ether? Was there an 'ether wind' blowing by the earth or by the sun? Did the earth or sun drag some of the ether with it as it moved through space?

The experiment that really got people's attention was done by Albert Michelson and Edward Morley in 1887. They were motivated by issues about the nature of light and the velocity of light, but especially by a particular phenomenon called the "aberration" of light. This was an important discovery in itself, so let us take a moment to understand it.

The Aberration of Light

Here is the idea: Consider a star very far from the earth. Suppose we look at this star through a telescope. Suppose that the star is "straight ahead" but the earth is moving sideways. Then, we will not in fact see the star as straight ahead.

Note that, because of the finite speed of light, if we point a long thin telescope straight at the star, the light will not make it all the way down the telescope but will instead hit the side because of the motion of the earth. A bit of light entering the telescope and moving straight down, will be smacked into by the rapidly approaching right wall of the telescope, even if it entered on the far left side of the opening.

The effect is the same as if the telescope was at rest and the light had been coming in at a slight angle so that the light moved a bit to the right. The only light that actually makes it to the bottom is light that is moving at an angle so that it runs away from the oncoming right wall as it moves down the telescope tube.

If we want light from the straight star in front of us to make it all the way down, we have to tilt the telescope. In other words, what we do see though the telescope is not the region of space straight in front of the telescope opening, but a bit of space slightly to the right.

Light Ray hits side instead of reaching bottom

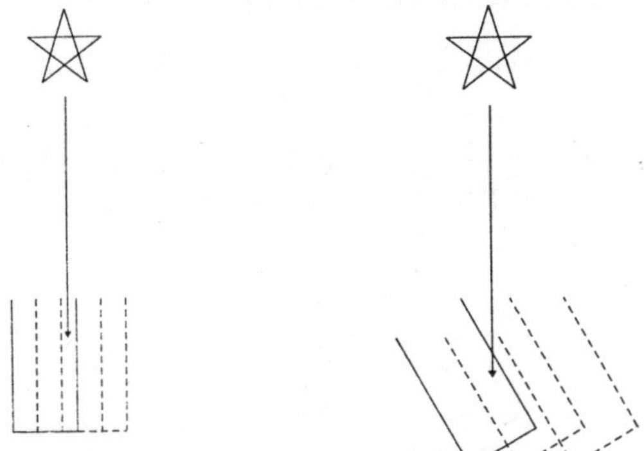

Fig. Telescope moves Through Ether Must Tilt Telescope to See Star

This phenomenon had been measured, using the fact that the earth first moves in one direction around the sun and then, six months later, it moves in the opposite direction. In fact, someone else (Fizeau) had measured the effect again using telescopes filled with water.

The light moves more slowly through water than through the air, so this should change the angle of aberration in a predictable way. While the details of the results were actually quite confusing, the fact that the effect occurred at all seemed to verify that the earth did move through the ether and, moreover, that the earth did not drag very much of the ether along with it.

You might wonder how Fizeau could reach such a conclusion. After all, as you can see from the diagram below, there is also and effect if the ether is dragged along by the earth.

In the region far from the earth where the ether is not being dragged, it still provides a 'current' that affects the path of the light. The point, however, is that the telescope on the Earth must now point at the place where the light ray enters the region of ether being dragged by the earth.

Note that this point does not depend on whether the telescope is filled with air or with water!

So, Fizeau's observation that filling the telescope with water increases the stellar aberration tells us that the ether is not strongly dragged along by the earth.

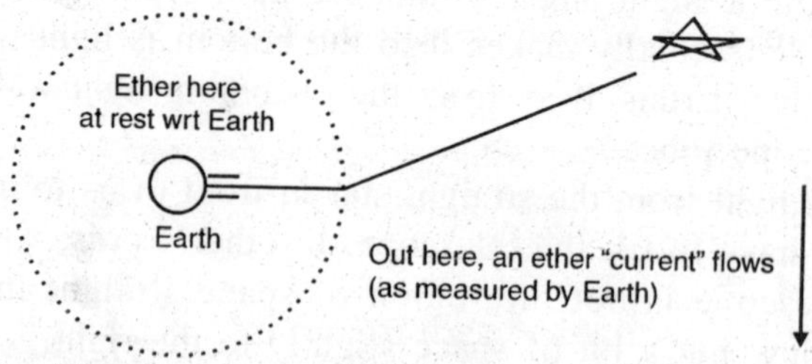

Michelson, Morely, and their Experiment

Because of the confusion surrounding the details of Fizeau's results, it seemed that the matter deserved further investigation.

Michelson and Morely thought that they might get a handle on things by measuring the velocity of the ether with respect to the earth in a different way. Have a look at their original paper to see what they did in their own words.

Michelson and Morely used a device called an interferometer, which looks like the picture below.

The idea is that they would shine light (an electromagnetic wave) down each arm of the interferometer where it would bounce off a mirror at the end and return.

The two beams are then recombined and viewed by the experimenters. Both arms are the same length, say L.

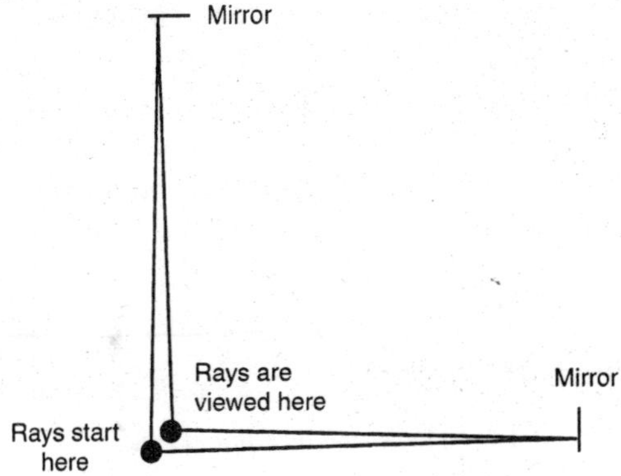

Fig. Light Rays Bounce off Both Mirrors

What do the experimenters see? Well, if the earth was at rest in the ether, the light would take the same amount of time to travel down each arm and return. Now, when the two beams left they were synchronized ("in phase"), meaning that wave crests and wave troughs start down each arm at the same time. Since each beam takes the same time to travel, this means that wave crests emerge at the same time from each arm and similarly with wave troughs. Waves add together as shown below8, with two crests combining to make a big crest, and two troughs combining to make a big trough. The result is therefore a a bright beam of light emerging from the device. This is what the experimenters should see.

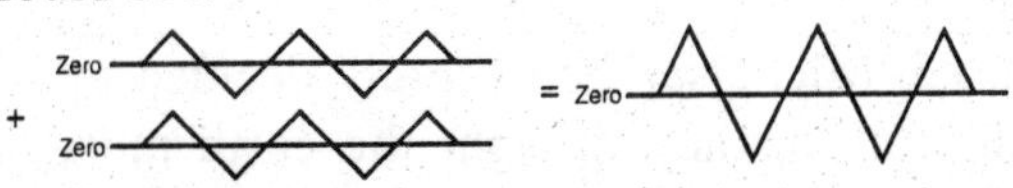

On the other hand, if the earth is moving through the ether (say, to the right), then the right mirror runs away from the light beam and it takes the light longer to go down the right arm than down the top arm. On the way back though, it takes less time to travel the right arm because of the opposite effect.*A* detailed calculation is required to see which effect is greater (and to properly take into account that the top beam actually moves at an angle as shown below).

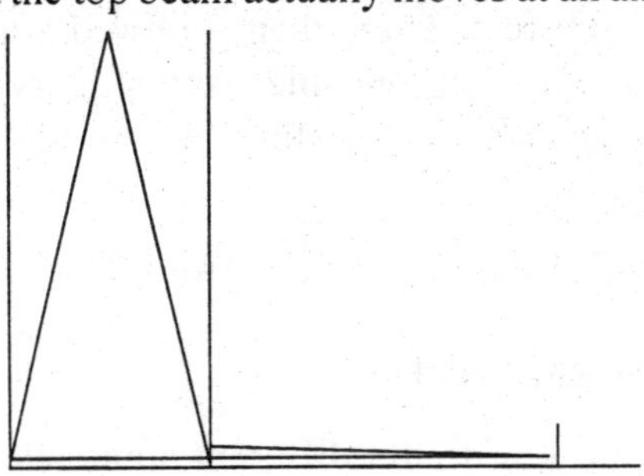

Fig. This one takes Less Time.

After doing this calculation one finds9 that the light beam in the right arm comes back faster than light beam in the top arm. The two signals would no longer be in phase, and the light would not be so bright. In fact, if the difference were great enough that a crest came back in one arm when a trough came back in the other, then the waves would cancel out completely and they would see nothing at all! Michelson and Morely planned to use this effect to measure the speed of the earth with respect to the ether.

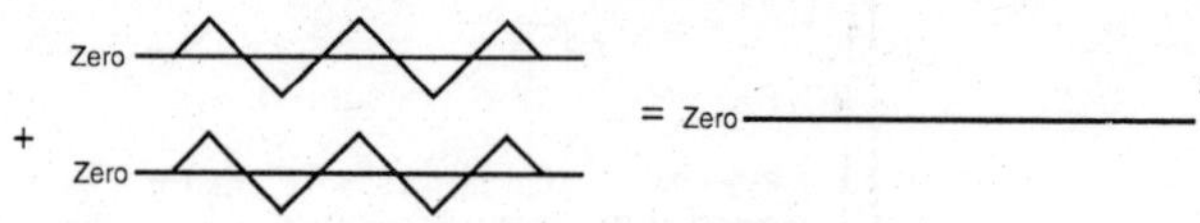

Fig. These waves Cancel out

However, they saw no effect whatsoever! No matter which direction they pointed their device, the light seemed to take the same time to travel down each arm. Clearly, they thought, the earth just happens to be moving with the ether right now (i.e., bad timing).

So, they waited six months until the earth was moving around the sun in the opposite direction, expecting a relative velocity between earth and ether equal to twice the speed of the earth around the sun. However, they still found that the light took the same amount of time to travel down both arms of the interferometer!

So, what did they conclude? They thought that maybe the ether is dragged along by the earth... But then, how would we explain the stellar aberration effects?

Deeply confused, Michelson and Morely decided to gather more data. Despite stellar aberration, they thought the earth must drag some ether along with it. After all, as we mentioned, the details of the aberration experiments were a little weird, so maybe the conclusion that the earth did not drag the ether was not really justified.....

If the earth did drag the ether along, they thought there might be less of this effect up high, like on a mountain top. So, they repeated their experiment at the top of a mountain.

Still, they found no effect. There then followed a long search trying to find the ether, but no luck. Some people were still trying to find an ether 'dragged along very efficiently by everything' in the 1920's and 1930's. They never had any luck.

EINSTEIN AND INERTIAL FRAMES

THE POSTULATES OF RELATIVITY

In 1905, Albert Einstein tried a different approach. He asked "What if there is no ether?" What if the speed of light in a vacuum really is the same in every inertial reference frame? He soon realized, as we have done, that this means that we must abandon *T* and *S*, our Newtonian assumptions about space and time.

Hopefully, you are sufficiently confused by the Michelson and Morely and stellar aberration results that you will agree to play along with Einstein for awhile. This is what we want to do in the next few sections. We will explore the consequences of Einstein's idea.

Surprisingly, one can use this idea to build a consistent picture of what is going on that explains both the Michelson-Morely and stellar aberration.

It turns out that this idea makes a number of other weird and ridiculous-sounding predictions as well. Perhaps even more surprisingly, these predictions have actually been confirmed by countless experiments over the last 100 years.

We are about to embark on a very strange path, one that runs counter to the intuition that we accumulate in our daily lives. We will have to tread carefully, taking the greatest care with our logical reasoning. Careful logical reasoning can only proceed from clearly stated assumptions (a.k.a. 'axioms' or 'postulates'). We're throwing out almost everything that we thought we understood about space and time. So then, what should we keep?

We'll keep the bare minimum consistent with Einstein's idea. We will take our postulates to be:

- The laws of physics are the same in every inertial frame.
- The speed of light in an inertial frame is always $c = 2.99.. \times 10^8$m/s.

We also keep Newton's first law, which is just the definition of an inertial frame: There exists a class of reference frames (called inertial frames) in which an object moves in a straight line at constant speed if and only if zero net force acts on that object.

Finally, we will need a few properties of inertial frames. We therefore postulate the following familiar statement.

Object A is in an inertial frame $\Leftrightarrow$ Object A experiences zero force $\Leftrightarrow$ Object A moves at constant velocity in any other inertial frame.

Since we no longer have *S* and T, we can no longer derive this last statement. It turns out that this statement does in fact follow from even more elementary (albeit technical) assumptions that we could introduce and use to derive it. This is essentially what Einstein did. However, in practice it is easiest just to assume that the result is true and go from there.

Finally, it will be convenient to introduce a new term:

Definition: An "observer" is a person or apparatus that makes measurements.

Using this term, assumption II becomes: The speed of light is always $c = 2.9979 \times 10^8$m/s as measured by any inertial observer.

By the way, it will be convenient to be a little sloppy in our language and to say that two observers with zero relative velocity are in the same reference frame, even if they are separated in space.

TIME AND POSITION, TAKE II

We used the old assumptions T and S to show that our previous notions of time and position were well-defined. Thus, we can no longer rely even on the definitions of 'time and position of some event in some reference frame'. We will need new definitions based on our new postulates.

For the moment, let us stick to inertial reference frames. What tools can we use? We don't have much to work with.

The only assumption that deals with time or space at all is postulate II, which sets the speed of light. Thus, we're going to somehow base out definitions on the speed of light.

We will use the following: To define position in a given inertial frame: Build a framework of measuring rods and make sure that the zero mark always stays with the object that defines the (inertial) reference frame. Note that, once we set it up, this framework will move with the inertial observer without us having to apply any forces.

The measuring rods will move with the reference frame. An observer (say,*a* friend of ours who rides with the framework) at an event can read o the position (in this reference frame) of the event from the mark on the rod that passes through that event.

To define time in a given inertial frame: Put an ideal clock at each mark on the framework of measuring rods above. Keep the clocks there, moving with the reference frame. The clocks can be synchronized with a pulse of light emitted, for example, from $t = 0$. A clock at x knows that, when it receives the pulse, it should read $| x |/c$.

These notions are manifestly well-defined. We do not need to make the same kind of checks as before as to whether replacing one clock with another would lead to the same time measurements. This is because the rules just given do not in fact allow us to use any other clocks, but only the particular set of clocks which are bolted to our framework of measuring rods.

Whether other clocks yield the same values is still an interesting question, but not one that a ects whether the above notions of time and position of some event in a given reference frame are well defined.

Significantly, we have used a different method here to synchronize clocks. The new method based on a pulse of light is available now that we have assumption II, which guarantees that it is an accurate way to synchronize clocks in an inertial frame. This synchronization process is shown in the spacetime diagram below.

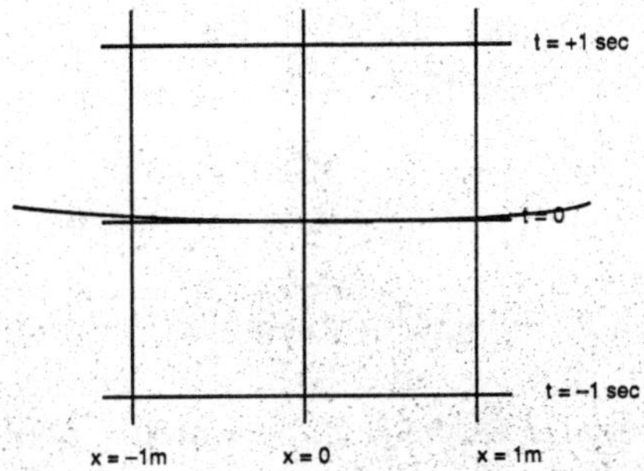

Note that the diagram is really hard to read if we use meters and seconds as

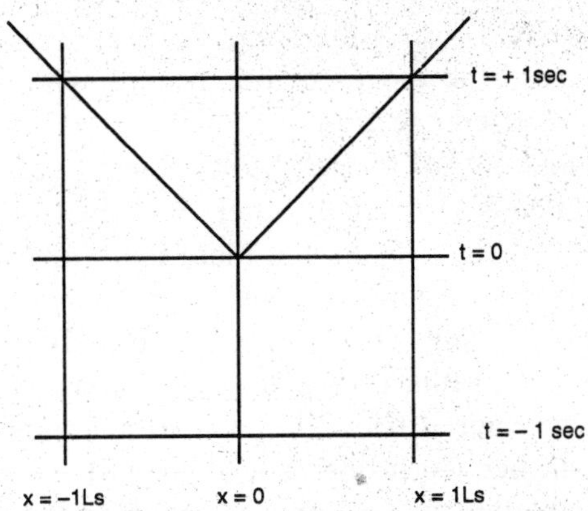

Therefore, it is convenient to use units of seconds and light-seconds: 1Ls = (1 sec)c = 3x × 10^8m = 3 × 10^5km. This is the distance that light can travel in one second, roughly 7 times around the equator. Working in such units is often called "choosing c = 1," since light travels at 1Ls/sec. We will make this choice for the rest of the course, so that light rays will always appear on our diagrams as lines at a 45? angle with respect to the vertical; i.e. slope = 1.

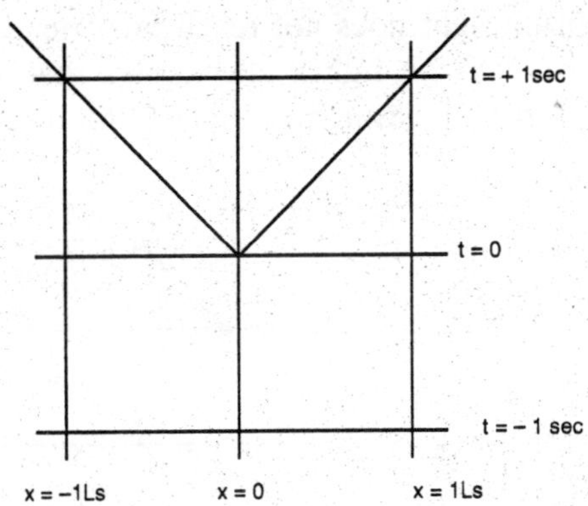

SIMULTANEITY: OUR FIRST DEPARTURE FROM GALILEO AND NEWTON

The above rules allow us to construct spacetime diagrams in various reference frames. An interesting question then becomes just how these diagrams are related.

Let us start with an important example. We went to some trouble to show that the notion that two events happen 'at the same time' does not depend on which reference frame (i.e., on which synchronized set of clocks) we used to measure these times. Now that we have thrown out *T* and *S*, will this statement still be true?

Let us try to find an operational definition of whether two events occur 'simul-taneously' (i.e., at the same time) in some reference frame. We can of course read the clocks of our friends who are at those events and who are in our ref-erence frame.

However, it turns out to be useful to find a way of determining which events are simultaneous with each other directly from postulate II, the one about the speed of light. Note that there is no problem in determining whether or not two things happen (like a door closing and a firecracker going o) at the same event. The question is merely whether two things that occur at different events take place simultaneously.

Suppose that we have a friend in an inertial frame and that she emit a flash of light from her worldline. The light will travel outward both to the left and the right, always moving at speed c. Suppose that some of this light is reflected back to her from event

A on the left and from event *B* on the right. The diagram below makes it clear that the two reflected pulses of light reach her at the same time if and only if A and *B* are simultaneous. So, if event C (where the reflected pulses cross) lies on her worldline, she knows that A and *B* are in fact simultaneous in our frame of reference.

Note: Although the light does not reach our friend until event C (at $t = 2$ sec., where she 'sees' the light), she knows that the light has taken some time to travel and he measurements place the reflections at $t = 1$ sec.

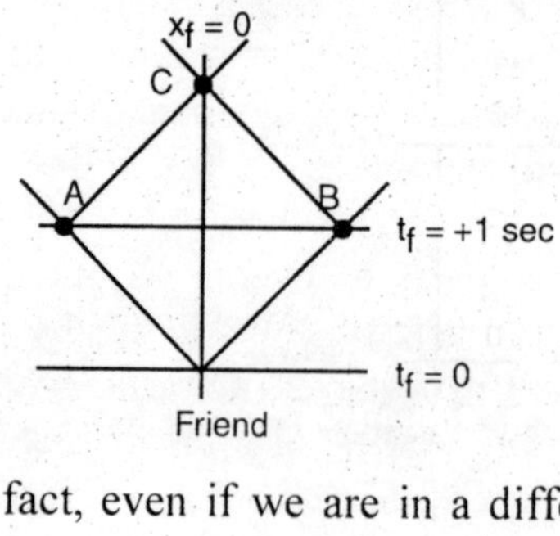

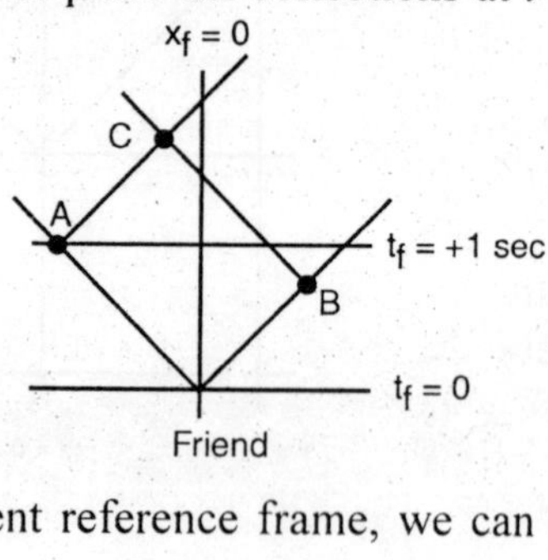

In fact, even if we are in a different reference frame, we can tell that

A and *B* are simultaneous in our friend's frame if event C lies on her worldline. Suppose that we are also inertial observers who meet our friend at the origin event and then move on. What does the above experiment look like in our frame?

Let's start by drawing our friend's worldline and marking event C. We don't really know where event C should appear, but it doesn't make much difference since we have drawn no scale. All that matters is that event C is on our friend's worldline ($x_f = 0$).

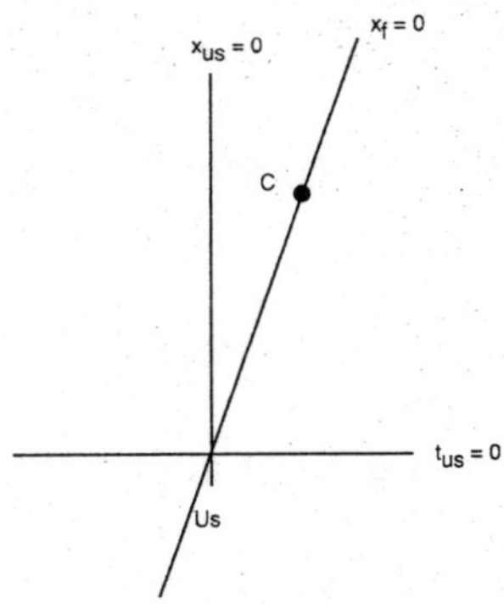

Now let's add the light rays from the origin and from event C. The events where these lines cross must be A and *B,* as shown below.

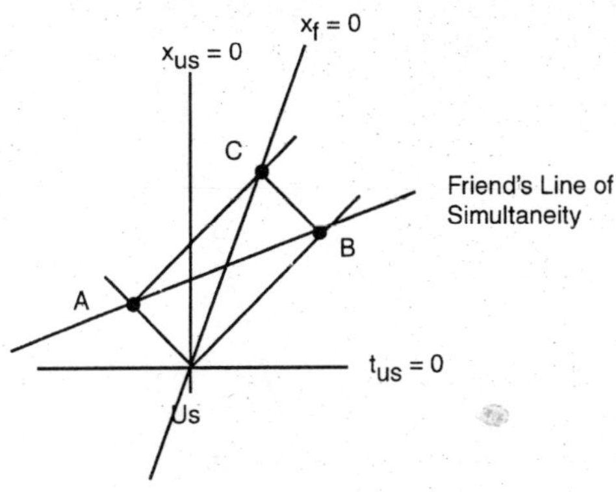

Note that, on either diagram, the worldline xf = 0 makes the same angle with the light cone as the line of simultaneity tf = const. That is, the angles α and β below are equal. You will in fact derive this in one of your homework problems.

By the way, we also can find other pairs of events on our diagram that are simultaneous in our friend's reference frame. We do this by sending out light signals from another observer in the moving frame. For example, the diagram below shows another event *(D)* that is also simultaneous with A and *B* in our friend's frame of reference.

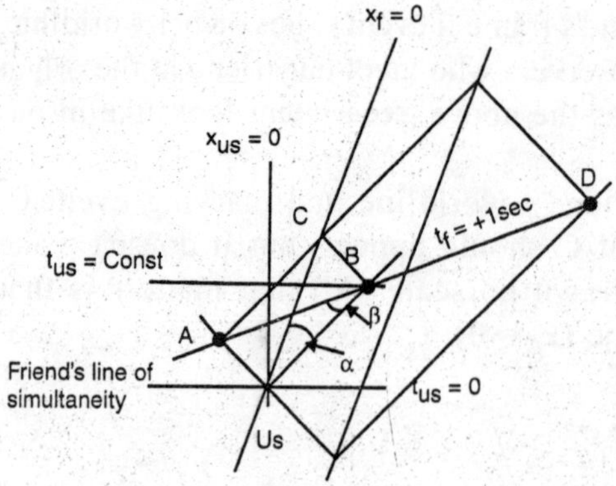

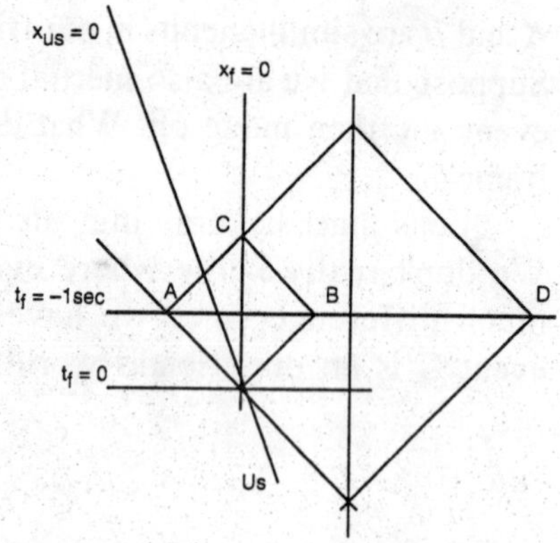

In this way we can map out our friend's entire line of simultaneity - the set of all events that are simultaneous with each other in her reference frame. The result is that the line of simultaneity for the moving frame does indeed appear as a straight line on our spacetime diagram. This property will be very important in what is to come.

Before moving on, let us get just a bit more practice and ask what set of events our friend (the moving observer) finds to be simultaneous with the origin (the event where the her worldline crosses ours)? We can use light signals to find this line as well. Let's label that line tf = 0 under the assumption that our friend chooses to set her watch to zero at the event where the worldlines cross. Drawing in a carefully chosen box of light rays, we arrive at the diagram below.

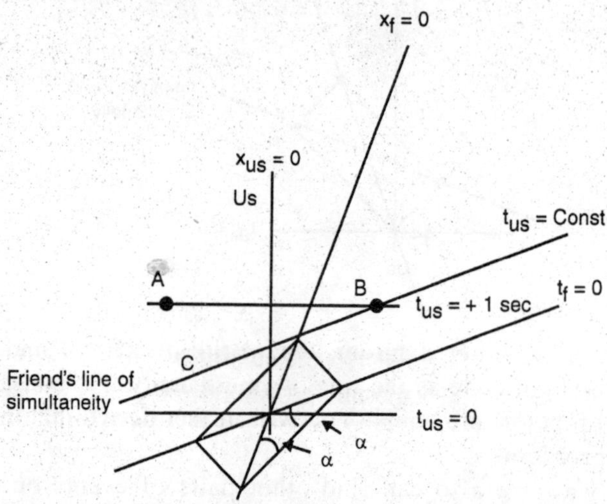

Note that we could also have used the rule noticed above: that the worldline and any line of simultaneity make equal angles with the light cone.

The line of simultaneity drawn above (tf = const) represents some constant time in the moving frame, we do not yet know which time that is! In particular, we do not yet know whether it represents a time greater than one second or a time less than on second. We were able to label the tf = 0 line with an actual value only because we explicitly assumed that our friend would measure time from the event (on that line) where our worldlines crossed. We will explore the question of how to assign actual time values to other lines of simultaneity shortly.

We have learned that events which are simultaneous in one inertial reference frame are not in fact simultaneous in a different inertial frame. We used light signals and postulate II to determine which events were simultaneous in which frame of reference.

RELATIONS BETWEEN EVENTS IN SPACETIME

It will take some time to absorb the implications, but let us begin with an interesting observation. A pair of events which is separated by "pure space" in one inertial frame (i.e., is simultaneous in that frame) is separated by both space and time in another. Similarly, a pair of events that is separated by "pure time" in one frame (occurring at the same location in that frame) is separated by both space and time in any other frame. This may remind you a bit of our discussion of electric and magnetic fields, where a field that was purely magnetic in one frame involved both electric and magnetic parts in another frame. In that case we decided that is was best to combine the two and to speak simply of a single "electromagnetic" field. Similarly here, it is best not to speak of space and time separately, but instead only of "spacetime" as a whole. The spacetime separation is fixed, but the decomposition into space and time depends on the frame of reference.

Note the analogy to what happens when you turn around in space. The notions of Forward/Backward vs. Right/Left get mixed up when you turn (rotate) your body. If you face one way, you may say that the Hall of Languages is "straight ahead." If you turn a bit, you might say that the Hall of Languages is "somewhat ahead and somewhat to the left." However, the separation between you and the Hall of Languages is the same no matter which way you are facing. As a result, Forward/Backward and Right/Left are not strictly speaking separate, but rather fit together to form two-dimensional space.

This is exactly what is meant by the phrase "space and time are not separate, but fit together to form four-dimensional spacetime." As a result, "time is the fourth dimension of spacetime." So then, how do we understand the way that events are related in this spacetime?

In particular, we have seen that simultaneity is not an absolute concept in spacetime itself. There is no meaning to whether two events occur at the same time unless we state which reference frame is being used.

If there is no absolute meaning to the word 'simultaneous,' what about 'before' and 'after' or 'past' and 'future?' Let's start o slowly. We have seen that if A and *B* are simultaneous in your (inertial) frame of reference (but are not located at the same place), then there is another inertial frame in which A occurs before *B.* A similar argument (considering a new inertial observer moving in the other direction) shows that there is another inertial frame in which *B* occurs before A. Looking back at our diagrams, the same is true if A occurs just slightly before *B* in your frame of reference.

However, this does not happen if *B* is on the light cone emitted from A, or if *B* is inside the light cone of A. To see this, remember that since the speed of light is c = 1 in any inertial frame, the light cone looks the same on everyone's spacetime diagram. A line more horizontal than the light cone therefore represents a 'speed' greater than c, while a line more vertical than the light cone represents a speed less than c. Because light rays look the same on everyone's spacetime diagram, the distinction between these three classes of lines must also be the same in all reference frames.

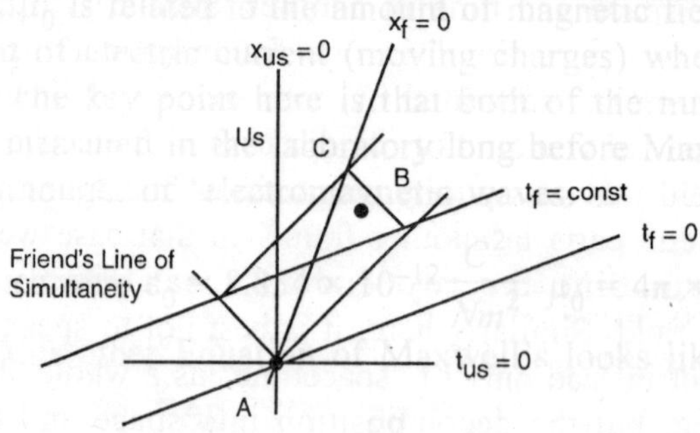

Thus, it is worthwhile to distinguish three classes of relationships that pairs of events can have. These classes and some of their properties are described below. Note that in describing these properties we limit ourselves to inertial reference frames that have a relative speed less than that of light.

Case 1: A and *B* are outside each other's light cones.

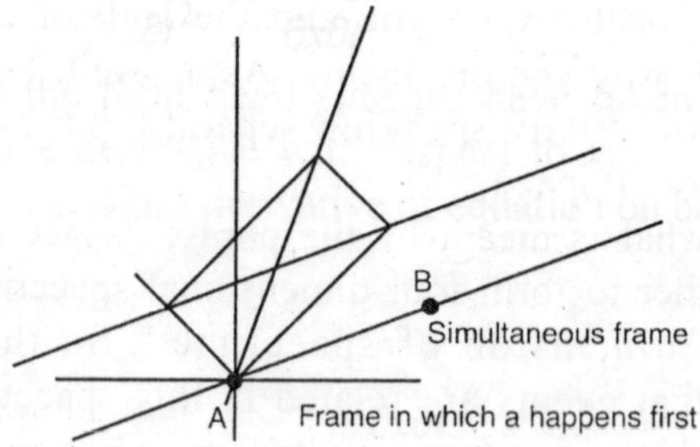

In this case, we say that they are spacelike related. Note that the following things are true in this case:

- There is an inertial frame in which A and *B* are simultaneous.
- There are also inertial frames in which event A happens first as well as frames in which event *B* happens first (even more tilted than the simultaneous frame shown above). However, A and *B* remain outside of each other's light cones in all inertial frames.

Case 2: A and *B* are inside each other's light cones in all inertial frames.

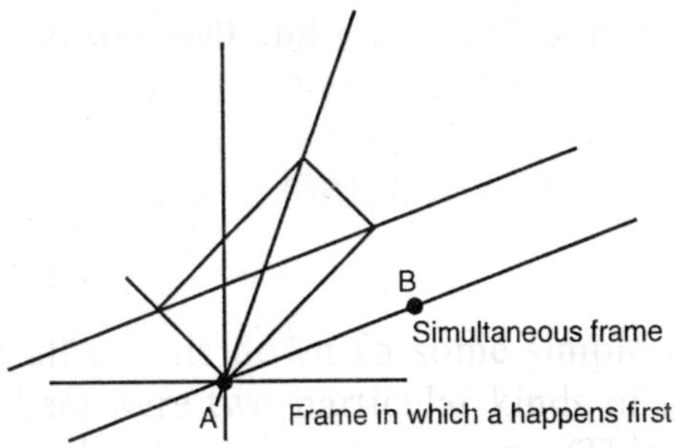

In this case we say that they are timelike related. Note that the following things are true in this case:

- There is an inertial observer who moves through both events and whose speed in the original frame is less than that of light.
- All inertial observers agree on which event (A or *B*) happened first.
- As a result, we can meaningfully speak of, say, event A being to the past of event *B*.

Case 3: A and *B* are on each other's light cones. In this case we say that they are lightlike related. Again, all inertial observers agree on which event happened first and we can meaningfully speak of one of them being to the past of the other.

Now, why did we consider only inertial frames with relative speeds less than c? Suppose for the moment that our busy friend (the inertial observer) could in fact travel at $v > c$ (i.e., faster than light).

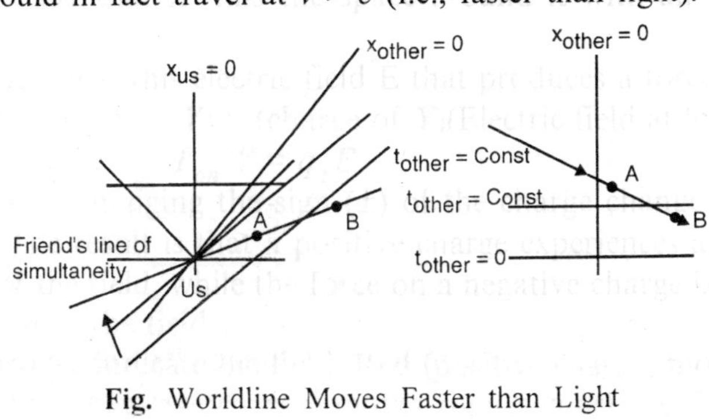

Fig. Worldline Moves Faster than Light

We have marked two events, A and *B* that occur on her worldline. In our frame event A occurs first.

However, the two events are spacelike related. Thus, there is another inertial frame (tother, xother) in which *B* occurs before. This means that there is some inertial observer (the one whose frame is drawn at right) who would see her traveling backwards in time.

This was too weird even for Einstein. After all, if she could turn around, our faster-than-light friend could even carry a message from some observer's future into that observer's past. This raises all of the famous 'what if you killed your grandparents' scenarios from science fiction fame.

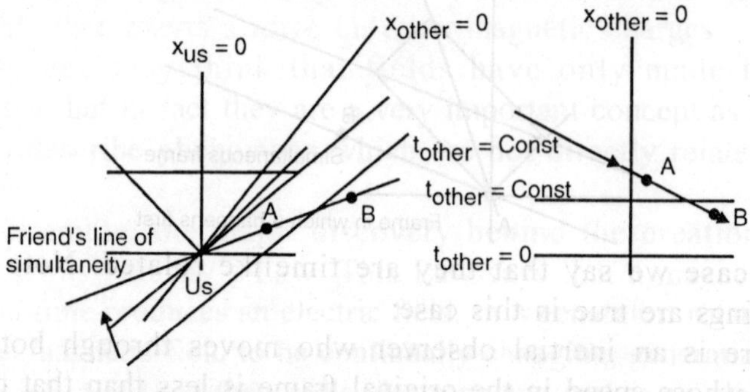

Fig. Worldline Moves Faster than Light

The point is that, in relativity, travel faster than light is travel backwards in time. For this reason, let us simply ignore the possibility of such observers for awhile. In fact, we will assume that no information of any kind can be transmitted faster than c.

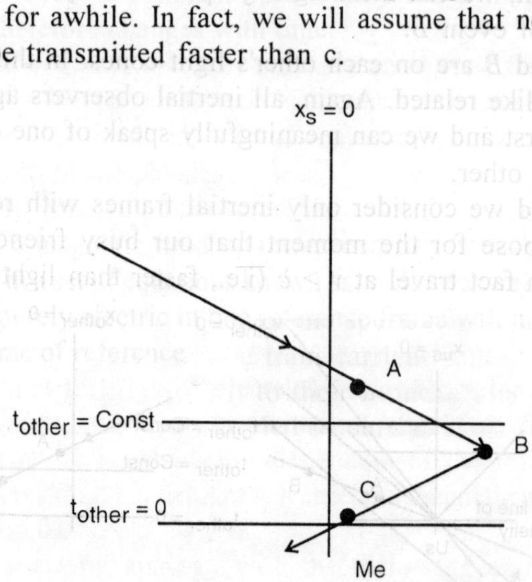

We are beginning to come to terms with simultaneity but, as pointed out earlier, we are still missing important information about how different inertial frames match up. In particular, we still do not know just what value of constant tf the line marked "friend's line of simultaneity" below actually represents.

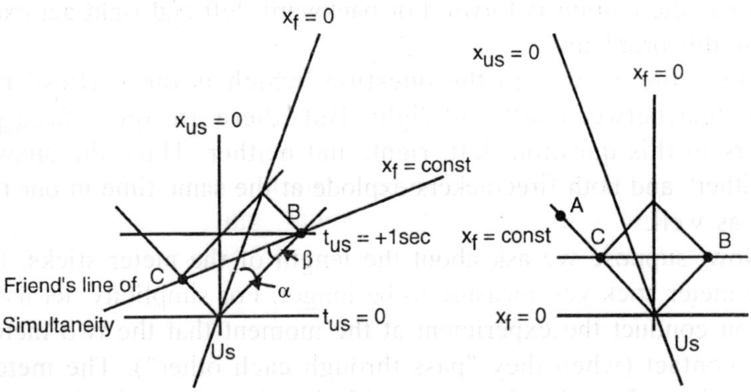

In other words, we do not yet understand the rate at which some observer's clock ticks in another observer's reference frame. That is, we should somehow make a clock out of light. For example, we can bounce a beam of light back and forth between two mirrors separated by a known distance.

Perhaps we imagine the mirrors being attached to a rod of fixed length *L*. Since we know how far apart the mirrors are, we know how long it takes a pulse of light to travel up and down and we can use this to mark the passage of time. We have a clock.

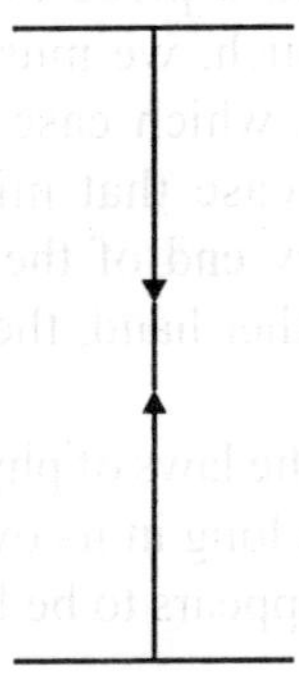

Rods in the Perpendicular Direction

A useful trick is to think about what happens when this 'light clock' is held perpendicular to the direction of relative motion. This direction is simpler than the direction of relative motion itself.

For example, two inertial observers actually do agree on which events are simultaneous in that direction. Suppose that you have two firecrackers, one placed one light second to your left and one placed one light second to your right. Suppose that both explode at the same time in your frame of reference. Does one of them explode earlier in mine? No, and the easiest way to see this is to argue by symmetry: the only difference between the two firecrackers is that one is on the right and the other is on the left.

Since the motion is forward or backward, left and right act exactly the same in this problem.

Thus, the answer to the question 'which is the earliest' must not distinguish between left and right. But, there are only three possible answers to this question: left, right, and neither. Thus, the answer must be 'neither', and both firecrackers explode at the same time in our reference frame as well.

Now, suppose we ask about the length of the meter sticks. Let's ask whose meter stick you measure to be longer. For simplicity, let us suppose that you conduct the experiment at the moment that the two meter sticks are in contact (when they "pass through each other"). The meter sticks passing through each other, since this involves only simultaneity in the direction along the meter sticks and, in the present case, this direction is perpendicular to our relative velocity.

On the one hand, since we both agree that we are discussing the same set of events, we must also agree on which meter stick is longer. This is merely a question of whether the event at the end of your meter stick is inside or outside of the line of events representing my meter stick.

Said more physically, suppose that we put a piece of blue chalk on the end of my meter stick, and a piece of red chalk on the end of yours. Then, after the meter sticks touch, we must agree on whether there is now a blue mark on your stick (in which case yours in longer), there is a red mark on my stick (in which case that mine is longer), or whether each piece of chalk marked the very end of the other stick (in which case they are the same length). On the other hand, the laws of physics are the same in all inertial frames.

In particular, suppose that the laws of physics say that, if you (as an inertial observer) take a meter stick 1m long in its own rest frame and move it toward you, then that that meter stick appears to be longer than a meter stick that is at rest in your frame of reference.

Here we assume that it does not matter in which direction (forward or backward) the meter stick is moving, as all direction in space are the same.

In that case, the laws of physics must also say that, if we (as an inertial observer) take a meter stick 1m long in its own rest frame and move it toward me, that that meter stick again appears to be longer than a meter stick that is at rest in my of reference.

Thus, if you find my stick to be longer, we must find your stick to be longer. If you find my stick to be shorter, then we must find your stick to be shorter. Consistency requires both of us find the two meter sticks to be of the same length.

We conclude that the length of a meter stick is the same in two inertial frames for the case where the stick points in the direction perpendicular to the relative motion.

LIGHT CLOCKS AND REFERENCE FRAMES

The property just derived makes it convenient to use such meter sticks to build clocks. We have given up most of our beliefs about physics for the moment, so that in particular we need to think about how to build a reliable clock.

The one thing that we have chosen to build our new framework upon is the constancy of the speed of light. Therefore, it makes sense to use light to build our clocks. We will do this by sending light signals out to the end of our meter stick and back.

For convenience, let us assume that the meter stick is one light-second long. This means that it will take the light one second to travel out to the end of the stick and then one second to come back. A simple model of such a light clock would be a device in which we put mirrors on each end of the meter stick and let a short pulse of light bounce back and forth. Each time the light returns to the first mirror, the clock goes 'tick' and two seconds have passed.

Now, suppose we look at our light clock from the side. Let's say that the rod in the clock is oriented in the vertical direction. The path taken by the light looks like this:

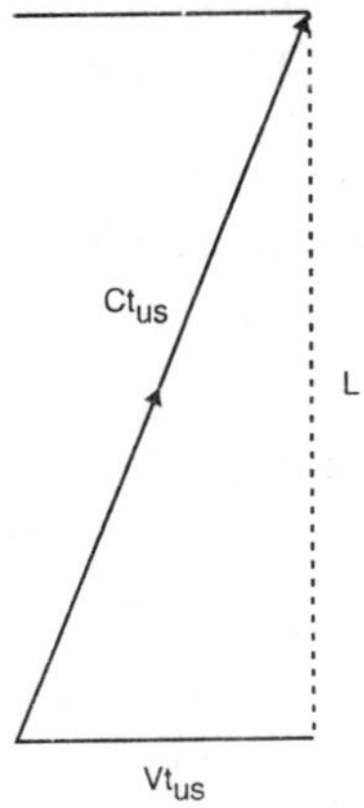

However, what if we look at a light clock carried by our inertial friend who is moving by at speed v?

Suppose that the rod in her clock is also oriented vertically, with the relative motion in the horizontal direction. Since the light goes straight up and down in her reference frame, the light pulse moves up and forward (and then down and forward) in our reference frame.

This should be clear from thinking about the path you see a basketball follow if someone lifts the basketball above their head while they are walking past you. The length of each side of the triangle is marked on the diagram above.

Here, L is the length of her rod and tus is the time that it takes the light to move from one end of the stick to the other. To compute two of the lengths, we have used the fact that, in our reference frame, the light moves at speed c while our friend moves at speed v.

The interesting question, of course, is just how long is this time tus. We know that the light takes 1 second to travel between the tips of the rod as measured in our friend's reference frame, but what about in ours? It turns out that we can calculate the answer by considering the length of the path traced out by the light pulse.

Using the Pythagorean theorem, the distance that we measure the light to travel is $\sqrt{(vt_{us})^2 + L^2}$.

However, we know that it covers this distance in a time tus at speed c. Therefore, we have

$$c^2t^2_{\text{us}} = v^2\, t^2_{\text{us}} + L^2,$$

or,

$$L^2/c^2 = t^2_{us} - (v/c)^2\, t^2_{us} = (1 - [v/c]^2)t^2_{us}.$$

Thus, we measure a time $t_{us} = \dfrac{L}{c\sqrt{1-(v/c)^2}}$ between when the light leaves one mirror and when it hits the next! This is in contrast to the time $t_{\text{friend}} = L/c = 1\text{second}$ measured by our friend between these same to events. Since this will be true for each tick of our friend's clock, we can conclude that:

Between any two events where our friend's clock ticks, the time t_{us} that we measure is related to the time t_{friend} measured by our friend by through

$$t_{us} = \frac{t_{friend}}{\sqrt{1-(v/c)^2}}$$

Finally, we have learned how to label another line on our diagram above:

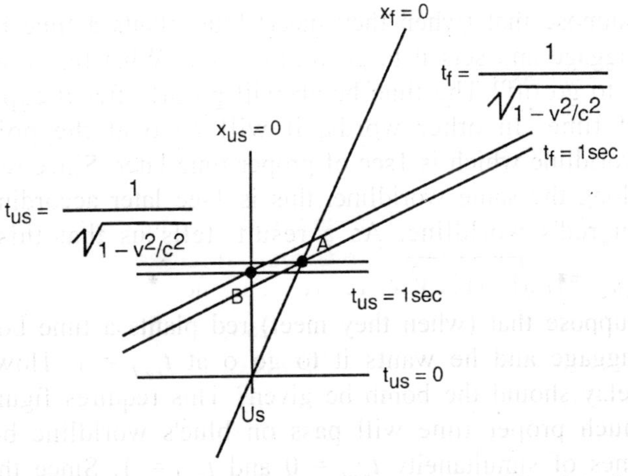

The dot labeled A is the event where the moving (friend's) clock ticks t = 1 second. It is an event on the friend's worldline. The dot labeled B is the event where our clock ticks t = 1 second. It is an event on our worldline.

PROPER TIME

We have seen that3 different observers in different inertial frames measure dif-ferent amounts of time to pass between two given events. We might ask if any one of these is a "better" answer than another? Well, in some sense the answer must be 'no,' since the principle of relativity tells us that all inertial frames are equally valid. However, there can be a distinguished answer. Note that, if one inertial observer actually experiences both events, then inertial observers in other frames have different worldlines and so cannot pass through both of these events. It is useful to use the term proper time between two events to refer to the time measured by an inertial observer who actually moves between the two events. Note that this concept exists only for timelike separated events.

Let's work through at a few cases to make sure that we understand what is going on. Consider two observers, red and blue. The worldlines of the two observers intersect at an event, where both set their clocks to read $t = 0$.

- Suppose that red sets a firecracker to go off on red's worldline a t t_{red} = 1. At what time does blue find it to go off? Our result tells us that $t_{blue} = 1/\sqrt{1-(v/c)^2}$.
- Suppose now that blue sets a firecracker to go off on blue's worldline at t_{blue} = 1. At what time does red find it to go off? From we now have $t_{red} = 1/\sqrt{1-(v/c)^2}$.

- Suppose that (when they meet) blue plants a time bomb in red's luggage and sets it to go o after 1sec. What times does blue find it to go off? The time bomb will go off after it experiences 1sec of time. In other words, it will go o at the point along its worldline which is 1sec of proper time later. Since red is traveling along the same worldline, this is 1sec later according to red and on red's worldline. As a result, tells us that this happens at $t_{blue} = 1/\sqrt{1-(v/c)^2}$.
- Suppose that (when they meet) red plants a time bomb in blue's luggage and he wants it to go o at $t_{red} = 1$. How much time delay should the bomb be given? This requires figuring out how much proper time will pass on blue's worldline between red's lines of simultaneity $t_{red} = 0$ and $t_{red} = 1$. Since the events are on blue's worldline, blue plays the role of the moving friend. As a result, the time until the explosion as measured by blue should be $t_{blue} = \sqrt{1-(v/c)^2}$, and this is the delay to set.

Why should you believe all of this ? So far, we have just been working out consequences of Einstein's idea. We have said little about whether you should actually believe that this represents reality.

In particular, the idea that clocks in different reference frames measure different amounts of time to pass blatantly contradicts your experience, doesn't it? Just because you go and fly around in an airplane does not mean that your watch becomes unsynchronized with the Cartoon Network's broadcast schedule, does it?

Well, let's start thinking about this by figuring out how big the time dilation effect would be in everyday life. Commercial airplanes move at about 300m/*s*.

So, $v/c \approx 10^{-6}$ for an airplane. Now, $\sqrt{1-(v/c)^2} \approx 1 - \frac{1}{2}(v/c)^2 + \ldots \approx 1 - 5 \times 10^{-13}$ for the airplane. This is less than 1 part in a trillion.

Tiny, eh? You'd never notice this by checking your watch against the Cartoon Network. However, physics is a very precise science. It turns out that it is in fact possible to measure time to better than one part in a trillion. A nice form of this experiment was first done in the 1960's. Some physicists got two identical atomic clocks, brought them together, and checked that they agreed to much better than 1 part in a trillion. Then, they left one in the lab and put the other on an airplane (such clocks were big, they bought a seat for the clock on a commercial airplane flight) and flew around for awhile. When they brought the clocks back together at the end of the experiment, the moving clock had in fact 'ticked' less times, measuring less time to pass in precise accord with our calculations above and

Einstein's prediction. We were merrily exploring Einstein's crazy idea. While Einstein's suggestion clearly fits with the Michelson-Morely experiment, we still have not figured out how it fits with the stellar aberration experiments. So, we were just exploring the suggestion to see where it leads.

It led to a (ridiculous) prediction that clocks in different reference frames measure different amounts of time to pass. This prediction has in fact been experimentally tested, and that Einstein's idea passed with flying colors. Now, you should begin to believe that all of this crazy stuff really is true. Oh, and there will be plenty more weird predictions and experimental verifications to come.

Another lovely example of this kind of thing comes from small subatomic parti-cles called muons (pronounced moo-ons). Muons are "unstable," meaning that they exist only for a short time and then turn into something else involving a burst of radiation. You can think of them like little time bombs. They live (on average) about 106 seconds. Now, muons are created in the upper atmosphere when a cosmic ray collides with the nucleus of some atom in the air (say, oxygen or nitrogen). In the 1930's, people noticed that these particles were traveling down through the atmosphere and appearing in their physics labs. Now, the atmosphere is about 30,000m tall, and these muons are created near the top.

The muons then travel downward at something close to the speed of light. Note that, if they traveled at the speed of light 3×10^8m/ *s*, it would take them a time $t = 3 \times 10^4\text{m}/(3 \times 10^8\text{m}/s) = 10^{-4}$ sec. to reach the earth. But, they are only supposed to live for 10^{-6} seconds! So, they should only make it 1/100 of they way down before they explode.

The point is that the birth and death of a muon are like the ticks of its clock and should be separated by 10^{-6} seconds as measured in the rest frame of the muon. In other words, the relevant concept here is 10^{-6} seconds of proper time. In our rest frame, we will measure a time $10^{-6}\text{sec}/\sqrt{1-(v/c)^2}$ to pass. For v close enough to c, this can be as large (or larger than) 10^{-4} seconds.

This concludes our first look at time dilation. We turn our attention to measurements of position and distance. However, there remain several subtleties involving time dilation that we have not yet explored.

LENGTH CONTRACTION

We learned how to relate times measured in different inertial frames. Clearly, the next thing to understand is distance. While we had to work fairly hard to compute the amount of time dilation that occurs, we will see that the effect on distances follow quickly from our results for time.

Let's suppose that two inertial observers both have measuring rods that are at rest in their respective inertial frames. Each rod has length L in the frame in which it is at rest (it's "rest frame").

We saw that distances in the direction perpendicular to the relative motion are not affected. So, to finish things o, this time we must consider the case where the measuring rods are aligned with the direction of the relative velocity.

For definiteness, let us suppose that the two observers each hold their meter stick at the leftmost end. The relevant spacetime diagram is shown below. As usual, we assume that the two observers clocks both read $t = 0$ at the event where their worldlines cross. We will call our observers 'student' and 'professor.' We begin by drawing the diagram in the student's rest frame and with the professor moving by at relative velocity v.

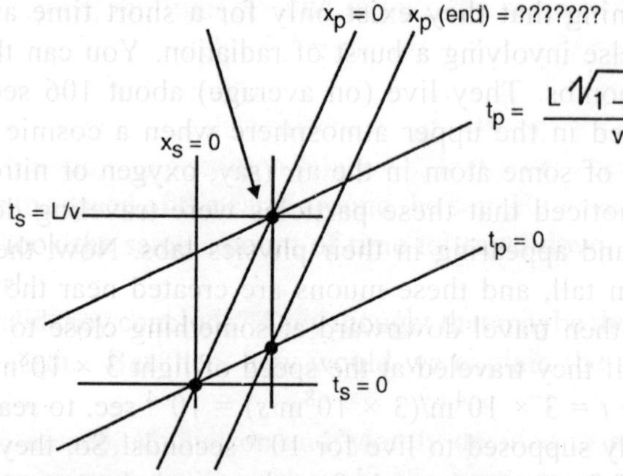

Now, the student must find that the professor takes a time L/v to traverse the length of the student's measuring rod. Let us refer to the event (marked in magenta) where the moving professor arrives at the right end of the student's measuring rod as "event A."

Since this event has $t_s = L/v$, we can use our knowledge of time dilation to conclude that the professor assigns a time $t_P = (L/v)\sqrt{1-(v/c)^2}$ to this event. Our goal is to determine the length of the student's measuring rod in the professor's frame of reference. That is, we wish to know what position xP (end) the professor assigns to the rightmost end of the students rod when this end crosses the professor's line of simultaneity $t_P = 0$.

To find this out, note that from the professor's perspective it is the student's rod that moves past him at speed v. It takes the rod a time $t_P = (L/v)\sqrt{1-(v/c)^2}$ to pass by. Thus, the student's rod must have a length $L_P = \sqrt{1-(v/c)^2}$ in the professor's frame of reference. The professor's rod, of course, will similarly be shortened in the student's frame of reference. So, we see that distance measurements also depend on the observer's frame of reference. Note however, that given any inertial object, there is a special inertial frame in which the object is at rest. The length of an object in its own rest frame is known as its proper length. The length of the object in any other inertial frame will be shorter than the object's proper length. We can summarize what we have learned by stating:

An object of proper length L moving through an inertial frame at speed v has length $L\sqrt{1-v^2/c^2}$ as measured in that inertial frame.

There is an important subtlety that we should explore. Note that the above statement refers to an object. However, we can also talk about proper distance between two events. When two events are spacelike related, there is a special frame of reference in which the events are simultaneous and the separation is "pure space" (with no separation in time). The distance between them in this frame is called the proper distance between the events. It turns out that this distance is in fact longer in any other frame of reference.

Why longer? To understand this, look back at the above diagram and compare the two events at either end of the students' rod that are simultaneous in the professor's frame of reference. Note that the proper distance is the distance measured in the professor's reference frame, which we just concluded is shorter than the distance measured by the student. The difference here is that we are now talking about events (points on the diagram) where as before we were talking about objects (whose ends appear as worldlines on the spacetime dia-gram). The point is that, when we talk about measuring the length of an object, different observers are actually measuring the distance between different pairs of events.

THE TRAIN PARADOX

Let us now test our new skills and work through some subtleties by considering an age-old parable known as the train paradox. It goes like this: Once upon a time there was a really fast Japanese bullet train that ran at 80% of the speed of light. The train was 100m long in its own rest frame. The train carried as cargo the profits of SONY corporation from Tokyo out to their headquarters in the countryside. The profits were, of course, carried in pure gold.

Now, some less than reputable characters found out about this and devised an elaborate scheme to rob the train. They knew that the train would pass through a 100m long tunnel on its route. Watching the train go by, they measured the train to be quite a bit less than 100m long and so figured that they could easily trap it in the tunnel.

Of course, the people on the train found that, when the train was in motion, it was the train that was 100m long while the tunnel was significantly shorter. As a result, they had no fear of being trapped in the tunnel by train robbers. Now, do you think the robbers managed to catch the train?

Let's draw a spacetime diagram using the tunnel's frame of reference. We can let E represent the tunnel entrance and *X* represent the tunnel exit. Similarly, we let *B* represent the back of the train and F represent the front of the train. Let event 1 be the event where the back of the train finally reaches the tunnel and let event 2 be the event where the front of the train reaches the exit.

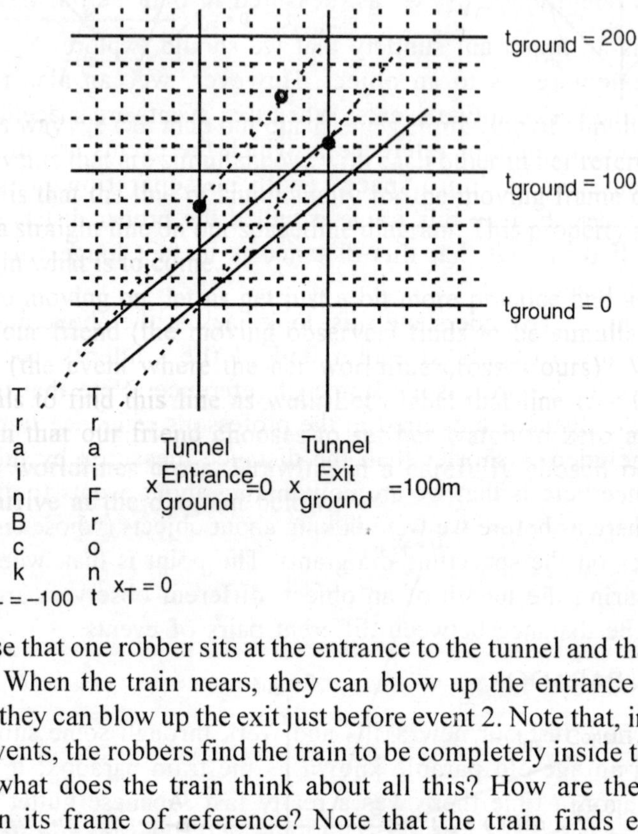

Suppose that one robber sits at the entrance to the tunnel and that one sits at the exit. When the train nears, they can blow up the entrance just after event 1 and they can blow up the exit just before event 2. Note that, in between these two events, the robbers find the train to be completely inside the tunnel.

Now, what does the train think about all this? How are these events described in its frame of reference? Note that the train finds event 2 to occur long before event 1. So, can the train escape?

Let's think about what the train would need to do to escape. At event 2, the exit to the tunnel is blocked, and (from the train's perspective) the debris blocking the exit is rushing toward the train at 80% the speed of light. The only way the train could escape would be to turn around and back out of the tunnel. The train finds that the entrance is still open at the time of event 2.

Of course, both the front and back of the train must turn around. How does the back of the train know that it should do this? It could find out via a phone call from an engineer at the front to an engineer at the back of the train, or it could be via a shock wave that travels through the metal of the train as the front of the train throws on its brakes and reverses its engines. The point is though that some signal must pass from event 2 to the back of the train, possibly relayed along the way by something at the front of the train. Sticking to our assumption that signals can only be sent at speed c or slower, the earliest possible time that the back of the train could discover the exit explosion is at the event marked *D* on the diagram. Note that, at event *D,* the back of the train does find itself inside the tunnel and also finds that event 1 has already occurred. The entrance is closed and the train cannot escape.

There are two things that deserve more explanation. The first is the above comment about the shock wave. Normally we think of objects like trains as being perfectly stiff. Also, it takes a (small but finite) amount of time for each atom to respond to the push it has been given on one side and to move over and begin to push the atom on the other side. The result is known as a "shock wave" that travels at finite speed down the object. Note that an important part of the shock wave are the electric forces that two atoms use to push each other around. Thus, the shock wave can certainly not propagate faster than an electromagnetic disturbance can. As a result, it must move at less than the speed of light.

For the other point, let's suppose that the people at the front of the train step on the brakes and stop immediately. Stopping the atoms at the front of the train will make them push on the atoms behind them, stopping them, etc. The shock wave results from the fact that atoms just behind the front slam into atoms right at the front; the whole system compresses a bit and then may try to reexpand, pushing some of the atoms farther back.

What we saw above is that the shock wave cannot reach the back of the train until event *D.* Suppose that it does indeed stop the back of the train there. The train has now come to rest in the tunnel's frame of reference. Thus, after event *D,* the proper length of the train is less than 100m!!!!

In fact, suppose that we use the lines of simultaneity in the train's original frame of reference (before it tries to stop) to measure the proper length of the train. Then, immediately after event 2 the front of the train changes its motion, but the back of the train keeps going. As a result, in this sense the proper length of the train starts to shrink immediately after event 2. This is how it manages to fit itself into a tunnel that, in this frame, is less than 100m long.

What has happened? The answer is in the compression that generates the shock wave. The train really has been physically compressed by the wall of debris at the exit slamming into it at half the speed of light6! This compression is of course accompanied by tearing of metal, shattering of glass, death screams of passengers, and the like, just as you would expect in a crash. The train is completely and utterly destroyed. The robbers will be lucky if the gold they wish to steal has not been completely vaporized in the carnage.

Now, you might want to get one more perspective on this by analyzing the problem again in a frame of reference or the equivalent damage inflicted through the use of the train's brakes. that moves with the train at all times, even slowing down and stopping as the train slows down and stops. However, we do not know enough to do this yet since such a frame is not inertial.

Chapter 4

Minkowskian Geometry

We were faced with the baffling results of the Michelson-Morely experiment and the stellar aberration experiments. In the end, we decided to follow Einstein and to allow the possibility that space and time simply do not work in the way that our intuition predicts. In particular, we took our cue from the Michelson-Morely experiment which seems to say that the speed of light in a vacuum is the same in all inertial frames and, therefore, that velocities do not add together in the Newtonian way. We wondered "How can this be possible?"

We then spent the last chapter working out "how this can be possible." That is, we have worked out what the rules governing time and space must actually be in order for the speed of light in a vacuum to be the same in all inertial reference frames. In this way, we discovered that different observers have different notions of simultaneity, and we also discovered time dilation and length contraction. Finally, we learned that some of these strange predictions are actually correct and have been well verified experimentally.

It takes awhile to really absorb what is going on here. The process does take time, though at this stage of the course the students who regularly come to my o ce hours are typically moving along well. There are lots of levels at which one might try to "understand" the various effects. Some examples are:

Logical Necessity: Do see that the chain of reasoning leading to these con-clusions is correct? If so, and if you believe the results of Michelson and Morely that the speed of light is constant in all inertial frames, then you must believe the conclusions.

External Consistency: Understanding at this level involves determin-ing how big the a ects actually would be in your everyday life. You will quickly find that they are seldom more than one part in a billion or a trillion. At this level, it is no wonder that you never noticed.

Internal Consistency: How can these various effects possibly be self-consistent? How can a train 100m long get stuck inside a tunnel that, in it's initial frame of reference, is less than 100m long?

Step Outside the old Structure: When people ask this question, what they mean is "Can you explain why these strange things occur in terms of things that are familiar to my experience, or which are reasonable to my intu-ition?" It is important to realise that, in relativity, this is most definitely not possible in a direct way.

This is because all of your experience has built up an intuition that believes in the Newtonian assumptions about space and time and, as we have seen, these cannot possibly be true! Therefore, you must remove your old intuition, remodel it completely, and then put a new kind of intuition back in your head.

Finding the new logic: If we have thrown out all of our intuition and experience, what does it mean to "understand" relativity? We will see that relativity has a certain logic of its own. What we need to do is to uncover the lovely structure that space and time really do have, and not the one that we want them to have. In physics as in life, this is often necessary.

Typically, when one understands a subject deeply enough, one finds that the subject really does have an intrinsic logic and an intrinsic sense that are all its own. This is the level at which finally see "what is actually going on." This is also the level at which people finally begin to "like" the new rules for space and time.

MINKOWSKIAN GEOMETRY

Minkowski was a mathematician, and he is usually credited with emphasizing the fact that time and space are part of the same "spacetime" whole in relativity. He also who emphasized the fact that this spacetime has a special kind of geometry. It is this geometry which is the underlying structure and the new logic of relativity.

Understanding this geometry will provide both insight and useful technical tools. For this reason, we now pursue what at first sight will seem like a technical aside in which we first recall how the familiar Euclidean geometry relates quantities in different coordinate systems. We can then build an analogous technology in which Minkowskian geometry relates different inertial frames.

Invariants: Distance vs. the Interval

A fundamental part of familiar Euclidean geometry is the Pythagorean theorem. One way to express this result is to say that

$$(\text{distance})^2 = \Delta x^2 + \Delta y^2,$$

where distance is the distance between two points and Δx, Δy are respectively the differences between the x coordinates and between the y coordinates of these points. Here the notation Δx^2 means $(\Delta x)^2$ and not the change in x^2. Note that this relation holds in either of the two coordinate systems drawn below.

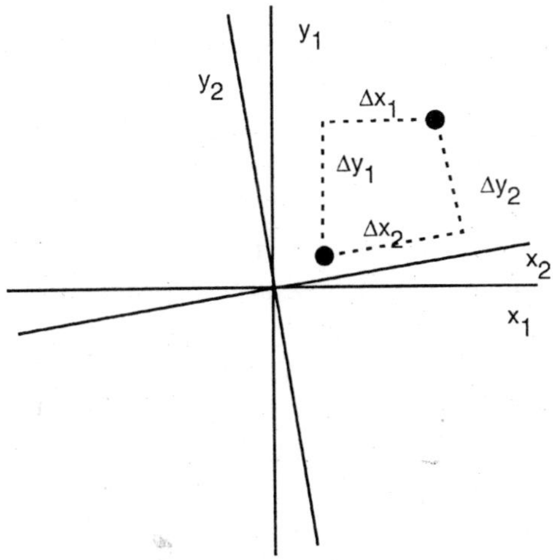

We compare coordinate systems (with one rotated relative to the other) *I* find

$$\Delta x_1^2 + \Delta y_1^2 = \Delta x_2^2 + \Delta y_2^2.$$

Let's think about an analogous issue involving changing inertial frames. Consider, for example, two inertial observers. Suppose that our friend flies by at speed v. For simplicity, let us both choose the event where our worldlines intersect to be $t = 0$. Let us now consider the event (on his worldline) where his clock 'ticks' $t_f = T$. Note that our friend assigns this event the position $x_f = 0$ since she passes through it.

What coordinates do we assign? Our knowledge of time dilation tells us that we assign a longer time: $t_{us} = T/\sqrt{1-v^2/c^2}$. For position, recall that at $t_{us} = 0$ our friend was at the same place that we are ($x_{us} = 0$). Therefore, after moving at a speed v for a time $t_{us} = T/\sqrt{1-v^2/c^2}$, our friend is at $x_{us} = vt_{us} = Tv/\sqrt{1-v^2/c^2}$.

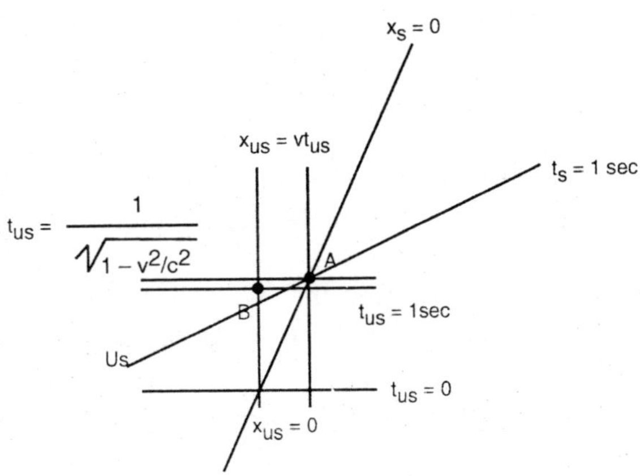

Now, we'd like to examine a Pythagorean-like relation. Of course we can't just mix x and t in an algebraic expression since they have different units. But, we have seen that x and ct do mix well! Thinking of the marked event where our friend's clock ticks, is it true that $x^2 + (ct)^2$ is the same in both reference frames? Clearly no, since both of these terms are larger in our reference frame than in our friend's ($x_{us} > 0$ and $ct_{us} > ct_f$)!

What we have just observed is that whenever an inertial observer passes through two events and measures a proper time T between them, any inertial observer finds $\Delta x^2 - c^2\Delta t^2 = -c^2T^2$. But, given any two timelike separated events, an inertial observer could in fact pass through them. So, we conclude that the quantity $\Delta x^2 - c^2\Delta t^2$ computed for a pair of timelike separated events is the same in all inertial frames of reference. Any quantity with this property is called an 'invariant' because it does not vary when we change reference frames.

A quick check shows that the same is true for spacelike separated events. For lightlike separated events, the quantity $\Delta x^2 - c^2\Delta t^2$ is actually zero in all reference frames. We see that for any pair of events the quantity $\Delta x^2 - c^2\Delta t^2$ is completely independent of the inertial frame that you use to compute it. This quantity is known as the (interval)^2.

$$(interval)^2 = \Delta x^2 - c^2\Delta t^2$$

The language here is a bit difficult since this can be negative. The way that physicists solve this in modern times is that we always discuss the (interval)2 and never (except in the abstract) just "the interval" (so that we don't have to deal with the square root). The interval functions like 'distance,' but in spacetime, not in space.

Let us now explore a few properties of the interval. As usual, there are three cases to discuss depending on the nature of the separation between the two events.

Timelike separation: In this case the squared interval is negative, for two timelike separated events there is (or could be) some inertial observer who actually passes through both events, experiencing them both. One might think that her notion of the amount of time between the two events is the most interesting and indeed we have given it a special name, the "proper time" ($\Delta\tau$; "delta tau") between the events. Note that, for this observer the events occur at the same place. Since the squared interval is the same in all inertial frames of reference, we therefore have:

$$0^2 - c^2(\Delta\tau)^2 = \Delta x^2 - c^2\Delta t^2.$$

Solving this equation, we find that we can calculate the proper time D in terms of the distance Δx and time Δt in any inertial frame using:

$$\Delta\tau = \sqrt{\Delta t^2 - \Delta x^2/c^2} = \Delta t\sqrt{1 - \frac{\Delta x^2}{c^2\Delta t^2}} = \Delta t\sqrt{1 - v^2/c^2}$$

We see that $\Delta\tau \leq \Delta t$.

Spacelike separation: Similarly, if the events are spacelike separated, there is an inertial frame in which the two are simultaneous - that is, in which $\Delta t = 0$. The distance between two events measured in such a reference frame is called the proper distance d. Much as above,

$$d = \sqrt{\Delta x^2 - c^2 \Delta t^2} \leq \Delta x \cdot$$

Note that this seems to "go the opposite way" from the length contraction effect. That is because here we consider the proper distance between two particular events. In contrast, in measuring the length of an object, different observers do NOT use the same pair of events to determine length.

Lightlike separation: Two events that are along the same light ray satisfy $\Delta x = \pm c\Delta t$. It follows that they are separated by zero interval in all reference frames. One can say that they are separated by both zero proper time and zero proper distance.

Curved Lines and Accelerated Objects

Thinking of things in terms of proper time and proper distance makes it easier to deal with, say, accelerated objects. Suppose we want to compute, for example, the amount of time experienced by a clock that is not in an inertial frame. Perhaps it quickly changes from one inertial frame to another, shown in the blue worldline (marked B) below. This worldline (B) is similar in nature to the worldline of the muon in part (b).

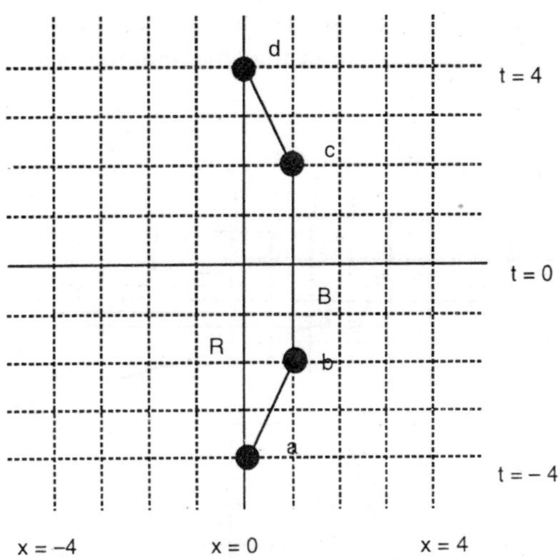

Note that the time experienced by the blue clock between events (a) and *(b)* is equal to the proper time between these events since, on that segment, the clock could be in an inertial frame. Surely the time measured by an ideal clock between (a) and *(b)* cannot depend on what it was doing before (a) or on what it does after *(b)*. Similarly, the time experienced by the blue clock between events *(b)* and (c) should be the same as that experienced by a truly inertial clock moving between these events; i.e. the proper time between these events. Thus, we can find the total proper time experienced by the clock by adding the proper time between (a) and *(b)* to the proper time between *(b)* and (c) and between (c) and *(d)*.

We also refer to this as the total proper time along the clock's worldline between (a) and *(d)*. A red observer (R) is also shown above moving between events (0, –4) and (0, +4). Let $\Delta\tau^{R}{}_{ad}$ and $\Delta\tau^{B}{}_{ad}$ be the proper time experienced by the red and the blue observer respectively between times t = –4 and t = +4; that is, between a, $d > D$?Ba, d and similarly for the other time intervals. Thus we see that the proper time along the broken line is less than the proper time along the straight line.

Since proper time (i.e., the interval) is analogous to distance in Euclidean ge-ometry, we also talk about the total proper time along a curved worldline in much the same way that we talk about the length of a curved line in space. We obtain this total proper time much as we did for the blue worldline above by adding up the proper times associated with each short piece of the curve. This is just the usual calculus trick in which we approximate a curved line by a sequence of lines made entirely from straight line segments. One simply replaces any Δx (or Δt) denoting a difference between two points with dx or dt which denotes the difference between two infinitesimally close points.

The rationale here, of course, is that if you look at a small enough (infinitesimal) piece of a curve, then that piece actually looks like a straight line segment. Thus we have

$$d\tau = dt\sqrt{1-v^2/c^2} < dt\text{, or } \int dt = \int\sqrt{1-v^2/c^2}\, dt < \Delta t$$

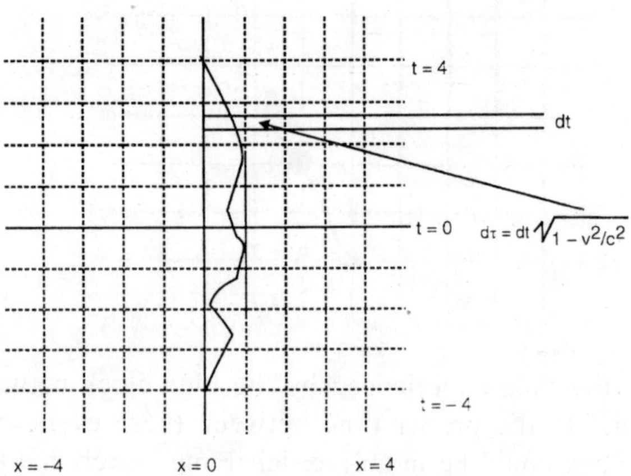

Again we see that a straight (inertial) line in spacetime has the longest proper time between two events. In other words, in Minkowskian geometry the longest line between two events is a straight line.

THE TWIN PARADOX

That's enough technical stuff for the moment. "The twin paradox." Using the notions of proper time and proper distance turns out to simplify the discussion significantly compared.

Let's think about two identical twins who, for obscure historical reasons are named Alphonse and Gaston. Alphonse is in an inertial reference frame floating in space somewhere near our solar system. Gaston, on the other hand, will travel to the nearest star (Alpha Centauri) and back at.8c.

Alpha Centauri is (more or less) at rest relative to our solar system and is four light years away. During the trip, Alphonse finds Gaston to be aging slowly because he is traveling at 8c. On the other hand, Gaston finds Alphonse to be aging slowly because, relative to Gaston, Alphonse is traveling at.8c.

During the trip out there is no blatant contradiction, since we have seen that the twins will not agree on which event (birthday) on Gaston's worldline they should compare with which event (birthday) on Alphonse's worldline in order to decide who is older. But, who is older when they meet again and Alphonse returns to earth?

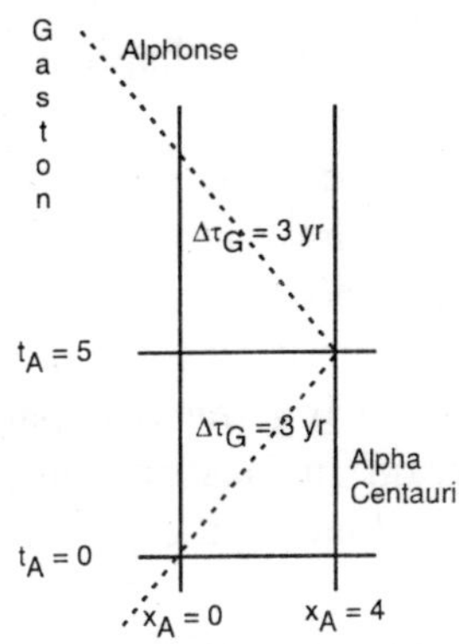

The above diagram shows the trip in a spacetime diagram in Alphonse's frame of reference. Let's work out the proper time experienced by each observer. For Alphonse, $\Delta x = 0$. How about Δt? Well, the amount of time that passes is long enough for Gaston to travel 8 light-years (there and back) at.8c. That is, Δt = 8lyr/(.8c) = 10yr. So, the proper time $\Delta\tau$A experienced by Alphonse is ten years.

On the other hand, we see that on the first half of his trip Gaston travels 4 light proper time of $\sqrt{5^2 - 4^2}$ = 3years. The same occurs on the trip back. So, the total proper time experienced by Gaston is $\Delta\tau_G$ = 6years. Is Gaston really younger then when they get back together? Couldn't we draw the same picture in Gaston's frame of reference and reach the opposite conclusion? NO, we cannot.

The reason is that Gaston's frame of reference is not an inertial frame! Gaston does not always move in a straight line at constant speed with respect to Alphonse. In order to turn around and come back, Gaston must experience some force which makes him non-inertial. Most importantly, Gaston knows this! When, say, his rocket engine fires, he will feel the force acting on him and he will know that he is no longer in an inertial reference frame.

The point here is not that the process is impossible to describe in Gaston's frame of reference. Gaston experiences what he experiences, so there must be such a description.

The point is, however, that so far we have not worked out the rules to understand frames of reference that are not inertial. Therefore, we cannot simply blindly apply the time dilation/length contraction rules for inertial frames to Gaston's frame of reference.

Thus, we should not expect our results so far to directly explain what is happening from Gaston's point of view.

But, you might say, Gaston is almost always in an inertial reference frame. He is in one inertial frame on the trip out, and he is in another inertial frame on the trip back. What happens if we just put these two frames of reference together?

Let's do this, but we must do it carefully since we are now treading new ground. First, we should draw in Gaston's lines of simultaneity on Alphonse's spacetime diagram above. His lines of simultaneity will match2 simultaneity in one inertial frame during the trip out, but they will match those of a different frame during the trip back. Then, those lines of simultaneity draw a diagram in Gaston's not-quite-inertial frame of reference, much as we have done in the past in going from one inertial frame to another.

Since Gaston is in a different inertial reference frame on the way out than on the way back, to draw two sets of lines of simultaneity and each set will have a different slope. Now, two lines with different slopes must intersect.

The lines of simultaneity with Gaston's proper time at the events where he crosses those lines. Note that there are two lines of simultaneity marked t_G = 3years!. One of these 3^- (which is "just before" Gaston turns around) and one 3^+ (which is "just after" Gaston turns around).

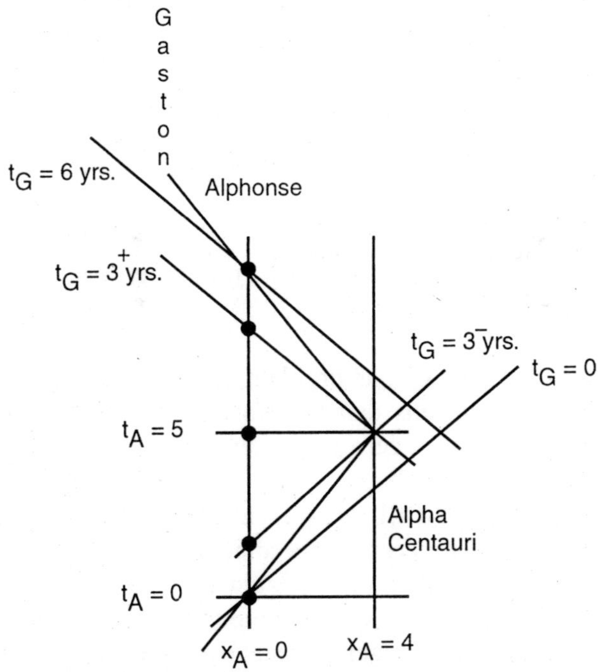

Simply knit together Gaston's lines of simultaneity and copy the events from the diagram above, the following diagram in Gaston's frame of reference. Note that it is safe to use the standard length contraction result to find that in the inertial frame of Gaston on his trip out and in the inertial frame of Gaston on his way back the distance between Alphonse and Alpha Centauri is $4\text{Lyr}\sqrt{1-(4/5)^2} = (12/5)\text{Lyr}$.

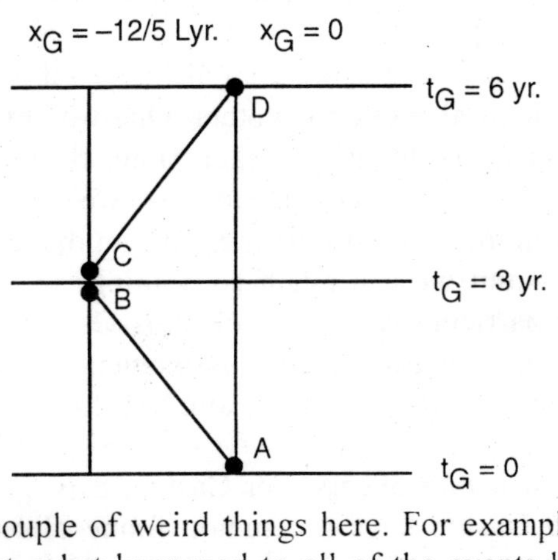

There are a couple of weird things here. For example, what happened to event E? In fact, what happened to all of the events between *B* and C?

By the way, how old is Alphonse at event *B*? In Gaston's frame of reference (which is inertial before $t_G = 3$, so we can safely calculate things that are confined to this region of time),

Alphonse has traveled (12/5)Lyr. in three years.

So, Alphonse must experience a proper time of $\sqrt{3^2 - (12/5)^2}$ = $\sqrt{9 - 144/25}$ = $\sqrt{81/25}$ (9/5)years.

Similarly, Alphonse experiences (9/5)years between events C and *D*. This means that there are 10 – 18/5 = (32/5)years of Alphonse's life missing from the diagram.

It turns out that part of our problem is the sharp corner in Gaston's worldline. The corner means that Gaston's acceleration is infinite there, since he changes velocity in zero time. Let's smooth it out a little and see what happens.

Suppose that Gaston still turns around quickly, but not so quickly that we cannot see this process on the diagram.

If the turn-around is short, this should not change any of our proper times very much (proper time is a continuous function of the curve!!!), so Gaston will still experience roughly 6 years over the whole trip, and roughly 3 years over half. Let's say that he begins to slow down (and therefore ceases to be inertial) after 2.9 years so that after 3 years he is momentarily at rest with respect to Alphonse.

Then, his acceleration begins to send him back home. A tenth of a year later (3.1 years into the trip) he reaches.8 c, his rockets shut o, and he coasts home as an inertial observer.

We have already worked out what is going on during the periods where Gaston is inertial. But, what about during the acceleration? Note that, at each instant,

Gaston is in fact at rest in some inertial frame - it is just that he keeps changing from one inertial frame to another. One way to draw a spacetime diagram for Gaston is try to use, at each time, the inertial frame with respect to which he is at rest. This means that we would use the inertial frames to draw in more of Gaston's lines of simultaneity on Alphonse's diagram, at which point we can again copy things to Gaston's diagram.

A line that is particularly easy to draw is Gaston's t_G = 3year line. This is because, at t_G = 3years, Gaston is momentarily at rest relative to Alphonse. This means that Gaston and Alphonse share a line of simultaneity.

For Alphonse, it it t_A = 5years. For Gaston, it is t_G = 3years. On that line, Alphonse and Gaston have a common frame of reference and their measurements agree.

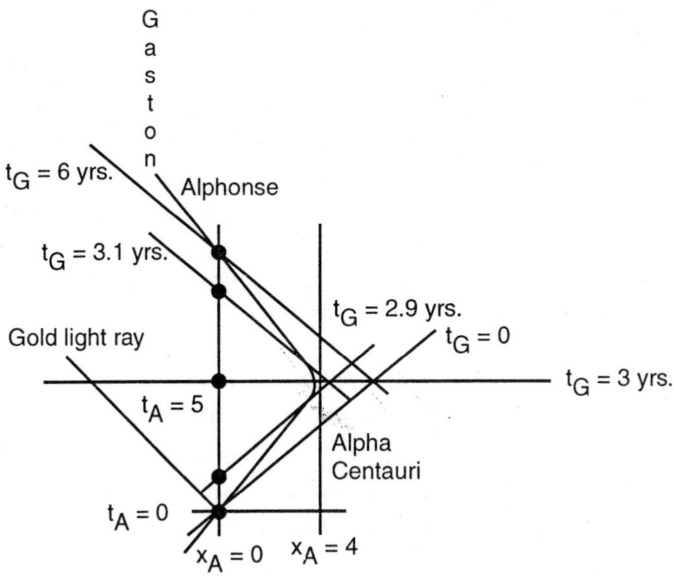

Note that we finally have a line of simultaneity for Gaston that passes through event E So, event E really does belong on Gaston's t_G = 3year line after all. By the way, just "for fun" added to our diagram an light ray moving to the left from the origin.

We are almost ready to copy the events onto Gaston's diagram. But, to properly place event R, we must figure out just where it is in Gaston's frame. In other words, how far away is it from Gaston along the line t_G = 3years? Gaston and Alphonse measure things in the same way. Therefore they agree that, along that line, event E is four light years away from Alphonse. Placing event E onto Gaston's diagram connecting the dots to get Alphonse's worldline, we find:

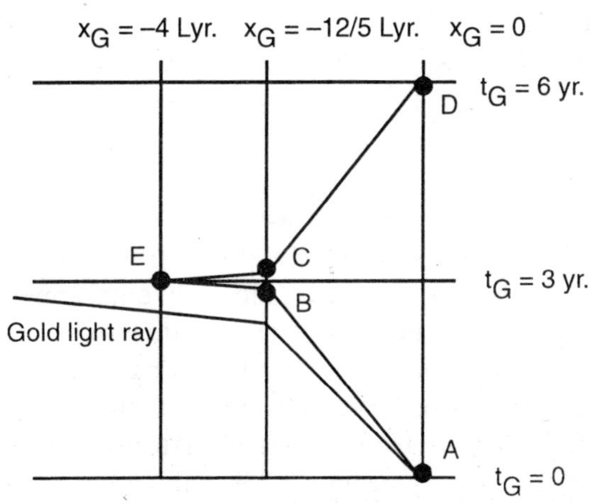

There is something interesting about Alphonse's worldline between *B* and E. It is almost horizontal, and has speed much greater than one light-year per year! What is happening?

The gold light ray, and that it too moves at more than one light-year per year in this frame. We see that Alphonse is in fact moving more slowly than the light ray, which is good. However, we also see that the speed of a light ray is not in general equal to c in an accelerated reference frame! In fact, it is not even constant since the gold light ray appears 'bent' on Gaston's diagram. Thus, it is only in inertial frames that light moves at a constant speed of 3 × 108 meters per second. This is one reason to avoid drawing diagrams in non-inertial frames whenever you can.

Actually, though, things are even worse than they may seem at first glance... Suppose, for example, that Alphonse has a friend Zelda who is an inertial observer at rest with respect to Alphonse, but located four light years on the other side of Alpha Centauri. We can then draw the following diagram in Alphonse's frame of reference:

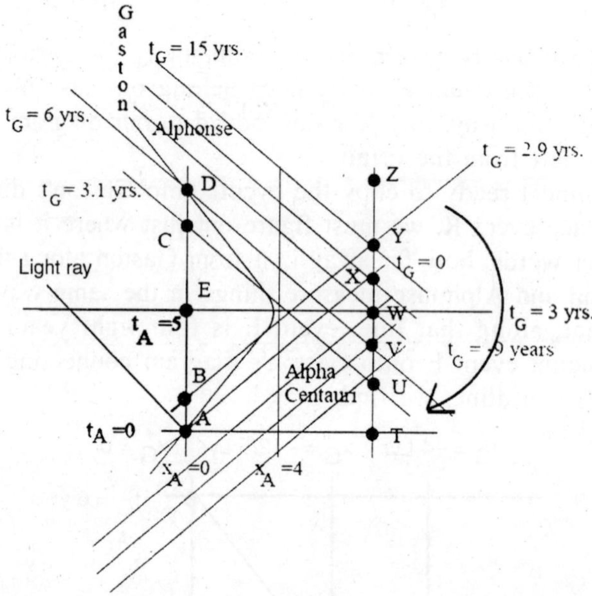

Once again, we simply can use Gaston's lines of simultaneity to mark the events (T,U,V,W, *X*,Y,Z) in Zelda's life on Gaston's diagram. In doing so, however, we find that some of Zelda's events appear on TWO of Gaston's lines of simultaneity - a (magenta) one from before the turnaround and a (green) one from after the turnaround! In fact, many of them (like event W) appear on three lines of simultaneity, as they are caught by a third 'during' the turnaround when Gaston's line of simultaneity sweeps downward from the magenta $t = 2.9$ to the green $t = 3.1$ as indicated by the big blue arrow! Marking all of these events on Gaston's diagram (taking the time to first calculate the corresponding positions) yields something like this:

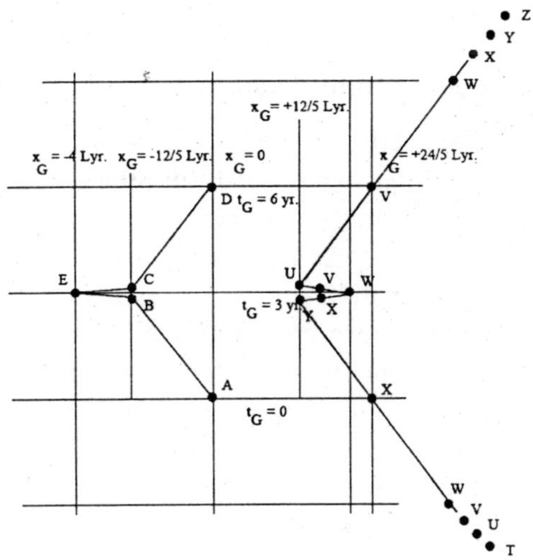

The events T, U,V, W at the very bottom and W, *X*, Y, Z are not drawn to scale, but they indicate that Zelda's worldline is reproduced in that region of the diagram in a more or less normal fashion.

Let us quickly run though Gaston's description of Zelda's life: Zelda merrily experiences events T, U, V, W, *X*, and Y. Then, Zelda is described as "moving backwards in time" through events Y, *X*, W, V, and U. During most of this period she is also described as moving faster than one light-year per year. After Gaston's tG = 3.1year line, Zelda is again described as moving forward in time (at a speed of 4 light-years per 5 years), experiencing events V, W, *X*, Y, for the third time and finally experiencing event Z.

The moral here is that non-inertial reference frames are "all screwed up." Ob-servers in such reference frames are likely to describe the world in a very funny way. To figure out what happens to them, it is certainly best to work in an iner-tial frame of reference and use it to carefully construct the non-inertial spacetime diagram. By the way, there is also the issue of what Gaston would see if he watched Alphonse and Zelda through a telescope. This has to do with the sequence in which light rays reach him, and with the rate at which they reach him.

MORE ON MINKOWSKIAN GEOMETRY

Now that we've ironed out the twin paradox, it's time to talk more about Minkowskian Geometry (a.k.a. "why you should like relativity").

We will shortly see that understanding this geometry makes relativity much simpler. Or, perhaps it is better to say that relativity is in fact simple but that we so far been viewing it through a confusing "filter" of trying to separate space and time. Understanding Minkowskian geometry removes this filter, as we realise that space and time are really part of the same object.

Drawing Proper time and Proper Distance

The notion of the spacetime interval, the interval was a quantity built from both time and space, but which had the interesting property of being the same in all reference frames. We write it as:

$$(\text{interval})^2 = \Delta x^2 - c^2 \Delta t^2.$$

This quantity has two different manifestations: proper time, and proper distance. In essence these are much the same concept. However, it is convenient to use one term (proper time) when the squared interval is negative and another (proper distance) when the squared interval is positive.

Let's draw some pictures to better understand these concepts. The set of all events that are one second of proper time ($\Delta\tau$ = 1sec) to the future of some event (x_0, t_0). We have

$$-(1\text{sec})^2 = -\Delta\tau^2 = \Delta t^2 - \Delta x^2/c^2.$$

Suppose that we take $x_0 = 0$, $t_0 = 0$ for simplicity. Then we have just $x_2/c^2 - t^2 = -(1\text{sec})^2$.

You may recognize this as the equation of a hyperbola with focus at the origin and asymptotes $x = \pm ct$. In other words, the hyperbola asymptotes to the light cone. Since we want the events one second of proper time to the future, we draw just the top branch of this hyperbola:

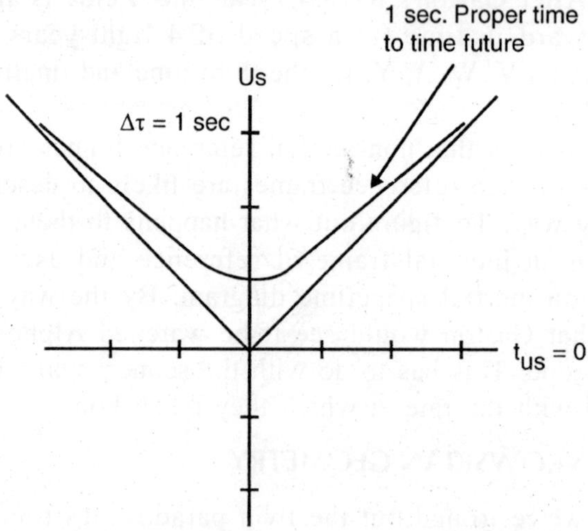

There are similar hyperbolae representing the events one second of proper time in the past, and the events one light-second of proper distance to the left and right.

We should also note in passing that the light light rays form the (somewhat degenerate) hyperbolae of zero proper time and zero proper distance.

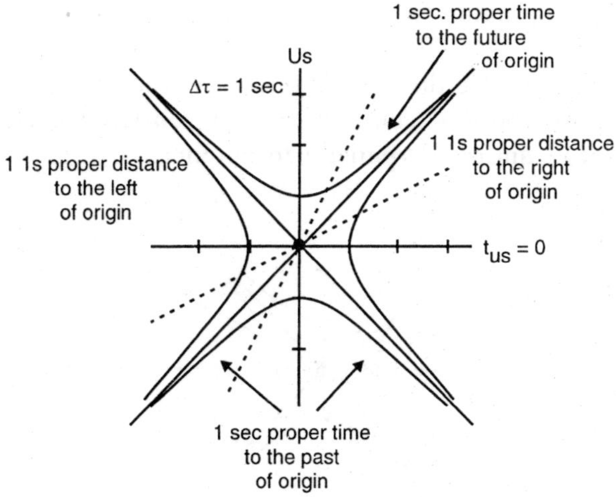

Changing Reference Frames

The worldline and a line of simultaneity for a second inertial observer moving at half the speed of light relative to the first. How would the curves of constant proper time and proper distance look if we re-drew the diagram in this new inertial frame? Stop reading and think about this for a minute.

Because the separation of two events in proper time and proper distance is invariant (i.e., independent of reference frame), these curves must look exactly the same in the new frame.

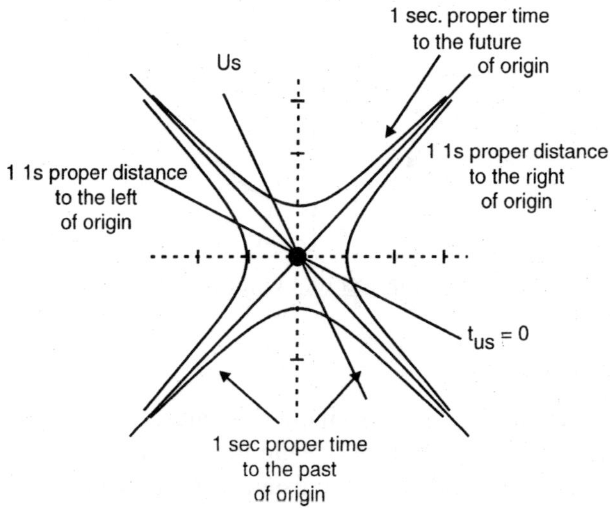

That is, any event which is one second of proper time to the future of some event A (say, the origin in the diagram above) in one inertial frame is also one second of proper time to the future of that event in any other inertial frame and therefore must lie on the same hyperbola $x^2 - c^2t^2 = -(1\,\mathrm{sec})^2$.

The same thing holds for the other proper time and proper distance hyperbolae.

We see that changing the inertial reference frame simply slides events along a given hyperbola of constant time or constant distance, but does not move events from one hyperbola to another.

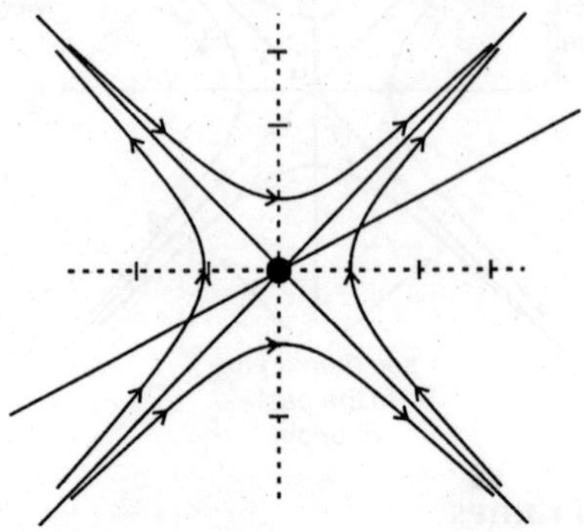

Remember our Euclidean geometry analogue from last time? The above observation is exactly analogous to what happens when we rotate an object4. The points of the object move along circles of constant radius from the axis, but do not hop from circle to circle.

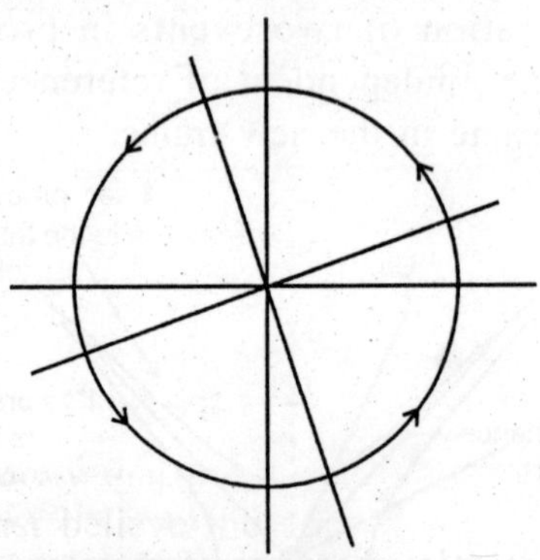

By the way, the transformation that changes reference frames is called a 'boost.'

Hyperbolae, Again

In order to extract the most from our diagrams, let's hit the analogy with circles one last time. If an arbitrary straight line through the centre of a circle, it always intersects the circle a given distance from the centre.

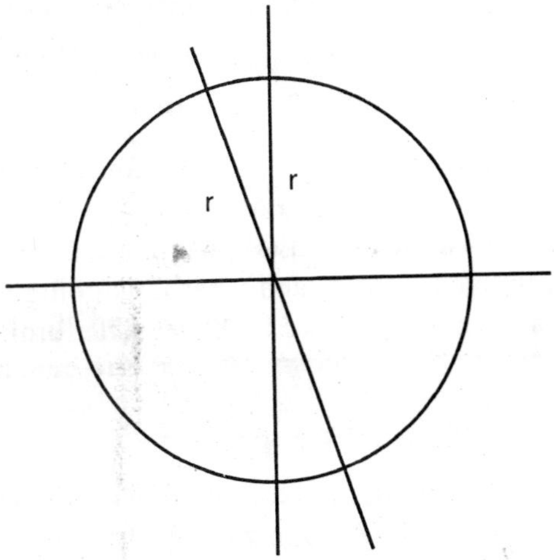

What happens if we draw an arbitrary straight line through the origin of our hyperbolae?

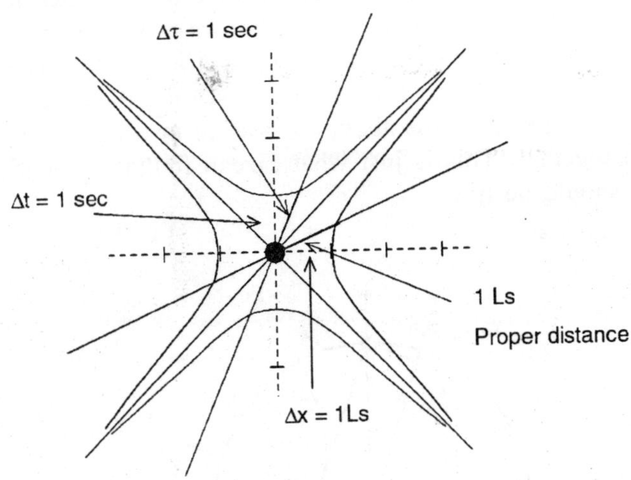

If it is a timelike line, it could represent the worldline of some inertial observer. Suppose that the observer's clock reads zero at the origin. Then the worldline intersects the future $\Delta\tau = 1\text{sec}$ hyperbola at the event where that observers clock reads one second.

Similarly, since a spacelike line is the line of simultaneity of some inertial observer. It intersects the $d = 1\text{Ls}$ curve at what that observer measures to be a distance of 1Ls from the origin.

What we have seen is that these hyperbolae encode the Minkowskian geometry of spacetime.

The hyperbolae of proper time and proper distance (which are different manifestations of the same concept: the interval) are the right way to think about how events are related in spacetime and make things much simpler than trying to think about time and space separately.

Boost Parameters and Hyperbolic Trigonometry

So, you might rightfully ask, what exactly can we do with this new way of looking at things? Let's go back and look at how velocities combine in relativity. This is the question of "why don't velocities just add?" Or, if you are going at 1/2 c relative to Alice, and Charlie is going at 1/2 c relative to you, how fast is Charlie going relative to Alice? The formula looks like

$$v_{AC}/c = \frac{v_{AB}/c + v_{BC}/c}{1 + v_{AB}v_{BC}/c^2}.$$

It is interesting to remark here that this odd effect was actually observed experimentally by Fizeau in the 1850's. He managed to get an effect big enough to see by looking at light moving through a moving fluid. The point is that, when it is moving through water, light does not in fact travel at speed c. Instead, it travels relative to the water at a speed *c/ n* where *n* is around 1.5. The quantity n is known as the 'index of refraction' of water. Thus, it is still moving at a good fraction of "the speed of light." Anyway, if the water is also flowing at a fast rate, then the speed of the light toward us is given by the above expression in which the velocities do not just add together. This is just what Fizeau found5, though he had no idea why it should be true.

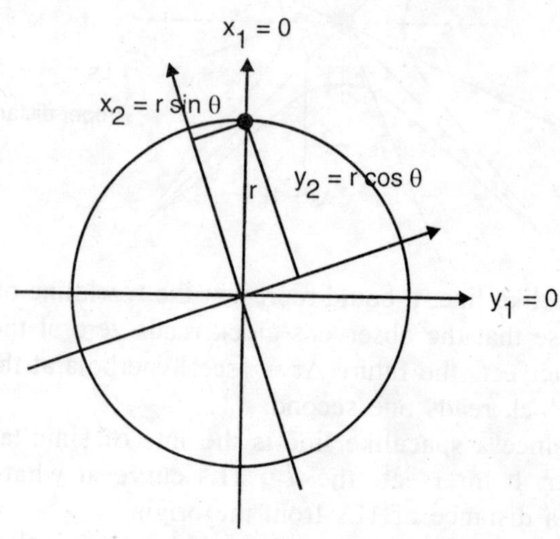

Now, the above formula looks like a mess. Why in the world should the composition of two velocities be such an awful thing? As with many questions, the answer is that the awfulness is not in the composition rule itself, but in the filter (the notion of velocity) through which we view it. We will now see that, when this filter is removed and we view it in terms native to Minkowskian geometry, the result is quite simple indeed.

The analogy between boosts and rotations. How do we describe rotations? We use an angle θ. That rotations mix x and y through the sine and cosine functions.

$$x_2 = r \sin \theta,$$
$$y_2 = r \cos \theta.$$

Note what happens when rotations combine. Well, they add of course. Combining rotations by θ_1 and θ_2 yields a rotation by an angle $\theta = \theta_1 + \theta_2$.

But we often measure things in terms of the slope $m = \frac{x}{y}$ (note the similarity to $v = x/t$). Now, each rotation θ_1, θ_2 is associated with a slope $m_1 = \tan \theta_1$, $m_2 = \tan \theta_2$. But the full rotation by θ is associated with a slope:

$$m = \tan \theta = \tan (\theta_1 + \theta_2)$$
$$= \frac{\tan\theta_1 + \tan\theta_2}{1 - \tan\theta_1 \tan\theta_2}$$
$$= \frac{m_1 + m_2}{1 - m_1 m_2}.$$

So, by expressing things in terms of the slope we have turned a simple addition rule into something much more complicated.

The point here is that the final result bears a strong resemblance to our formula for the addition of velocities. In units where $c = 1$, they differ only by the minus sign in the deominator above. This suggests that the addition of velocities can be simplified by using something similar to, but still different than, the trigonometry above. To get an idea of where to start, recall one of the basic facts associated with the relation of sine and cosine to circles is the relation:

$$\sin^2 \theta + \cos^2 \theta = 1.$$

It turns out that there are other natural mathematical functions called hyperbolic sine (sinh) and hyperbolic cosine (cosh) that satisfy a similar (but different!)

$$\cosh^2 \theta - \sinh^2 \theta = 1,$$

so that they are related to hyperbolae.

These functions can be defined in terms of the exponential function, ex:

$$\sinh \theta = \frac{e^{\theta} - e^{-\theta}}{2}$$

$$\cosh \theta = \frac{e^{\theta} + e^{-\theta}}{2}.$$

You can do the algebra to check for yourself that these satisfy relation above. By the way, although you may not recognize this form, these functions are actually very close to the usual sine and cosine functions. Introducing $i = \sqrt{-1}$, one can write sine and cosine as.

$$\sin\theta = \frac{e^{i\theta} - e^{-i\theta}}{2i}$$

$$\cos\theta = \frac{e^{i\theta} + e^{-i\theta}}{2}$$

Thus, the two sets of functions differ only by factors of i which, as you can imagine, are related to the minus sign that appears in the formula for the squared interval.

Now, consider any event (A) on the hyperbola that is a proper time τ to the future of the origin. Due to the relation 4.8, we can write the coordinates t, x of this event as:

$$t = \tau \cosh\theta,$$
$$x = c\tau \sinh\theta.$$

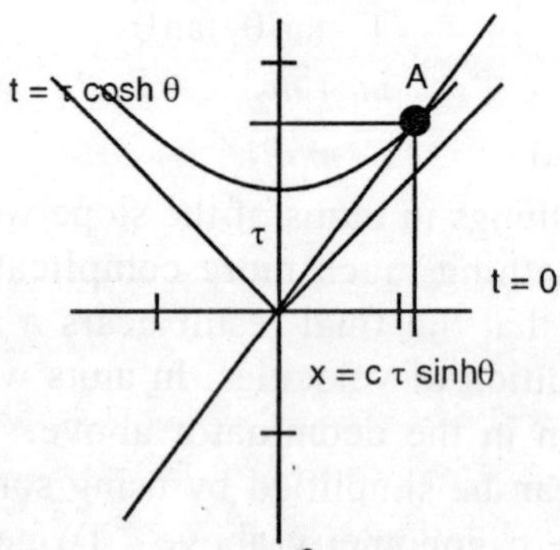

The worldline of an inertial observer that passes through both the origin and event A. Note that the parameter θ gives some notion of how different the two inertial frames (that of the moving observer and that of the stationary observer) actually are. For $\theta = 0$, event A is at $x = 0$ and the two frames are the same, while for large θ event A is far up the hyperbola and the two frames are very different.

We can parameterize the points that are a proper distance d from the origin in a similar way, though we need to 'flip x and t.'

$$t = d/c \sinh\theta,$$
$$x = d \cosh\theta.$$

If we choose the same value of θ, then we do in fact just interchange x and t, "flipping things about the light cone." Note that this will take the worldline of the above inertial observer into the corresponding line of simultaneity. In other words, a given worldline and the corresponding line of simultaneity have the same 'hyperbolic angle,' though we measure this angle from different reference lines ($x = 0$ vs. $t = 0$) in each case.

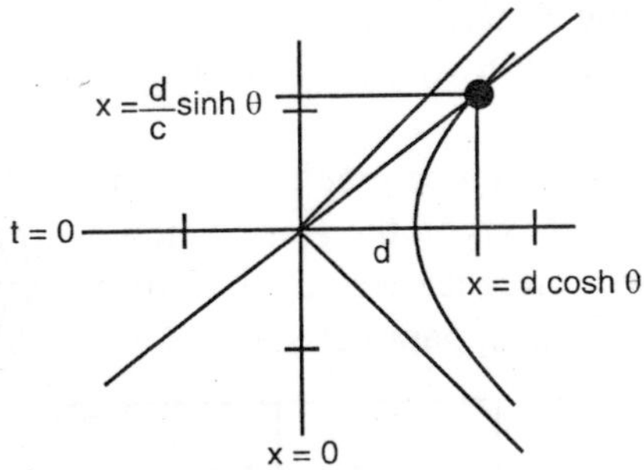

Again, we see that θ is really a measure of the separation of the two reference frames. In this context, we also refer to θ as the boost parameter relating the two frames. The boost parameter is another way to encode the information present in the relative velocity, and in particular it is a very natural way to do so from the viewpoint of Minkowskian geometry.

In what way is the relative velocity v of the reference frames related to the boost parameter θ? Let us again consider the inertial observer passing from the origin through event A on the hyperbola of constant proper time. This observer moves at speed:

$$v = \frac{x}{t} = \frac{c\tau \sinh\theta}{\tau\cosh\theta} = c\frac{\sinh\theta}{\cosh\theta} = c\tanh\theta,$$

and we have the desired relation. Here, we have introduced the hyperbolic tangent function in direct analogy to the more familiar tangent function of trigonometry. Note that we may also write this function as

$$\tanh\,\theta = \frac{e^{\theta} - e^{\theta}}{e^{\theta} + e^{-\theta}}$$

The hyperbolic tangent function may seem a little weird, but we can get a better feel for it by drawing a graph like the one below. The vertical axis is tanh θ and the horizontal axis is θ.

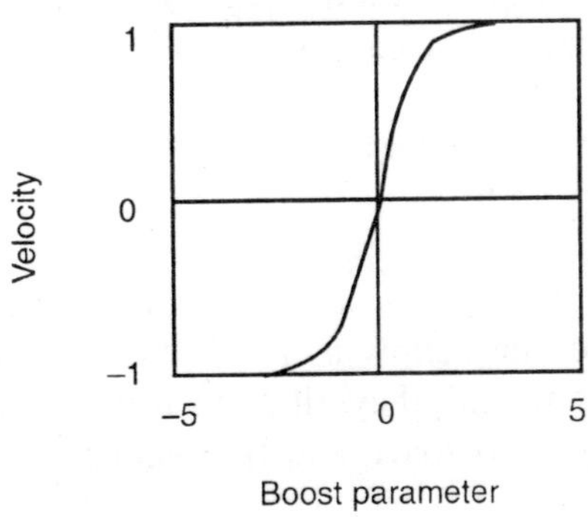

To go from velocity v to boost parameter θ, we just invert the relationship:

$$\theta = \tanh^{-1}(v/c).$$

Here, $\tanh^{-1}$ is the function such that $\tanh^{-1}(\tanh \theta) = \tanh(\tanh^{-1} \theta) = 1$. This one is difficult to write in terms of more elementary functions (though it can be done).

However, we can draw a nice graph simply by 'turning the above picture on its side.' The horizontal axis on the graph below is x and the vertical axis is $\tanh^{-1} x$. Note that two reference frames that differ by the speed of light in fact differ by an infinite boost parameter.

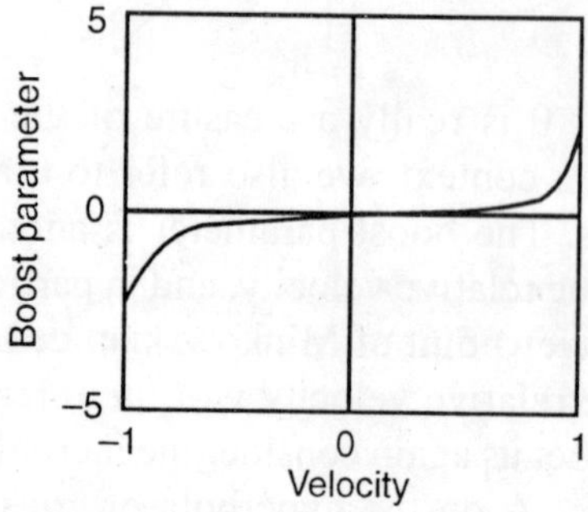

Now for the magic: Let's consider three inertial reference frames, Alice, Bob, and Charlie. Let Bob have boost parameter $\theta_{BC} = \tanh^{-1}(v_{BC}/c)$ relative to Charlie, and let Alice have boost parameter $\theta_{AB} = \tanh^{-1}(v_{AB}/c)$ relative to Bob. Then the relative velocity of Alice and Charlie is

$$v_{AC}/c = \frac{v_{AB}/c + v_{BC}/c}{1 + v_{AB}v_{BC}/c^2}.$$

Let's write this in terms of the boost parameter:

$$v_{AC}/c = \frac{\tanh(\theta_{AB}) + \tanh(\theta_{BC})}{1 + \tanh(\theta_{AB}) + \tanh(\theta_{BC}/c^2)}.$$

After a little algebra, one can show that this is in fact:

$$v_{AC}/c = \tanh(\theta_{AB} + \theta_{BC}).$$

In other words, the boost parameter θ_{AC} relating Alice to Charlie is just the

sum of the boost parameters θ_{AB} and θ_{BC}.

Boost parameters add:

$$\theta_{AC} = \theta_{AB} + \theta_{BC}!!$$

Because boost parameters are part of the native Minkowskian geometry of spacetime, they allow us to see the rule for combining boosts in a simple form. In particular, they allow us to avoid the confusion created by first splitting things into space and time and introducing the notion of "velocity."

2+1 and Higher Dimensional Effects: A re-turn to Stellar Aberration

So, we are beginning to understand how this relativity stuff works, and how it can be self-consistent. Although we now 'understand' the fact that the speed of light is the same in all inertial reference frames (and thus the Michelson-Morely experiment), recall that it was not just the Michelson-Morely experiment that compelled us to abandon the ether and to move to this new point of view. Another very important set of experiments involved stellar aberration (the tilting telescopes) - a subject to which we need to return.

One might think that assuming the speed of light to be constant in all reference frames would remove all effects of relative motion on light, in which case the stellar aberration experiments would contradict relativity. However, we will now see that this is not so.

Stellar Aberration in Relativity

The basic setup of the aberration experiments. Starlight hits the earth from the side, but the earth is "moving forward" so this somehow means that astronomers can't point their telescopes straight toward the star if they actually want to see it. This is shown in the diagram below.

Light Ray hits side instead of reaching bottom

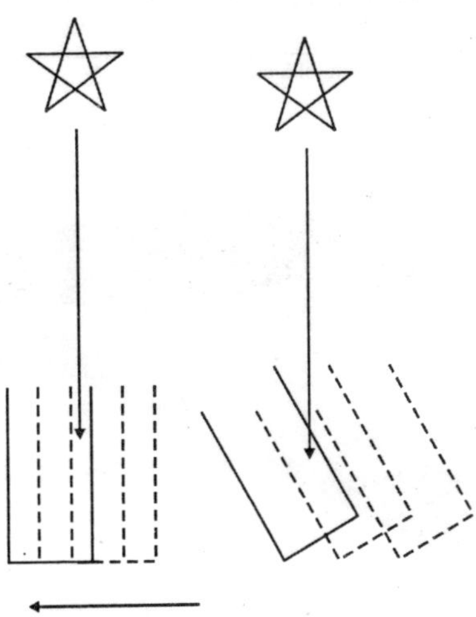

Fig. Telescope Moves Through Ether Must tilt Telescope to See Star

To reanalyze the situation using our new understanding of relativity we will have to deal the fact that the star light comes in from the side while the earth travels forward (relative to the star). Thus, we will need to use a spacetime diagram having three dimensions - two space, and one time. One often calls such diagrams "2+1 dimensional."

These are harder to draw than the 1+1 dimensional diagrams that we have been using so far, but are really not so much different. After all, we have already talked a little bit about the fact that, under a boost, things behave reasonably simply in the direction perpendicular to the action of the boost: neither simultaneity nor lengths are affected in that direction.

We'll try to draw 2+1 dimensional spacetime diagrams using our standard conventions: all light rays move at 45 degrees to the vertical. Thus, a light cone looks like this:

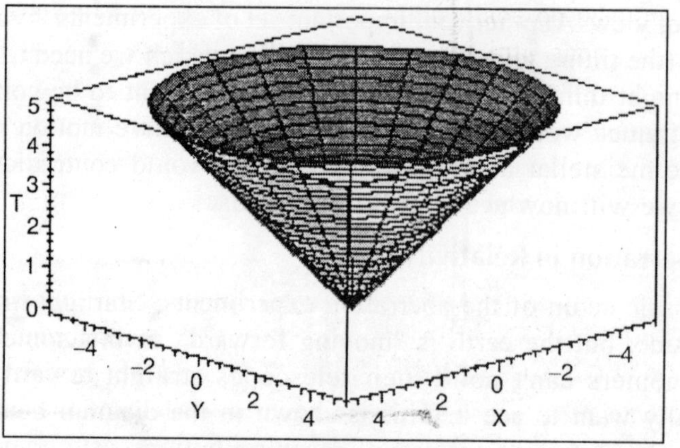

We can also draw an observer and their plane of simultaneity.

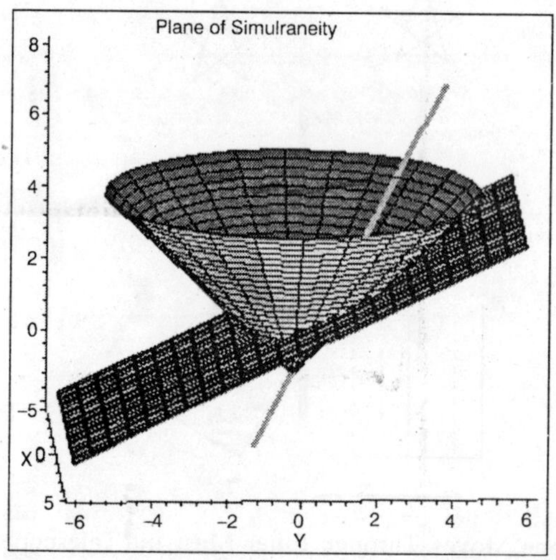

In the direction of the boost, this plane of simultaneity acts just like the lines of simultaneity that we have been drawing. However, in the direction perpendicular to the boost direction, the boosted plane of simultaneity is not tilted. This is the statement that simultaneity is not affeected in this direction.

The moving observer's idea of "right and left," so the plane of events that the moving observer finds to be straight to her right or to her left. Here, the observer is moving across the paper, so her "right and left" are more or less into and out of the paper.

Of course, we would like to know how this all looks when redrawn in the moving observer's reference frame. One thing that we know is that every ray of light must still be drawn along some line at 45 degrees from the vertical. Thus, it will remain on the light cone. However, it may not be located at the same place on this light cone. In particular, note that the light rays direct straight into and out of the page as seen in the original reference frame are 'left behind' by the motion of the moving observer.

That is to say that our friend is moving away from the plane containing these light rays. Thus, in the moving reference frame these two light rays do not travel straight into and out of the page, but instead move somewhat in the "backwards" direction!

This is how the aberration effect is described in relativity. Suppose that, in the reference frame of our sun, the star being viewed through the telescope is "straight into the page." Then, in the reference frame of the sun, the light from the star is a light ray coming straight out from the page. However, in the "moving" reference frame of the earth, this light ray appears to be moving a bit "backwards." Thus, astronomers must point their telescopes a bit forward in order to catch this light ray.

Qualitatively, the aberration effect is actually quite similar in Newtonian and post-Einstein physics. However, the actual amount of the aberration effect observed in the 1800's made no sense to physicists of the time.

This is because, at the quantitative level, the Newtonian and post-Einstein aberration effects are quite different. As usual, the post Einstein version gets the numbers exactly correct, finally tying up the loose ends of 19th century observations.

Einstein's idea that the speed of light is in fact the same in all inertial reference frames wins again.

MORE ON BOOSTS AND THE 2+1 LIGHT CONE: THE HEAD-LIGHT EFFECT

It is interesting to explore the effect of boosts on 2+1 light cones in more detail, as this turns out to uncover two more new effects. Instead of investigating this by drawing lots of three-dimensional pictures, it is useful to find a way to encode the information in terms of a two-dimensional picture that is easily drawn on the blackboard or on paper. We can do this by realizing that the light cone above can be thought of as being made up of a collection of light rays arrayed in a circle.

Some inertial reference frame, perhaps the one in which one of the above diagrams is drawn. That observer finds that light from an "explosion" at the origin moves outward along various rays of light. One light ray travels straight forward, one travels straight to the observer's left, one travels straight to the observer's right, and one travels straight backward.

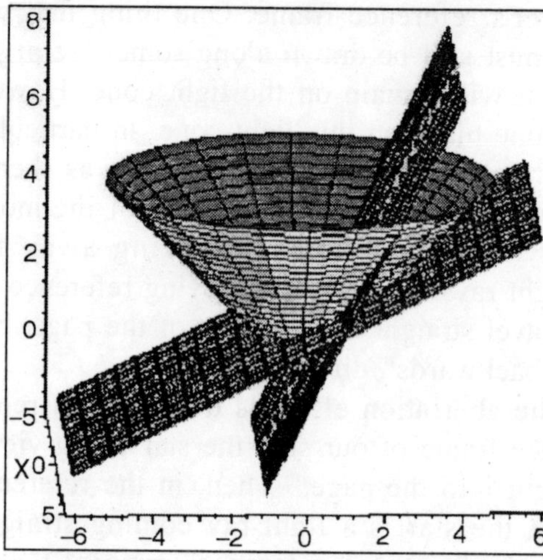

There is one light ray traveling outward in each direction, and of course the set of all directions (in two space dimensions) forms a circle. Thus, we may talk about the circle of light rays. It us convenient to dispense with all of the other parts of the diagram and just draw this circle of light rays. The picture below depicts the circle of light rays in the same reference frame used to draw the above diagram and uses the corresponding colored dots to depict the front, back, left, and right light rays.

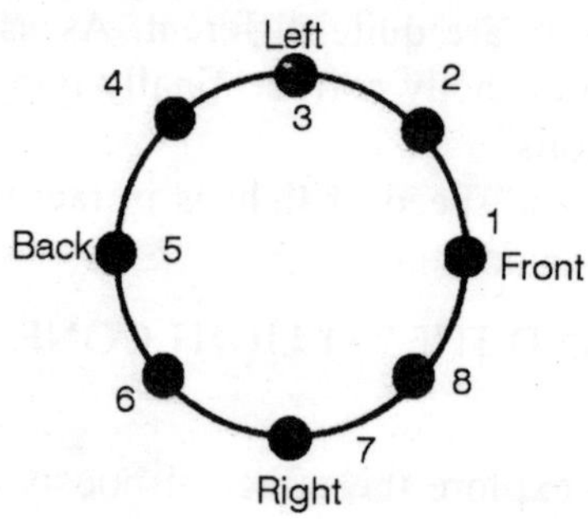

Fig. Light Circle in the Original Frame

Now let's draw the corresponding circle of light rays from the moving observer's perspective. A given light ray from one reference frame is still some light ray in the new reference frame. Therefore, the effect of the boost on the light cone can be described by simply moving the various dots to appropriate new locations on the circle. For example, the light rays that originally traveled straight into and out of the page now fall a bit 'behind' the moving observer. So, they are now moved a bit toward the back. Front, back, left, and right now refer to the new reference frame.

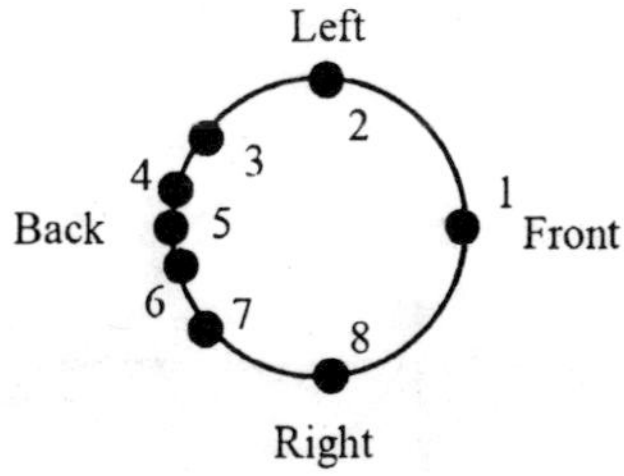

Fig. Light Circle in the Second Frame

Note that most of the dots have fallen toward our current observer's back side - the side which represents the direction of motion of the first observer! Suppose then that the first observer were actually, say, a star like the sun.

In it's own rest frame, a star shines more or less equally brightly in all directions - in other words, it emits the same number of rays of light in all directions. So, if we drew those rays as dots on a corresponding light-circle in the star's frame of reference, they would all be equally spread out as in the first light circle we drew above.

What we see, therefore, is that in another reference frame (with respect to which the star is moving) the light rays do not radiate symmetrically from the star. Instead, most of the light rays come out in one particular direction! In particular, they tend to come out in the direction that the star is moving. Thus, in this reference frame, the light emitted by the star is bright in the direction of motion and dim in the opposite direction and the star shines like a beacon in the direction it is moving. For this reason, this is known as the "headlight" effect.

By the way, this effect is seen all the time in high energy particle accelerators and has important applications in materials science and medicine. Charged particles whizzing around the accelerator emit radiation in all directions as described in their own rest frame. However, in the frame of reference of the laboratory, the radiation comes out in a tightly focussed beam in the direction of the particles' motion. This means that the radiation can be directed very precisely at materials to be studied or tumors to be destroyed.

Multiple Boosts in 2+1 Dimensions

The above two circles of light rays and notice that there is a certain symmetry about the direction of motion. So, suppose you are given a circle of light rays marked with dots which show, as above, the direction of motion of light rays in your reference frame. Suppose also that these light rays were emitted by a star, or by any other source that emits equally in all directions in its rest frame. Then you can tell which direction the star is moving relative to you by identifying the symmetry axis in the circle! There must always be such a symmetry axis. The result of the boost was to make the dots flow as shown below:

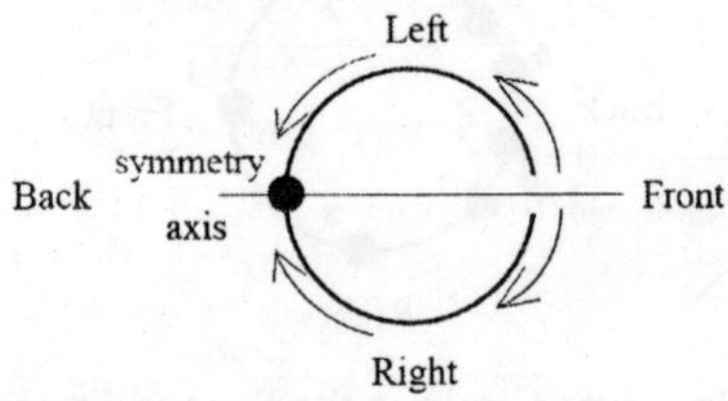

green (font and back) dots are on the symmetry axis, and so do not move at all.

So, just for fun, let's take the case above and consider another observer who is moving not in the forward/backward direction, but instead is moving in the direction that is "left/right" relative to the "moving" observer above. To find out what the dots looks like in the new frame of references, we just rotate the flow shown above by 90 degrees as shown below

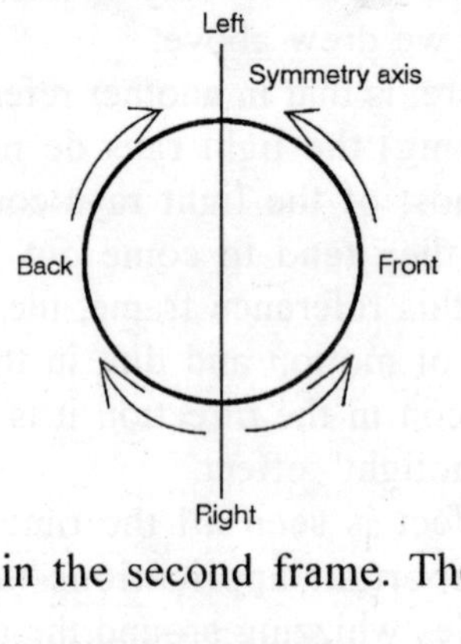

and apply it to the dots in the second frame. The result looks something like this:

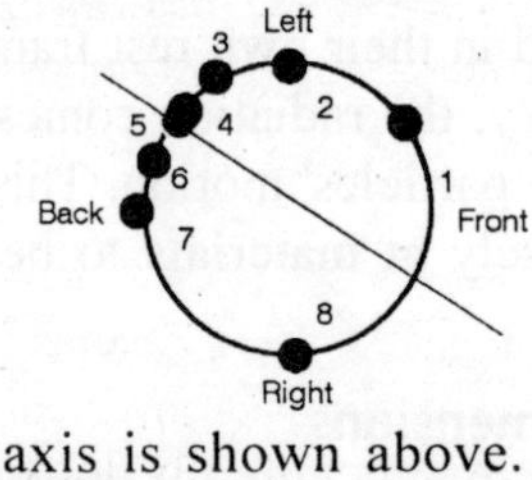

The new symmetry axis is shown above. Thus, with respect to the original observer, this new observer is not moving along a line straight to the right. Instead, the new observer is moving somewhat in the forward direction as well. But wait.... something else interesting is going on here.... the light rays don't line up right. Note that if we copied the above symmetry axis onto the light circle in the original frame, it would sit exactly on top of rays 4 and 8. However, in the figure above the symmetry axis sits half-way between 1 and 8 and 4 and 5. This is the equivalent of having first rotated the light circle in the original frame by 1/16 of a revolution before performing a boost along the new symmetry axis! The new observer differs from the original one not just by a boost, but by a rotation as well!

In fact, by considering two further boost transformations as above (one acting only backward, and then one acting to the right), one can obtain the following circle of light rays, which are again evenly distributed around the circle. You should work through this for yourself, pushing the dots around the circle with care.

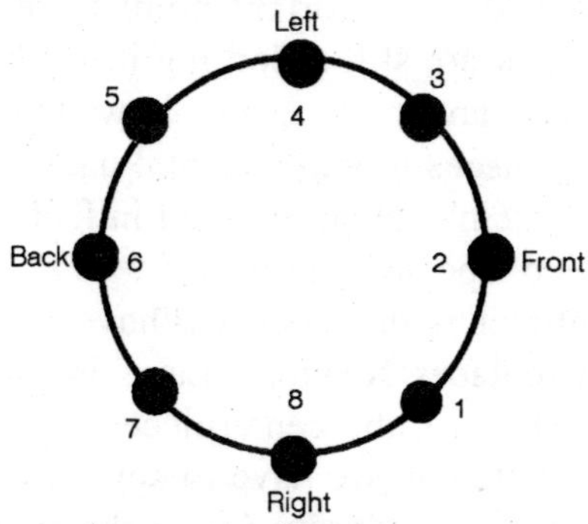

Thus, by a series of boosts, one can arrive at a frame of reference which, while it is not moving with respect to the original fame, is in fact rotated with respect to the original frame. By applying only boost transformations, we have managed to turn our observer by 45 degrees in space. This just goes to show again that time and space are completely mixed together in relativity, and that boost transformations are even more closely related to rotations than you might have thought. A boost transformation can often be thought of as a "rotation of time into space." In this sense the above effect may be more familiar: Consider three perpendicular axes, *x*, y, and z. By performing only rotations about the *x* and y axes, one can achieve the same result as any rotation about the z axes.

Other Effects

Boosts in 3+1 dimensions and higher works pretty much like it they do in 2+1 dimensions, which as we have seen has only a few new effects beyond the 1+1 case on which we spent most of our time. This has to do with how rapidly moving objects actually look; that is, they have to do with how light rays actually reach your eyes to be processed by your brain.

Chapter 5

Accelerating Reference Frames

We have now reached an important point in our study of relativity. Although that many of you are still absorbing it, we have learned the basic structure of the new ideas about spacetime, how they developed, and how they fit with the various pieces of experimental data. We have also finished all of the material in Einstein's Relativity (and in fact in most introductions) associated with so-called 'special relativity.'

One important subject with which we have not yet dealt is that of "dynamics," or, "what replaces Newton's Laws in post-Einstein physics?" Newton's second Law (F=ma), the centerpiece of pre-relativistic physics, in-volves acceleration. Although we have to some extent been able to deal with accelerations in special relativity (as in the twin paradox), we have seen that accelerations produce further unexpected effects. We need to study these more carefully before continuing onward. So, we are going to carefully investigate the simple but illustrative special case known as 'uni-form' acceleration.

THE UNIFORMLY ACCELERATING WORLDLINE

One might at first think that this means that the acceleration a = dv/ dt of some object is constant, as measured in some inertial frame. However, this would imply that the velocity (relative to that frame) as a function of time is of the form v = v0 + at. One notes that this eventually exceed the speed of light. Given our experience to date, this would seem to be a bit odd.

Also, on further reflection, one realizes that this notion of acceleration depends strongly on the choice of inertial frame. The dv part of a involves subtracting velocities, and we have seen that plain old subtraction does not in fact give the relative velocity between two inertial frames. Also, the dt part involves time measurements, which we know to vary greatly between reference frames.

Thus, there is no guarantee that a constant acceleration a as measured in some inertial frame will be constant in any other inertial frame, or that it will in any way "feel" constant to the object that is being accelerated.

Defining Uniform Acceleration

What we have in mind for uniform acceleration is something that does in fact feel constant to the object being accelerated. In fact, we will take this as a definition of "uniform acceleration." We can in fact feel accelerations directly when an airplane takes off, a car goes around a corner, or an elevator begins to move upward we feel the forces associated with this acceleration (as in Newton's law F=ma). To get the idea of uniform acceleration, picture a large rocket in deep space that burns fuel at a constant rate. Here we have in mind that this rate should be constant as measured by a clock in the rocket ship. Presumably the astronauts on this rocket experience the same force at all times.

Newton's laws will need to be modified in relativity. However, we know that Newton's laws hold for objects small velocities (much less than the speed of light) relative to us. These laws are precisely correct in the limit of zero relative velocity.

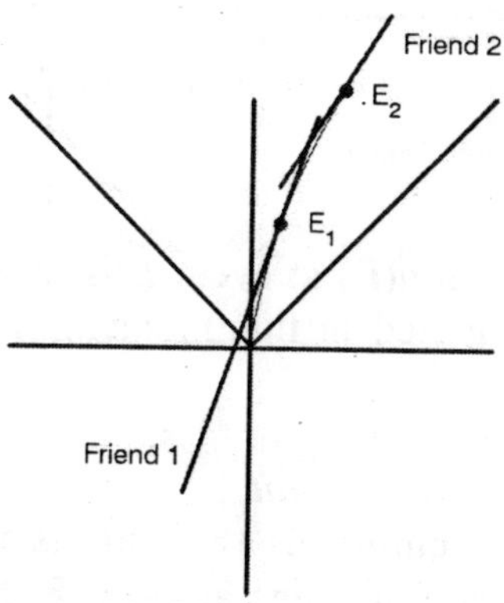

So, how can we keep the rocket "moving slowly" relative to us as it continues to accelerate? We can do so by continuously changing our own reference frame. Perhaps a better way to say this is that we should arrange for many of our friends to be inertial observers, but with a wide range of velocities relative to us. During the short time that the rocket moves slowly relative to us, we use our reference frame to describe the motion. Then, at event E_1 (after the rocket has sped up a bit), we'll use the reference frame of one of our inertial friends whose velocity relative to us matches that of the rocket at event E_1. Then the rocket will be at rest relative to our friend. Our friend's reference frame is known as the momentarily co-moving inertial frame at event E_1. A bit later (at event E_2), we will switch to another friend, and so on.

In fact, to do this properly, we should switch friends (and reference frames) fast enough so that we are always using a reference frame in which the rocket is moving only infinitesimally slowly. Then the relativistic effects will be of zero size. In other words, we wish to borrow techniques from calculus and take the limit in which we switch reference frames continually, always using the momentarily co-moving inertial frame.

Anyway, the thing that we want to be constant in uniform acceleration is called the "proper acceleration." Of course, it can change along the rocket's worldline (depending on how fast the rocket decides to burn fuel), so we should talk about the proper acceleration 'at some event (E) on the rocket's worldline.' To find the proper acceleration (α) at event E, first consider an inertial reference frame in which the rocket is at rest at event E.

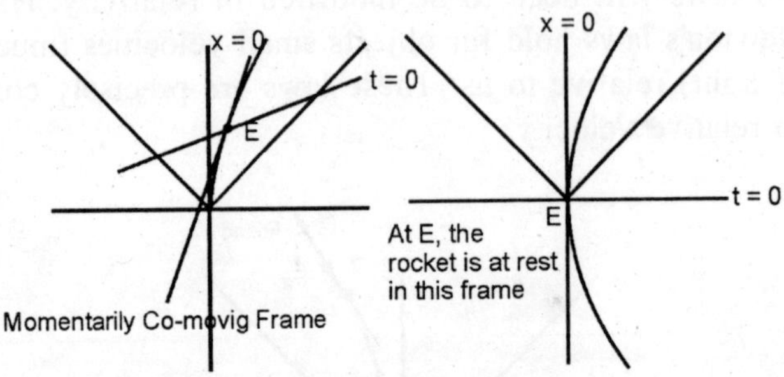

The proper acceleration α(E) at event E is just the acceleration of the rocket at event E as computed in this momentarily co-moving reference frame.

Thus we have

$$\alpha(E) = dv_E/dt_E,$$

where the E -subscripts remind us that this is to be computed in the momentarily co-moving inertial frame at event E. Notice the analogy with the definition of proper time along a worldline, which says that the proper time is the time as measured in a co-moving inertial frame (i.e., a frame in which the worldline is at rest).

An important point is that, although our computation of α(E) involves a discussion of certain reference frames, α(E) is a quantity that is intrinsic to themotion of the rocket and does not depend on choosing of some particular inertial frame from which to measure it. Thus, it is not necessary to specify an inertial frame in which α(E) is measured, or to talk about α(E) "relative" to some frame. As with proper time, we use a Greek letter (α) to distinguish proper acceleration from the more familiar frame-dependent acceleration a.

We should also point out that the notion of proper acceleration is also just how the rocket would naturally measure its own acceleration (relative to inertial frames). For example, a person in the rocket might decide to drop a rock out the window at event E. If the rock is gently released at event E, it will initially have no velocity relative to the rocket - its frame of reference will be the momentarily co-moving inertial frame at event E.

If the observer in the rocket measures the relative acceleration between the rock and the rocket, this will be the same size (though in the opposite direction) as the acceleration of the rocket as measured by the (inertial and momentarily co-moving) rock. In other words, it will be the proper acceleration of the rocket.

Uniform Acceleration and Boost Parameters

So, now we know what we mean by uniform acceleration. But, it would be useful to know how to draw this kind of motion on a spacetime diagram (in some inertial frame). In other words, we'd like to know what sort of worldline this rocket actually follows through spacetime.

There are several ways to approach this question, to use some of the tools that we've been developing. Uniform acceleration is a very natural notion that is not tied to any particular reference frame. We also know that, in some sense, it involves a change in velocity and a change in time. One might expect the discussion to be simplest if we measure each of these in the most natural way possible, without referring to any particular reference frame.

The natural way to describe velocity (Minkowskian geometry) is in terms of the associated boost parameter θ. Boost parameters really do add together in the simple, natural way. This means that when we consider a difference of two boost parameters (like, say, in $\Delta\theta$ or $d\,\theta$), this difference is in fact independent of the reference frame in which it is computed. The boost parameter of the reference frame itself just cancels out.

What about measuring time? The 'natural' measure of time along a worldline is the proper time. The proper time is again independent of any choice of reference frame. Let's again think about computing the proper acceleration $\theta(E)$ at some event E using the momentarily co-moving inertial frame. We have

$$\alpha(E) = \frac{dv_E}{dt_E} \, .$$

What we want to do is to write dvE and dtE in terms of the boost parameter (θ) and the proper time (τ). Let's start with the time part. The proper time τ along the rocket's worldline is just the time that is measured by a clock on the rocket.

Thus, the question is just "How would a small time interval dτ measured by this clock (at event E) compare to the corresponding time interval dtE measured in the momentarily co-moving inertial frame?" But we are interested only in the infinitesimal time around event E where there is negligible relative velocity between these two clocks. Clocks with no relative velocity measure time intervals in exactly the same way. So, we have dt_E = dτ.

Now let's work in the boost parameter, using dθ to replace the dvE in equation. The boost parameter θ is just a function of the velocity v/c = tanh θ. So, let's try to compute dvE/dtE using the chain rule. You can use the definition of tanh to check that

$$\frac{dv}{d\theta} = \frac{c}{\cosh^2\theta}$$

$$\frac{dv}{d\tau} = \frac{dv}{d\theta}\frac{dv}{d\tau} = \frac{c}{\cosh^2\theta}\frac{d\theta}{d\tau}.$$

Thus, we have

Finally, note that at event E, the boost parameter θ of the rocket relative to the momentarily co-moving inertial frame is zero. So, if we want $\frac{dv_E}{d\tau}$ we should substitute θ = 0 into the above equation:

$$\frac{dv}{d\tau} = \frac{dv_E}{d\tau} = \frac{c}{\cosh^2\theta}\bigg|_{\theta=0}\frac{d\theta}{d\tau} = c\frac{d\theta}{d\tau}$$

In other words,

$$\frac{d\theta}{d\tau} = \alpha/c.$$

d θ and *dr* do not in fact depend on a choice of inertial reference frame. The relation holds whether or not we are in the momentarily co-moving inertial frame.

If we translate equation into words, it will come as no surprise: "An object that experiences uniform acceleration gains the same amount of boost parameter for every second of proper time; that is, for every second of time measured by a clock on the rocket."

It will be useful to solve for the case of uniform α and in which the boost parameter (and thus the relative velocity) vanishes at τ = 0. For this case, yields the relation:

$$\theta = \alpha\tau/c.$$

This statement encodes a particularly deep bit of physics. In particular. it turns out to answer the question "Why can't an object go faster than the speed of light?" Here, we have considered the simple case of a rocket that tries to continually accelerate by burning fuel at a constant rate.

What we see is that it gains equal boost parameter in every interval of proper time. So, will it ever reach the speed of light ? No. After a very long (but finite) proper time has elapsed the rocket will merely have a large (but finite) boost parameter.

Since any finite boost parameter (no matter how large) corresponds to some v less than c, the rocket never reaches the speed of light. Similarly, it turns out that whether or not the acceleration is uniform, any rocket must burn an infinite amount of fuel to reach the speed of light. Thus, the speed of light (infinite boost parameter) plays the same role in relativity that was played by infinite velocity in Newtonian physics.

Finding the Worldline

We worked out the relation between the proper acceler-ation α of an object, the boost parameter θ that describes the object's motion, and the proper time τ along the object's worldline. This relation was encoded in equation $\frac{d\theta}{d\tau} = \alpha/c$.

This results told us quite a bit, and in particular let to insight into the "why things don't go faster than light" issue. However, we still don't know exactly what worldline a uniformly accelerating object actually follows in some inertial frame. This means that we don't yet really know how to draw the uniformly accelerating object on a spacetime diagram, so that we cannot yet apply our powerful diagrammatic tools to understanding the physics of uniform accelera-tion.

Let's start by drawing the rough qualitative shape of the worldline on a spacetime diagram.

The worldline will have $v = 0$ at $t = 0$, but the velocity will grow with t. The velocity will thus be nearly +c for large positive t and it will be nearly +c for large negative t.

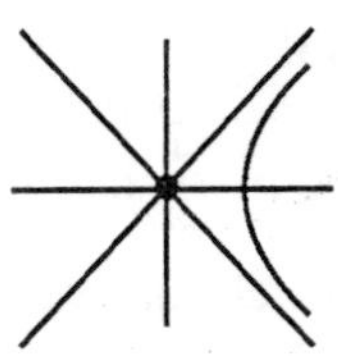

Uniform acceleration is in some sense invariant. When the uniformly accelerated rocket enters our frame of reference (i.e., when $v = 0$), no matter what inertial frame we are in! Thus, the curve should in some sense 'look the same' in every inertial frame.

So, any guesses? Can you think of a curve that looks something like the figure above that is 'the same' in all inertial frames?

How about the constant proper distance curve $x = d \cosh \theta$. Since θ

was a boost angle there, it is natural to guess that it is the same $\theta = \alpha\tau/c$ that we used above.

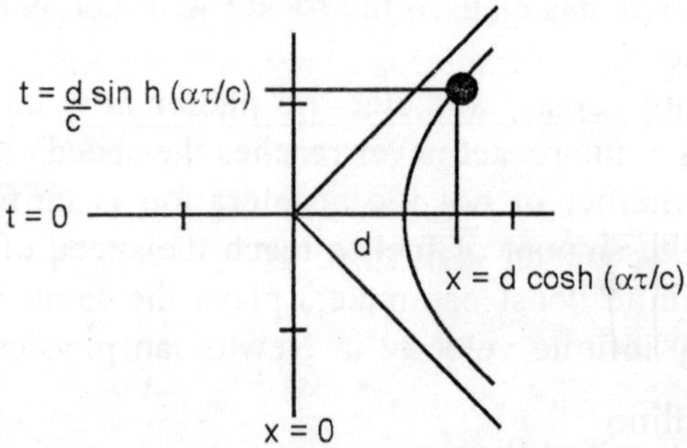

Let us check our guess to see that it is in fact correct. What we will do is to simply take the curve $x = d\cosh(\alpha\tau/c)$, $t = (d/c)\sinh(\alpha\tau/c)$ and show that, for the proper choice of the distance d, its velocity is $v = c\tanh(\alpha\tau/c)$, where τ is the proper time along the curve.

But we have seen that this relation between time is the defining property of a uniformly accelerated worldline with proper acceleration α, so this will indeed check our guess.

First, we simply calculate:

$$dx = \frac{\alpha d}{c}\sinh(\alpha\tau)d\tau$$

$$dt = \frac{\alpha d}{c^2}\cosh(\alpha\tau)d\tau.$$

Dividing these two equations we have

$$v = \frac{dx}{dt} = c\tanh(\alpha\tau/c);$$

i.e., θ is indeed $\alpha\tau/c$ along this curve. Now, we must show that τ is the proper time along the curve. But

$$d\text{propertime}^2 = dt^2 - \frac{1}{c^2}dx^2 = \frac{\alpha^2 d^2}{c^4}d\tau^2.$$

So, we need only choose d such that $\alpha d/c^2 = 1$ and we are done. Thus, $d = c^2/\alpha$. In summary,

If we start a uniformly accelerated object in the right place (c^2/α away from the origin), it follows a worldline that remains a constant proper distance (c^2/α) from the origin.

For a general choice of starting location (say, x_0), it follows a worldline that remains a constant proper distance $\frac{c^2}{\alpha}$ from some other event. Since it is some-times useful to have this more general equation, let us write it down here:

$$x - x_0 = \frac{c^2}{\alpha}\left(\cosh\left(\frac{\alpha\tau}{c}\right) - 1\right).$$

Exploring the Uniformly Accelerated Reference Frame

We have now found that a uniformly accelerating observer with proper acceleration α follows a worldline that remains a constant proper distance c^2/α away from some event.

Just which event this is depends on where and when the observer began to accelerate. For simplicity, let us consider the case where this special event is the origin. Let us now look more closely at the geometry of the situation.

Horizons and Simultaneity

The diagram below shows the uniformly accelerating worldline together with a few important light rays.

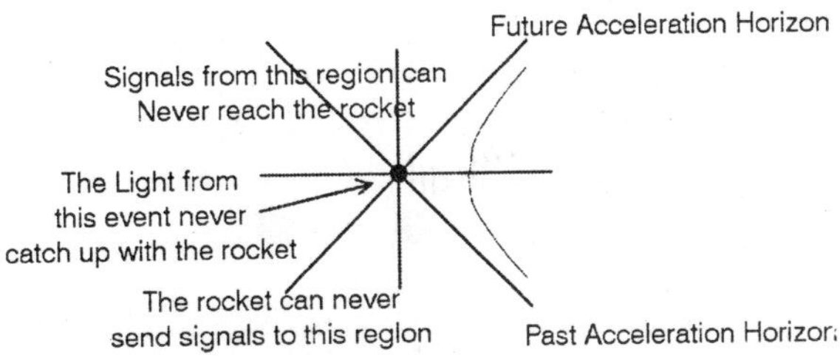

Note the existence of the light ray marked "future acceleration horizon." It marks the boundary of the region of spacetime from which the uniformly accel- erated observer can receive signals, since such signals cannot travel faster than c.

This is an interesting phenomenon in and of itself: merely by undergoing uniform acceleration, the rocket ship has cut itself o from communication with a large part of the spacetime. In general, the term 'horizon' is used whenever an object is cut o in this way. On the diagram above there is a light ray marked "past acceleration horizon" which is the boundary of the region of spacetime to which the uniformly accelerated observer can send signals.

When considering inertial observers, we found it very useful to know how to draw their lines of simultaneity and their lines of constant position. Presumably, we will learn equally interesting things from working this out for the uniformly accelerating rocket.

But, what notion of simultaneity should the rocket use? Let us define the rocket's lines of simultaneity to be those of the associated momentarily co-moving inertial frames. It turns out that these are easy to draw. Let us simply pick any event A on the uniformly accelerated worldline. A Z the event from which the worldline maintains a constant proper distance.

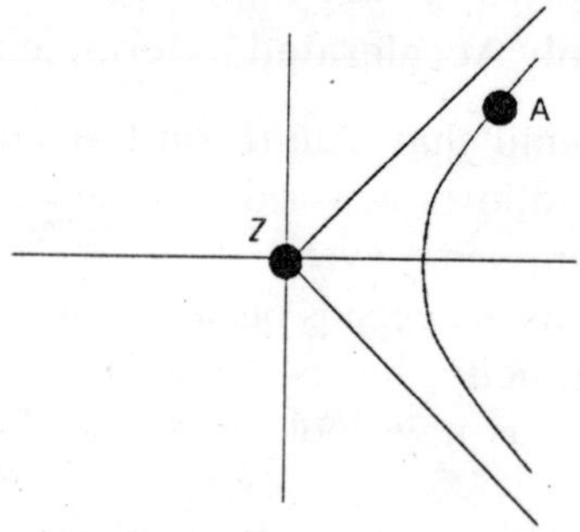

A boost transformation simply slides the events along the hyperbola.

This means that we can find an inertial frame in which the above picture looks like this:

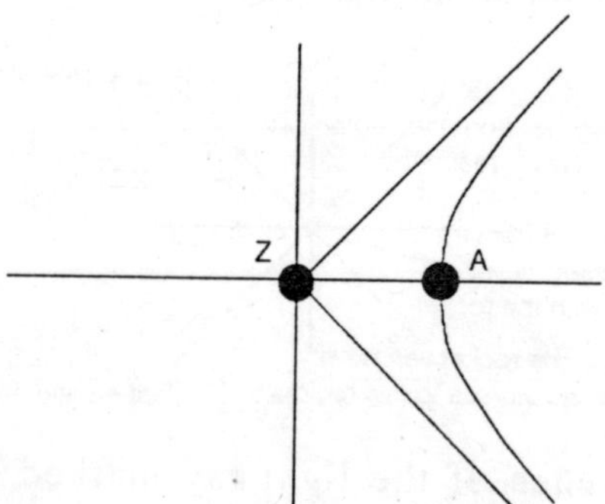

In the new frame of reference, the rocket is at rest at event A. Therefore, the rocket's line of simultaneity through A is a horizontal line. Note that this line passes through event Z.

This makes the line of simultaneity easy to draw on the original diagram. What we have just seen is that: Given a uniformly accelerating observer, there is an event Z from which it maintains proper distance. The observer's line of simultaneity through any event A on her worldline is the line that connects event A to event Z.

Thus, the diagram below shows the rocket's lines of simultaneity.

Let me quickly make one comment here on the passage of time. Suppose that events –2, – 1 above are separated by the same sized boost as events –1, 0, events 0, 1, and events 1, 2. From the relation $\theta = \alpha\tau/c^2$ it follows that each such pair of events is also separated by the same interval of proper time along the worldline.

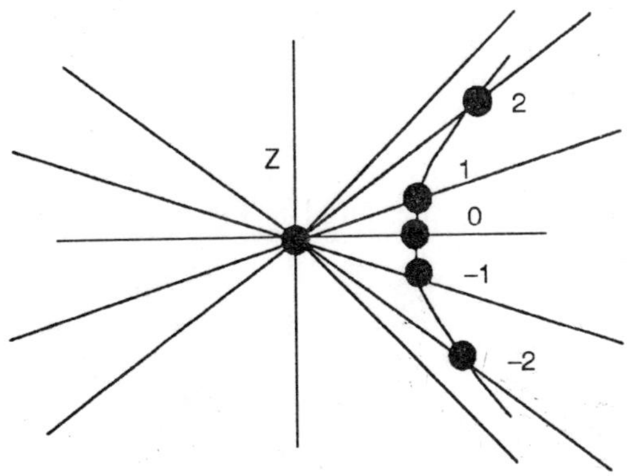

But now on to the more interesting features of the diagram above! Note that the acceleration horizons divide the spacetime into four regions. In the right-most region, the lines of simultaneity look more or less normal. However, in the top and bottom regions, there are no lines of simultaneity at all! The rocket's lines of simultaneity simply do not penetrate into these regions. Finally, in the left-most region things again look more or less normal except that the labels on the lines of simultaneity seem to go the wrong way, 'moving backward in time.'

And, of course, all of the lines of simultaneity pass through event Z where the horizons cross. These strange-sounding features of the diagram should remind you of the weird effects we found associated with Gaston's acceleration in our discussion of the twin paradox.

As with Gaston, one is tempted to ask "How can the rocket see things running backward in time in the left-most region?" In fact, the rocket does not see, or even know about, anything in this region. No signal of any kind from any event in this region can ever catch up to the rocket. This phenomenon of finding things to run backwards in time is a pure mathematical artifact and is not directly related to anything that observers on the rocket actually notice.

Friends on a Rope

We uncovered some odd effects associated with the the acceleration horizons. In particular, we found that there was a region in which the lines of simultaneity seemed to run backward. However, we also found that the rocket could neither signal this region nor receive a signal from it. The fact that the lines of simultaneity run backward here is purely a mathematical artifact.

Despite our discussion above, you might wonder if that funny part of the rocket's reference frame might somehow still be meaningful. It turns out to be productive to get another perspective on this, so let's think a bit about how we might actually construct a reference frame for the rocket.

We would like to know what happens to the ones that lie below the horizon. Let us begin by asking the question: what worldlines do these fellow observers follow?

Consider a friend who remains a constant distance Δ below us as measured by us; that is, as measured in the momentarily co-moving frame of reference. This means that this distance is measured along our line of simultaneity. But look at what this means on the diagram below:

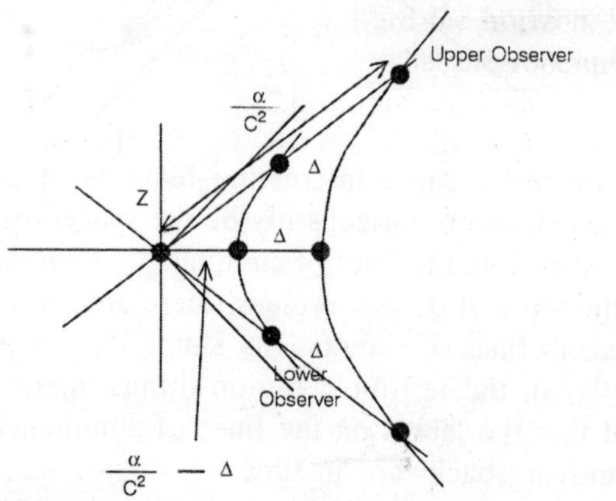

A distance (measured in some inertial frame) between two events on a given a line of simultaneity (associated with that same inertial frame) is in fact the proper distance between those events. Thus, on each line of simultaneity the proper distance between us and our friend is Δ. But, along each of these lines the proper distance between us and event *Z* is α/c^2. Thus, along each of these lines, the proper distance between our friend and event *Z* is α/c^2. Δ. In other words, the proper distance between our friend and event *Z* is again a constant and our friend's worldline must also be a hyperbola! Note, however, that the proper distance between our friend and event Z is less than the proper distance c^2/α between us and event Z. This means that our friend is again a uniformly accelerated observer, but with a different proper acceleration!

We can use the relations to find the proper acceleration α_L of our lower friend. The result is $\dfrac{c^2}{\alpha_L}$ = proper distance between *Z* and lower friend $=\dfrac{c^2}{\alpha}-\Delta$,

or
$$\frac{\alpha_L}{c^2}=\frac{1}{\dfrac{c^2}{\alpha}-\Delta}=\frac{\alpha}{c^2-\Delta\alpha},$$
so that our friend's proper acceleration is larger than our own.

In particular, let's look at what happens when our friend is sufficiently far below us that they reach the acceleration horizons. This is $\Delta = \frac{c^2}{\alpha}$. At this value, we find $\alpha_L = \infty$!! Note that this fits with the fact that they would have to travel right along a pair of light rays and switch between one ray and the other in zero time...

So then, suppose that we hung someone below us on a rope and slowly lowered them toward the horizon. The proper acceleration of the person (and thus the force that the rope must exert on them) becomes infinite as they get near the horizon. Similarly (by Newton's 3rd law) the force that they exert on the rope will become infinite as they near the horizon. Thus, no matter what it is made out of, the rope must break (or begin to stretch, or somehow fail to remain rigid such that the person falls away, never to be seen by us again) before the person is lowered across the horizon.

Again we see that, in the region beyond the horizon, the reference frame of a uniformly accelerating object is "unphysical" and could never in fact be constructed. There is no way to make one of our friends move along a worldline below the horizon that remains at a constant proper distance from us.

The Long Rocket

Suppose now that our rocket is long enough that we should draw separate world-lines for its front and back. If the rocket is 'rigid,' it will remain a constant proper length Δ as time passes. This is just like our 'friend on a rope' example. Thus, the back of the rocket also follows a uniformly accelerated worldline with a proper acceleration α_B which is related to the proper acceleration F of the front by:

$$\alpha_B = \frac{\alpha_F c^2}{c^2 - \Delta\alpha_F}.$$

Clearly, the back and front have different proper accelerations.

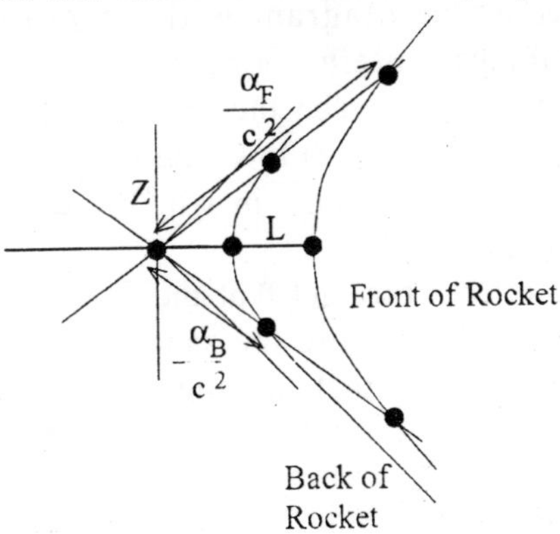

Note that the front and back of the rocket do in fact have the same lines of simul-taneity, so that they agree on which events happen "at the same time." But do they agree on how much time passes between events that are not simultaneous? Since they agree about lines of simultaneity it must be that, along any such line,both ends of the rocket have the same speed v and the same boost parameter θ.

However, because the proper acceleration of the back is greater than that of the front, the relation $\theta = \alpha\tau/c^2$ then tells us that more proper time τ passes at the front of the rocket than at the back. In other words, there is more proper time between the events A_F, B_F below than between events A_B, B_B. In fact, $\alpha_{top}\tau_{top} = \alpha_{bottom}\tau_{bottom}$.

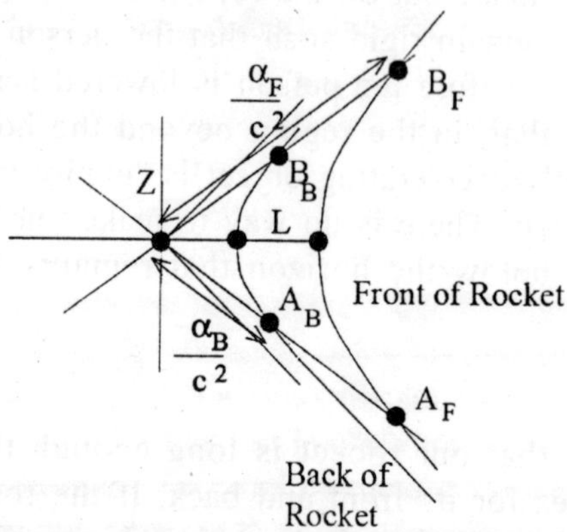

Here it is important to note that, since they use the same lines of simultaneity, both ends of the rocket agree that the front (top) clock runs faster! Thus, this effect is of a somewhat different nature than the time dilation associated with inertial observers. This, of course, is because all accelerated observers are not equivalent - some are more accelerated than others.

By the way, we could have read off the fact that $\Delta\tau_{Front}$ is bigger than $\Delta\tau_{Back}$ directly from our diagram without doing any calculations. (This way of doing things is useful for certain similar homework problems.)

To see this, note that between the two lines ($t = \pm t_0$) of simultaneity (for the inertial frame!!) drawn below, the back of the rocket is moving faster (relative to the inertial frame in which the diagram is drawn) than is the front of the rocket.

You can see this from the fact that the front and back have the same line of simultaneity (and therefore the same speed) at events B_F, B_B and at events A_F, A_B. This means that the speed of the back at B_B is greater than that of the front at D_F and that the speed of the back at A_B is greater than that of the front at C_F.

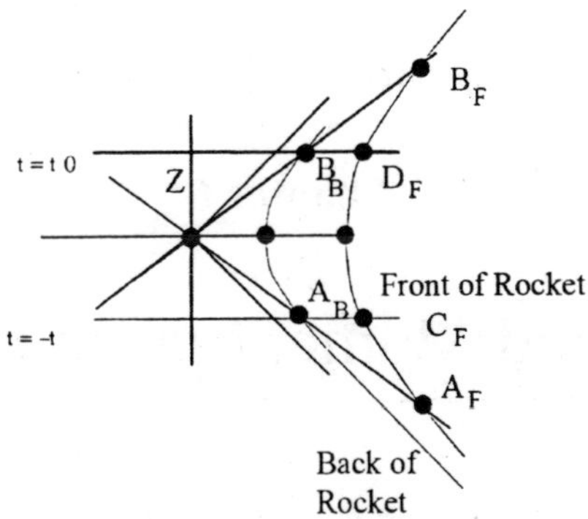

Thus, relative to the inertial frame in which the diagram is drawn, the back of the rocket experiences more time dilation in the interval ($-t_0$, t) and it's clock runs more slowly. Thus, the proper time along the back's worldline between events A_B and B_B is less than the proper time along the front's worldline between events C_F and D_F.

We now combine this with the fact that the proper time along the front's worldline between A_F and B_F is even greater than that between C_F and D_F. Thus, we see that the front clock records much more proper time between A_F and BF than does the back clock between A_B and B_B.

Chapter 6

Energy and Momentum in Relativity

Up until now, we have been concerned mostly with describing motion. We have asked how various situations appear in different reference frames, both inertial and accelerated.

However, we have largely ignored the question of what would make an object follow a given worldline ('dynamics'). The one exception was when we studied the uniformly accelerated rocket and realized that it must burn equal amounts of fuel in equal amounts of proper time. This realization came through using Newton's second law in the regime where we expect it to hold true: in the limit in which v/c is vanishingly small.

DYNAMICS, OR, "WHATEVER HAPPENED TO FORCES?"

That Newton's various laws used the old concepts of space and time. Before we can apply them to situations with finite relative velocity, they will have to be at least rewritten and perhaps greatly modified to accommodate our new understanding of relativity. This was also true for our uniformly accelerating rocket. A constant thrust does not provide a constant acceleration as measured from a fixed inertial reference frame but, instead, it produces a constant proper acceleration.

Now, a central feature of Newton's laws (of much of pre-Einstein physics) was the concept of force. It turns out that the concept of force is not as useful in relativistic physics.

This has something to do with our discovery that accelera-tion is now a frame-dependent concept (so that a statement like F = ma would be more complicated), but the main point actually involves Newton's third law: The third law of Newtonian Physics: When two objects (A and *B)* exert forces FA on *B* and FB on A on each other, these forces have the same size but act in opposite directions.

To understand why this is a problem, let's think about the gravitational forces between the Sun and the Earth.

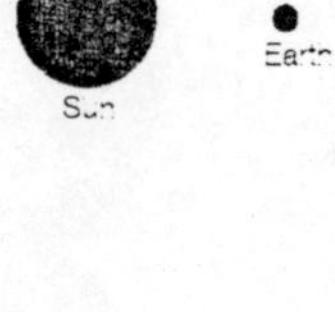

Newton said that the force between the earth and the sun is given by an inverse square law: $F = \dfrac{GM_{\text{earth}}M_{\text{sun}}}{d^2}$ where d is the distance between them. In particular, the force between the earth and sun decreases if they move farther apart. Let's draw a spacetime diagram showing the two objects moving apart.

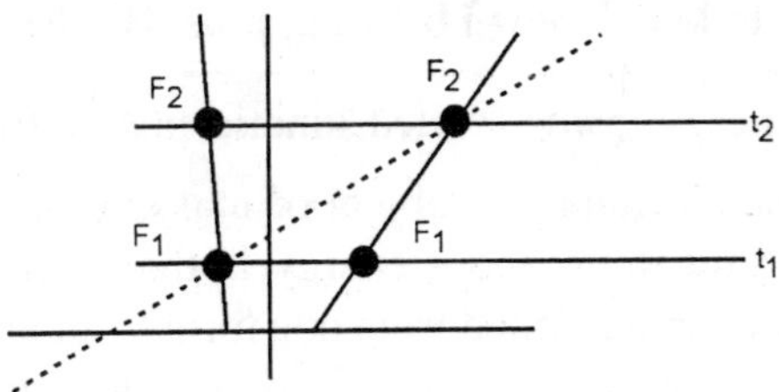

At some time t_1 when they are close together, there is some strong force F_1 acting on each object. Then, later, when they are farther apart, there is some weaker force F_2 acting on each object.

However, what happens if we consider this diagram in a moving reference frame? In a line of simultaneity (the dashed line) for a different reference frame above, and we can see that it passes through one point marked F_1 and one point marked F_2! This shows that Newton's third law as stated above cannot possibly hold1 in all reference frames.

So, Newton's third law has to go. But of course, Newton's third law is not completely wrong - it worked very well for several hundred years! So, as with the law of composition of velocities and Newton's second law, we may expect that it is an approximation to some other law, with this approximation being valid only for velocities that are very small compared to c.

It turns out that this was not such a shock to Einstein, as there had been a bit of trouble with Newton's third law even before relativity itself was understood. Again, the culprit was electromagnetism.

FIELDS, ENERGY, AND MOMENTUM

To see the point, consider an electron in an electric field. We have said that it is really the field that exerts a force on the electron.

Newton's third law would seem to say that the electron then exerts a force on the electric field. But what would this mean? Does an electric field have mass? Can it accelerate?

Luckily for Einstein, this problem had been solved. It was understood that the way out of this mess was to replace the notion of force with two somewhat more abstract notions: energy and momentum. Since not all of you are intimately familiar with these notions, let me say just a few words about them before we continue.

A word on Energy (E)

Most people have an intuitive concept of energy as "what comes out of a power plant" and this is almost good enough for our purposes. Anything which can do something has energy: batteries, light, gasoline, wood, coal (these three can be burned), radioactive substances, food, and so on. Also, any moving object has energy due to it's motion. For example, a moving bowling ball has energy that allows it to knock down bowling pins. By the way, in Newtonian physics, there in an energy $\frac{1}{2}mv^2$ (called 'kinetic energy' from the Greek word for motion) due to the motion of an object of mass m.

The most important thing about energy is that it cannot be created or destroyed; it can only be transformed from one form to another. As an example, in a power plant, coal is burned and electricity is generated. Burning coal is a process in which the chemical energy stored in the coal is turned into heat energy.

This heat energy boils water and creates a rising column of steam (which has energy due to its motion). The column of steam then turns a crank which turns a wire in a magnetic field. This motion converts the mechanical energy of motion of the wire into electrical energy.

Physicists say that Energy is "conserved," which means that the total energy E in the universe can not change.

A Few Words on Momentum (P)

Momentum is a bit less familiar, but it is like energy in that it cannot be created or destroyed: it can only be transferred from one object to another. Thus, momentum is also "conserved." Momentum is a quantity that describes in a certain sense "how much motion is taking place, and in what direction." If the total momentum is zero, we might say that there is "no net motion" of a system. Physicists say that the velocity of the "centre of mass" vanishes in this case.

Let's look back at the bowling ball example above. The energy of the bowling ball is a measure of how much mayhem the ball can cause when it strikes the pins. However, when the ball hits the pins, the pins do not fly about in an arbitrary way. In particular, the pins tend to fly away in more or less the same direction as the ball was moving originally. This is because, when the ball hits the pins, it gives up not only some of its energy to the pins, but also some of its momentum. The momentum is the thing that knows what direction the ball was traveling and makes the pins move in the same direction that the ball was going.

As an example, consider what happens if the bowling ball explodes when it reaches the pins.

This releases more energy (so that the pins fly around more) but will not change the momentum. As a result, the pins and ball shards will have the same net forward motion as would have happened without the explosion. Some pins and shards will now move more forward, but some other bits will also move more backward to cancel this effect.

In Newtonian physics, the momentum p of a moving object is given by the formula $p = mv$. This says that an object that moves very fast has more momentum than one that moves slowly, and an object that has a large mass has more momentum than one with a small mass. This second bit is why it is easier to knock over a bowling pin with a bowling ball than with a ping-pong ball.

By the way, the fact that momentum is a type of object which points in some direction makes it something called a vector. A vector is something that you can visualize as an arrow. The length of the arrow tells you how big the vector is (how much momentum) and the direction of the arrow tells you the direction of the momentum.

Now, in Newtonian physics, momentum conservation is closely associated with Newton's third law. One way to understand this is to realise that both rules (Newton's third law and momentum conservation) guarantee that an isolated system (say, a closed box o in deep space) that begins at rest cannot ever start to move.

In terms of Newton's third law, this is because, if we add up all of the forces between things inside the box, they will cancel in pairs: F_A on B + F_B on $A = 0$. In terms of momentum conservation, it is because the box at rest has zero momentum, whereas a moving box has a nonzero momentum. Momentum conservation says that the total momentum of the box cannot change from zero to non-zero.

In fact, in Newtonian physics, Newton's third law is equivalent to momentum conservation. To see this, consider two objects, A and B, with momenta $pA = mv_A$ and $p_B = mv_B$. Suppose for simplicity that the only forces on these objects are caused by each other. Note what happens when we take a time derivative:

$$\frac{dp_A}{dt} = m_A a_A = FB \text{ on } A,$$

$$= m_B a_B = F_A \text{ on } B.$$

The total momentum is $P_{\text{total}} = p_A + p_B$. We have

$$\frac{dP_{total}}{dt} = F_B \text{ on } A + F_A \text{ on } B = 0.$$

Thus, Newton's 3rd law is equivalent to momentum conservation. One holds if and only if the other does.

Anyway, physicists in the 1800's had understood that there was a problem with Newton's third law when one considered electric fields. It did not really seem to make sense to talk about an electron exerting a force on an electric field. However, it turns out that one can meaningfully talk about momentum carried by an electromagnetic field, and one can even compute the momentum of such a field - say, for the field representing a light wave or a radio wave. Furthermore, if one adds the momentum of the electro-magnetic field to the momentum of all other objects, Maxwell's equations tell us that the resulting total momentum is in fact conserved.

In this way, physicists had discovered that momentum conservation was a slightly more abstract principle that held true more generally than did Newton's third law.

In relativity, too, it turns out to be a good idea to think in terms of momentum and momentum conservation instead of thinking in terms of Newton's third law. For example, in the Sun-Earth, the field between the Sun and the Earth can carry momentum. As a result, momentum conservation does not have to fail if, on some slice of simultaneity, the momentum being gained by the Earth does not equal the momentum being lost by the Sun! Instead, the missing momentum is simply being stored or lost by the field in between the two objects.

ON TO RELATIVITY

Now, while the concepts of momentum and energy can make sense in relativistic physics, the detailed expressions for them in terms of mass and velocity should be somewhat different than in the Newtonian versions. However, as usual we expect that the Newtonian versions are correct in the particular limit in which all velocities are small compared to the speed of light.

There are a number of ways to figure out what the correct relativistic expressions are however, that way of getting at the answer is a bit technical. So, for the moment, we're going to approach the question from a different standpoint.

Einstein noticed that, even within electromagnetism, there was still something funny going on. Momentum was conserved, but this did not necessarily seem to keep isolated boxes (initially at rest) from running away! The example he had in mind was connected with the observation that light can exert pressure.

This was well known in Einstein's time and could even be measured. The measurements were made as early as 1900, while Einstein published his theory of special relativity in 1905. It was known, for example, that pressure caused by light from the sun was responsible for the long and lovely tails on comets: light pressure (also called radiation pressure or solar wind) pushed droplets of water and bits of dust and ice backwards from the comet making a long and highly reflective tail. Nowadays, we can use lasers to lift grains of sand, or to smash things together to induce nuclear fusion.

Lasers in a Box

Anyway, suppose that we start with a box having a powerful laser5 at one end.

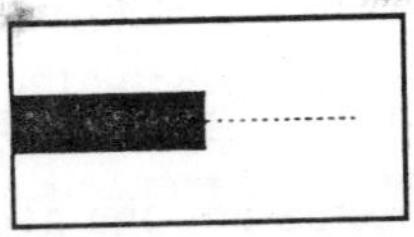

When the laser fires a pulse of light, the light is near the left end and pressure from this light pushes the box to the left. The box moves to the left while the pulse is traveling to the right. Then, when the pulse hits the far wall, its pressure stops the motion of the box. The light itself is absorbed by the wall and disappears.

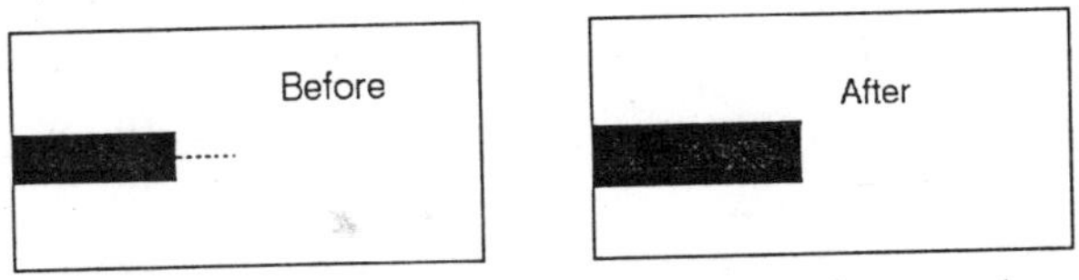

Now, momentum conservation says that the total momentum is always zero. Nevertheless, the entire box seems to have moved a bit to the left. With a large enough battery to power the laser, we could repeat this experiment many times and make the box end up very far to the left of where it started. Or, perhaps we do not even need a large battery: we can imagine recycling the energy used the laser.

If we could catch the energy at the right end and then put it back in the battery, we would only need a battery tiny enough for a single pulse. By simply recycling the energy many times, we could still move the box very far to the left. This is what really worried our friend Mr. Einstein.

Centre of Mass

The moving laser box worried him because of something called the centre of mass. Here's the idea: Imagine yourself in a canoe on a lake. You stand at one end of the canoe and then walk forward. However, while you walk forward, the canoe will slide backward a bit. A massive canoe slides only a little bit, but a light canoe will slide a lot. It turns out that in non-relativistic physics the average position of all the mass (including both you and the canoe) does not move. This average location is technically known as the 'center of mass'.

This follows from Newton's third law and momentum conservation. To understand the point suppose that in the above experiment we throw rocks from left to right instead of firing the laser beam. While most of the box would shift a bit to the left (due to the recoil) with each rock thrown, the rock in flight would travel quite a bit to the right. In this case, a sort of average location of all of the things in the box (including the rock) does not move.

Suppose now that we want to recycle the rock, taking it back to the left to be thrown again. We might, for example, try to throw it back. But this would make the rest of the box shift back to the right, just where it was before. It turns out that any other method of moving the rock back to the left side has the same effect.

To make a long story short, since the average position cannot change, a box can never move itself more than one box-length in any direction, and this can only be done by piling everything inside the box on one side. In fact, when there are no forces from outside the box, the centre of mass of the stuff in the box does not accelerate at all! In general, it is the centre of mass that responds directly to outside forces.

Mass vs. Energy

So, what's going on with our box? Let's look at the experiment more carefully.

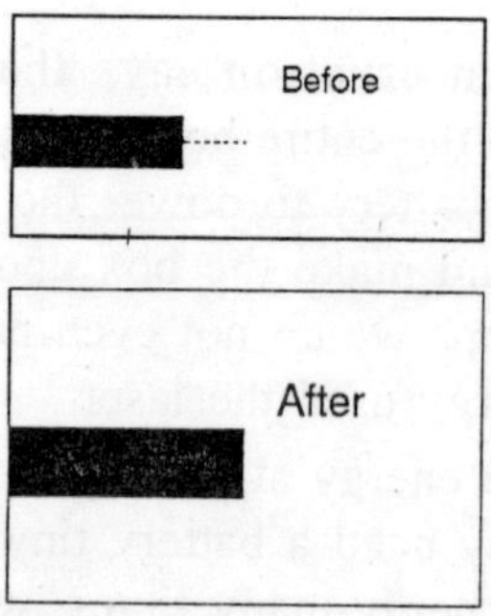

After the experiment, it is clear that the box has moved, and in fact that every single atom in the box has slid to the left. So, the centre of mass seems to have moved! But, Einstein asked, might something else have changed during the experiment which we need to take into account? Is the box after the experiment really identical to the one before the experiment began?

The answer is: "not quite." Before, the experiment, the battery that powers the laser is fully charged. After the experiment, the battery is not fully charged. What happened to the associated energy? It traveled across the box as a pulse of light. It was then absorbed by the right wall, causing the wall to become hot. The net result is that energy has been transported from one end of the box (where it was battery energy) to the other (where it became heat).

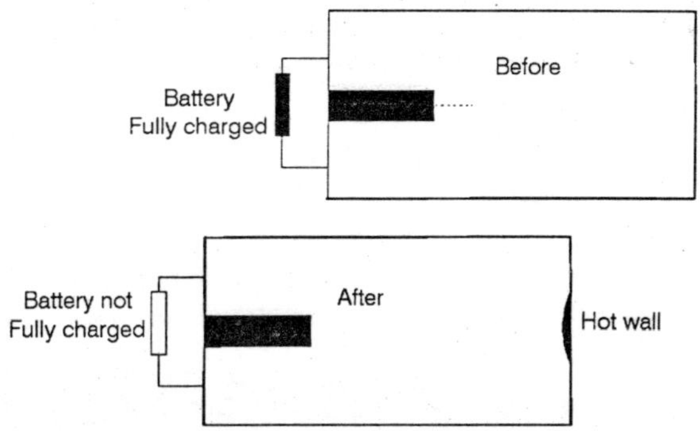

So, Einstein said, "perhaps we should think about something like the centre of energy as opposed to the centre of mass." But, of course, the mass must also contribute to the centre of energy... so is mass a form of energy?

Anyway, the relevant question here is "Suppose we want to calculate the centre of mass/energy. Just how much mass is a given amount of energy worth?" Or, said another way, how much energy is a given amount of mass worth?

Well, from Maxwell's equations, Einstein could figure out the energy transported.

He could also figure out the pressure exerted on the box so that he knew how far all of the atoms would slide. Assuming that the centre of mass-energy did not move, this allowed him to figure out how much energy the mass of the box was in fact worth. The computation is a bit complicated, so we won't do it here6. However, the result is that an object of mass m which is at rest is worth the energy:

$$E = mc^2$$

Note that, since $c^2 = 9 \times 10^{16} \mathrm{m}^2/\mathrm{s}^2$ is a big number, a small mass is worth a lot of energy. Or, a 'reasonable amount' of energy is in fact worth very little mass.

This is why the contribution of the energy to the 'center of mass-energy' had not been noticed in pre-Einstein experiments. Let's look at a few. We buy electricity in 'kilowatt-hours' (kWh) - roughly the amount of energy it takes to run a house for an hour. The mass equivalent of 1 kilowatt-hour is

$$m = \frac{1kWh}{c^2} = \frac{1kWh}{c^2}\frac{3600\sec}{hr.}\frac{1000W}{1kW} = \frac{3.6\times 10^6}{9\times 10^{16}} = 4\times 10^{-10} \text{ kg.}$$

In other words, not much.

By the way, one might ask whether the fact that both mass and energy contribute to the 'center of mass-energy' really means that mass and energy are convertible into one another. Let's think about what this really means. We have a fair idea of what energy is, but what is mass? We have not really talked about this yet in this course, but what Newtonian physicists meant by mass might be better known as 'inertia.' In other words, mass is defined through its presence in the formula $F = ma$ which tells us that the mass is what governs how diffcult an object is to accelerate.

Mass, Energy, and Inertia

So, then, what we really want to know is whether adding energy to an object increases its inertia. That is, is it harder to move a hot wall than a cold wall?

To get some perspective on this, recall that one way to add energy to an object is to speed it up. But we have already seen that rapidly moving objects are indeed hard to accelerate (e.g., a uniformly accelerating object never accelerates past the speed of light). But, this just means that you make the various atoms speed up and move around very fast in random ways. So, this example is really a lot like our uniformly accelerating rocket.

In fact, there is no question about the answer. We saw that heat enters into the calculation of the centre of mass. So, let's think back to the example of you walking in a canoe floating in water. If the canoe is hot, we have seen that it counts more in figuring the centre of mass than when it is cold. It acts like a heavier canoe and will not move as far. Why did it not move as far when you walked in it in the same way? It must have been harder to push; i.e., it had more inertia when it was hot. Thus we conclude that adding energy to a system (say, charging a battery) does in fact give it more inertia; i.e, more mass.

By the way, this explains something rather odd that became known through experiments in the 1920's and 30's, a while after Einstein published his theory of relativity. Atomic nuclei are made out of protons and neutrons. An example is the Helium nucleus (also called an α particle) which contains two neutrons and two protons. However, the masses of these objects are:

Proton mass: 1.675×10^{-27} kg.

Neutron mass: 1.673×10^{-27} kg.

α–Particle mass: 6.648×10^{-27} kg.

So, if we check carefully, we see that an α particle has less mass than the mass of two protons plus the mass of two neutrons. The difference is $m_\alpha - 2m_p - 2m_n = -.0477 \times 10^{-27}$ kg.

Why should this be the case? First note that, since these when these four particles stick together (i.e., the result is stable), they must have less energy when they are close together than when they are far apart. That is, it takes energy to rip them apart. But, if energy has inertia, this means exactly that the object you get by sticking them together will have less inertia (mass) than $2m_p + 2m_n$.

This, by the way, is how nuclear fusion works as a power source. For example, inside the sun, it often happens that two neutrons and two protons will be pressed close together. If they bind together to form an α particle then this releases an extra.0477 $\times 10^{-27}$ kg of energy that becomes heat and light.

Again, it is useful to have a look at the numbers. This amount of mass is worth an energy of $E = mc^2 = 5 \times 10^{-12}$Watt – seconds $\approx 1.4 \times 10^{-15}$ kWh. This may not seem like much, but we were talking about just 2 protons and 2 neutrons. What if we did this for one gram7 worth of stuff? Since four particles, each of which is about 1 Amu of mass, give the above result, one gram would produce the above energy multiplied by $\frac{1}{4}$ of Avagadro's number. In other words, we should multiply by 1.5×10^{23}. This yields roughly 2×10^8kWh = 5kW - years. In other words, fusion energy from 1 gram of material could power 5 houses for one year! Nuclear fission yields comparable results.

By the way, when we consider any other form of power generation (like burning coal or gasoline), the mass of the end products (the burned stu) is again less than the mass of the stuff we started with by an amount that is exactly c^{-2} times the energy released. However, for chemical processes this turns out to be an extremely tiny fraction of the total mass and is thus nearly impossible to detect.

MORE ON MASS, ENERGY, AND MOMENTUM

We saw that what we used to call mass and energy can be converted into each other - and in fact are converted into each other all of the time. Does this mean that mass and energy really are the same thing? Well, that depends on exactly how one defines mass and energy.... the point is that, as with most things in physics, the old (Newtonian) notions of mass and energy will no longer be appropriate.

So, we must extend both the old concept of mass and the old concept of energy before we can even start talking. There are various ways to extend these concepts.

Energy and Rest Mass

My notion of mass will be independent of reference frame. This is not the case for an older convention which has a closer tie to the old $F = ma$. This older convention then defines a mass that changes with velocity. However, for the moment, let me skirt around this issue by talking about the "rest mass" (m_0, by definition an invariant) of an object, which is just the mass (inertia) it has when it is at rest. In particular, an object at rest has inertia $m_0 c^2$.

In Newtonian physics, an object also has an energy $\frac{1}{2}m_0v^2$ due to its motion. Almost certainly, this expression will need to be modified in relativity, but it should be approximately correct for velocities small compared with the speed of light. Thus, for a slowly moving object we have

$$E = m_0c^2 + \frac{1}{2}m_0v^2 + \text{small corrections.}$$

Note that we can factor out an m_0c^2 to write this as:

$$E = m_0c^2(1 + \frac{1}{2}m_0v^2 + \text{small corrections}).$$

The precise form of these small corrections. However, this derivation is somewhat technical and relies on a more in-depth knowledge of energy and momentum in Newtonian physics. It came up there because it gives the first few terms in the Taylor's series expansion of the time-dilation factor,

$$\frac{1}{\sqrt{1 - v^2/c^2}} = 1 + \frac{1}{2}\frac{v^2}{c^2} + \text{small corrections,}$$

a factor which has appeared in almost every equation we have due to its connection to the interval and Minkowskian geometry.

It is therefore natural to guess that the correct relativistic formula for the total energy of a moving object is

$$E = \frac{m_0c^2}{\sqrt{1 - v^2/c^2}} = m_0c^2 \cosh \theta.$$

Momentum and Mass

Momentum is a little trickier, since we only have one term in the expansion so far: $p = mv$ + small corrections. Based on the analogy with energy, we expect that this is the expansion of something native to Minkowskian geometry - probably a hyperbolic trig function of the boost parameter θ. Unfortunately there are at least two natural candidates, m0 c sinh θ and m_0 c tanh θ (which is of course just m_0 v). The detailed derivation is given in 6.6, but it should come as no surprise that the answer is the sinh θ one that is simpler from the point of Minkowskian geometry and which is not the Newtonian answer. Thus the relativistic formula for momentum is:

$$p = \frac{m_0c^2}{\sqrt{1 - v^2/c^2}} = m_0c \sinh \theta.$$

If you don't really know what momentum is, don't worry too much about it. However, that the relativistic formulas for energy and momentum are very important for things you encounter everyday - like high resolution computer graphics.

The light from your computer monitor10 is generated by electrons traveling the speed of light and then hitting the screen. This is fast enough that, if engineers did not take into account the relativistic formula for momentum and tried to use just $p = mv$, the electrons would not land at the right places on the screen and the image would be all screwed up. There are some calculations about this in homework problem.

By the way, you may notice a certain similarity between the formulas for p and E in terms of rest mass m_0 and, say, the position x and the coordinate time t relative to the origin for a moving inertial object in terms of it's own proper time τ and boost parameter θ. In particular, we have

$$\frac{pc}{E} = \frac{v}{c} = \tanh\,\theta$$

We also have

$$E^2 - c^2p^2 = m_0c^4\,(\cosh^2\theta - \sinh^2\theta) = m_0^2\,c^4.$$

Since m0 does not depend on the reference frame, this is an invariant like, say, the interval. Hmm.... The above expression even looks kind of like the interval.... Perhaps it is a similar object? Here is what is going on: a displacement (like Δx, or the position relative to an origin) in general defines a vector - an object that can be thought of like an arrow. Now, an arrow that you draw on a spacetime diagram can point in a timelike direction as much as in a spacelike direction. Furthermore, an arrow that points in a 'purely spatial' direction as seen in one frame of reference points in a direction that is not purely spatial as seen in another frame.

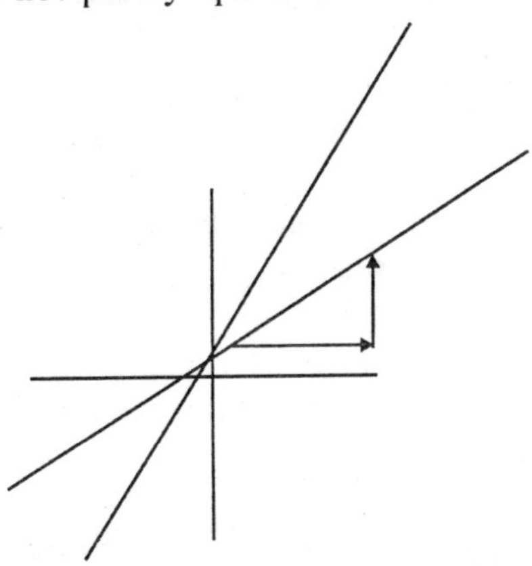

So, spacetime vectors have time parts (components) as well as space parts. A displacement in spacetime involves $c\Delta t$ as much as a Δx. The interval is actually something that computes the size of a given spacetime vector. For a displacement, it is $\Delta x^2 - c\Delta t^2$.Together, the momentum and the energy form a single spacetime vector. The momentum is already a vector in space, so it forms the space part of this vector.

It turns out that the energy forms the time part of this vector. So, the size of the energy-momentum vector is given by a formula much like the one above for displacements. This means that the rest mass m_0 is basically a measure of the size of the energy-momentum vector.

Furthermore, we see that this 'size' does not depend on the frame of reference and so does not depend on how fast the object is moving. However, for a rapidly moving object, both the time part (the energy) and the space part (the momentum) are large - it's just that the Minkowskian notion of the size of a vector involves a minus sign, and these two parts largely cancel against each other.

How About an Example?

As with many topics, a concrete example is useful to understand certain details of what is going on. The point that while energy and momentum are both conserved, mass is not conserved.

Let's suppose we take two electrons and places them in a box. Suppose that both electrons are moving at 4/5*c*, but in opposite directions. If me is the rest mass of an electron, each particle

$$|p| = \frac{m_e v}{\sqrt{1 - v^2/c^2}} = \frac{4}{3} m_e c$$

and an energy

$$E = \frac{m_e v}{\sqrt{1 - v^2/c^2}} = \frac{5}{3} m_e c^2$$

We also need to consider the box. For simplicity, let us suppose that the box also has mass me. But the box is not moving, so it has $p_{\text{Box}} = 0$ and $E_{\text{Box}} = m_e c^2$.

Now, what is the energy and momentum of the system as a whole? Well, the two electron momenta are of the same size, but they are in opposite directions. So, they cancel out. Since $p_{\text{Box}} = 0$, the total momentum is also zero. However, the energies are all positive (energy doesn't care about the direction of motion), so they add together. We find:

$$p_{\text{system}} = 0,$$

$$E_{\text{system}} = \frac{13}{3} m_e c^2.$$

So, what is the rest mass of the systemas a whole?

$$E^2 - p^2 c^2 = \frac{169}{9} m_e^2 c^4$$

So, the rest mass of the positronium system is given by dividing the right hand side by c^4. The result is $\frac{13}{3}m_e$, which is significantly greater than the rest mass of the Box plus twice the rest mass of the electron!

Similarly, two massless particles can in fact combine to make an object with a finite non-zero mass. For example, placing photons in a box adds to the mass of the box.

ENERGY AND MOMENTUM FOR LIGHT

At this point we have developed a good understanding of energy and momentum for objects. However, there has always been one other very important player in our discussions, which is of course light itself. We'll take a moment to explore the energy and momentum of light waves and to see what it has to teach us.

Light Speed and Massless Objects

"What happens if we try to get an object moving at a speed greater than c?" Let's look at the formulas for both energy and momentum. Notice that $E = \frac{m_0 c^2}{\sqrt{1 - v^2/c^2}}$ becomes infinitely large as v approaches the speed of light. Similarly, an object (with finite rest mass m_0) requires an infinite momentum to move at the speed of light. Again this tells us is that, much as with our uniformly accelerating rocket from last week, no finite effort will ever be able to make any object (with $m_0 > 0$) move at speed c. By the way, what happens if we try to talk about energy and momentum for light itself? Many of our formulas fail to make sense for $v = c$. However, some of them do. Consider, for example,

$$\frac{pc}{E} = \frac{v}{c}.$$

Since light moves at speed c through a vacuum, this would lead us to expect that for light we have $E = pc$. In fact, one can compute the energy and momentum of a light wave using Maxwell's equations. One finds that both the energy and the momentum of a light wave depend on several factors, like the wavelength and the size of the wave. However, in all cases the energy and momentum exactly satisfies the relation $E = pc$. We can consider a bit of light (a.k.a., a photon) with any energy E so long as we also assign it a corresponding momentum $p = E/c$. The energy and momentum of photons adds together in just the way. So, what is the rest mass of light? Well, if we compute $m_0^2 c^2 = E^2 - p^2 c^2 = 0$, we find $m_0 = 0$. Thus, light has no mass. This to some extent shows how light can move at speed c and have finite energy. The zero rest mass 'cancels' against the infinite factor coming from $1 - v^2/c^2$ in our formulas above. By the way, note that this also goes the other way: if $m_0 = 0$ then $E = \pm pc$ and so $\frac{v}{c}\frac{pc}{E} = \pm 1$. Such an object has no choice but to always move at the speed of light.

Another Look at the Doppler Effect

For a massive particle (i.e., with $m_0 > 0$), if we are in a frame that is moving rapidly toward the object, the object has a large energy and momentum as measured by us. One might ask if the same is true for light.

The easy way to discover that it is in fact true for light as well is to use the fact (which we have not yet discussed, and which really belongs to a separate subject called 'quantum mechanics,' but what the heck...) that light actually comes in small chunks called 'photons.'

The momentum and energy of a single photon are both proportional to its frequency f, which is the number of times that the corresponding wave shakes up and down every second. The frequency with which the light was emitted in, say, Alphonse's frame of reference was not the same as the frequency at which the light was received in the other frame (Gaston's).

The result was that if Gaston was moving toward Alphonse, the frequency was higher in Gaston's frame of reference. Using the relation between frequency and energy (and momentum), we see that for this case the energy and momentum of the light is indeed higher in Gaston's frame of reference than in Alphonse's frame of reference. So, moving toward a ray of light has a similar effect on how we measure its energy and momentum as does moving toward a massive object.

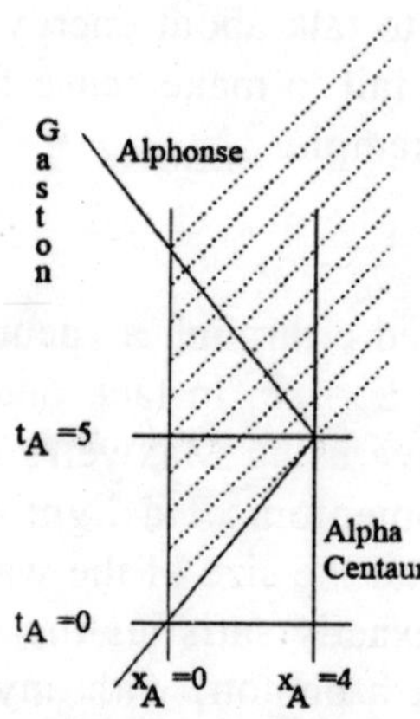

DERIVING THE RELATIVISTIC EXPRESSIONS FOR ENERGY AND MOMENTUM

Due to its more technical nature and the fact that this discussion requires a more solid understanding of energy and momentum in Newtonian physics.

Still, if you're inclined to see just how far logical reasoning can take you in this subject.

It turns out that the easiest way to do the derivation is by focusing on momentum11. The energy part will then emerge as a pleasant surprise. The argument has four basic inputs:

- We know that Newtonian physics is not exactly right, but it is a good approximation at small velocities. So, for an object that moves slowly, it's momentum is well approximated by p = mv.
- We will assume that, whatever the formula for momentum is, momentum in relativity is still conserved. That is, the total momentum does not change with time.
- We will use the principle of relativity; i.e., the idea that the laws o physics are the same in any inertial frame of reference.
- We choose a clever special case to study. We will look at a collisio of two objects and we will assume that this collision is 'reversible.' That is, we will assume that it is possible for two objects to collide in such a way that, if we filmed the collision and played the resulting movie backwards, what we see on the screen could also be a real collision. In Newtonian physics, such collisions are called elastic because energy is conserved.

Let us begin with the observation that momentum is a vector. In Newtonian relativity, the momentum points in the same direction as the velocity vector. This follows just from symmetry considerations. It must also be true in relativistic physics. The only special direction is the one along the velocity vector.

It turns out that to make our argument we will have to work with at least two dimensions of space. This is sort of like how we needed to think about sticks held perpendicular to the direction of motion when we worked out the time dilation effect. There is just not enough information if we stay with only one dimension of space.

So, let us suppose that we are in a long, rectangular room. The north and south walls are fairly close together, while the east and west walls are far apart:

Now, suppose that we have two particles that have the same rest mass m0, and which in fact are exactly the same when they are at rest. We will set things up so that the two particles are moving at the same speed relative to the room, but in opposite directions.

We will also set things up so that they collide exactly in the middle of the room, but are not moving exactly along either the north-south axis or the east-west axis. Also, the particles will not quite collide head-on, so that one scatters to each side after the collision. In the reference frame of the room, the collision will look like this:

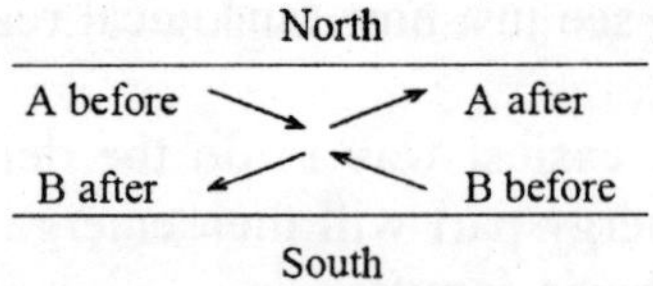

However, we will assume that the particles are nearly aligned with the east-west axis and that the collision is nearly head-on, so that their velocities in the northsouth direction are small.

To proceed, we will analyse the collision in a different reference frame. Suppose that one of our friends (say, Alice) is moving rapidly to the east through the room. If she travels at the right speed she will find that, before the collision and relative to her, particle A does not move east or west but only moves north and south.

We wish to set things up so that the motion of particle A in Alice's frame of reference is slow enough that we can use the Newtonian formula $p = mv$ for this particle in this frame of reference.

For symmetry purposes, we will have another friend Bob who travels to the right fast enough that, relative to him, particle *B* only moves in the north-south direction.

Now, suppose we set things up so that the collision is not only reversible, but in fact looks exactly the same if we run it in reverse. That is, we suppose that in Alice's frame of reference, the collision looks like:

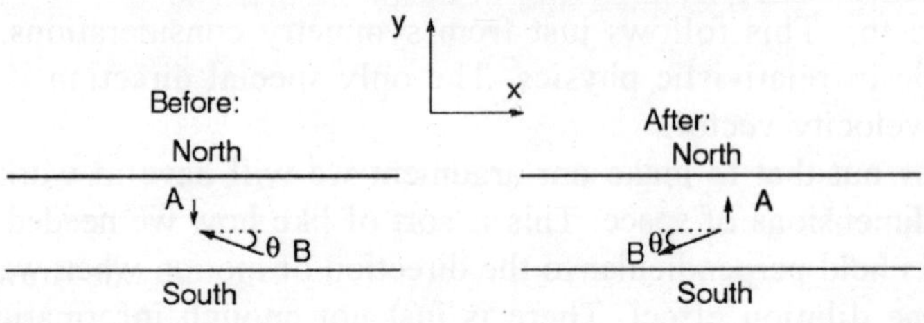

where particle A has the same speed before as it does after, as does particle *B*. Also, the angle is the same both before and after. Such a symmetric situation must be possible unless there is an inherent breaking of symmetry in spacetime.

Now, the velocity of particle A in this frame is to be slow enough that its momentum is given by the Newtonian formula $p_A = m_0 v_A$. For convenience, we take coordinate directions *x* and y on the diagram in Alice's reference frame. It's velocity in the *x* direction is zero, so its momentum in this direction must also be zero.

Thus, particle A only has momentum in the *y* direction. As a result, the change in the momentum of particle A is $2m_0 dy/dt$, where dy/dt denotes the velocity in the y direction after the collision.

If momentum is to be conserved, the total vector momentum must be the same before as after. That is to say, if (in Alice's frame of reference) we add the arrows corresponding to the momentum before, and the momentum after, we must get the same result:

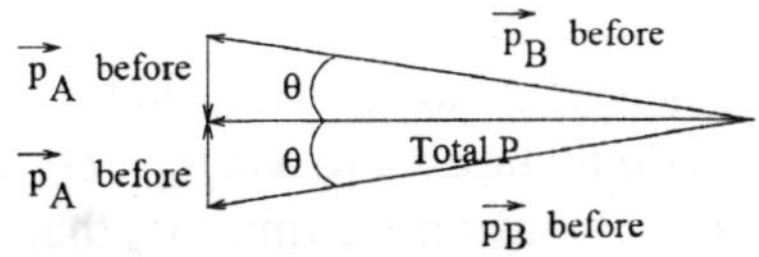

For the next part of the argument notice that if, after the collision, we observe particle *B* for awhile, it will eventually hit the south wall. Let us call this event *Y*, where *B* hits the south wall after the collision. The collision takes place in the middle of the box. Event *Y* and the collision will be separated by some period of time Δt_B (as measured by Alice) and some displacement vector $\Delta \vec{x}_B = (\Delta x_B, \Delta y_B)$ in space as measured by Alice. If the box has some length 2L in the north-south direction, then since the collision took place in the middle, $\Delta y_B = L$.

Also, if we trace particle *B* back in time before the collision, then there was some event before the collision when it was also at the south wall. Let us call this event *X*, when *B* was at the south wall before the collision. By symmetry, this event will be separated from the collision by the same Δt_B and by $-\Delta \vec{x}_B$. The displacement $\Delta \vec{x}_B$ points in the same direction as the momentum of particle *B,* since that is the direction in which *B* moves. Thus, we can draw another nice right triangle:

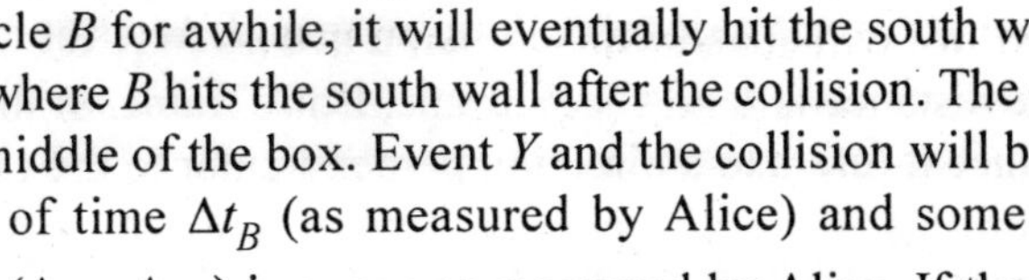

Note that this triangle has the same angle θ as the one drawn above. As a result, we have

$$\frac{L}{|\Delta \vec{x}_B|} = \sin\theta \frac{p_A}{|\vec{p}_B|}.$$

Note that, since p_A has no x component. If this notation bothers you, just replace all my p_A's with p_{Ay}. Here, $|\vec{p}_B|$ is the usual length of this vector, and similarly for $\vec{x}_B$.

Technically speaking, what we will do is to rearrange this formula as

$$\vec{p}_B = \frac{(p_A)(\Delta \vec{x}_B)}{L},$$

where now we have put the direction information back in. We will then compute $\vec{p}_B$ in the limit as the vertical velocity of particle A (and thus p_A, in

Alice's frame) goes to zero. In other words, we will use the idea that a slowly moving particle (in Alice's frame) could have collided with particle B to determine particle B 's momentum.

Let us now take a moment to calculate p_A. In the limit where the velocity of particle A is small, we should be able to use $p_A = m_0\, dy/dt$ after the collision. Now, we can calculate dy/dt by using the time Δt_A that it takes particle A to go from the collision site (in the centre of the box) to the north wall.

In this time, it travels a distance L, so $p_A = m_0 L/\Delta t_A$.

Again, the distance is L in Alice's frame, Bob's frame, or the Box's frame of reference since it refers to a direction perpendicular to the relative motion of the frames.

Thus we have

$$\vec{p}_B = \lim_{v_A \to 0} \frac{m_0(\Delta \vec{x}_B)}{\Delta t_A}$$

This is a somewhat funny formula as two bits (p_B and $\Delta\vec{x}_B$) are measured in the lab frame while another bit t_A is measured in Alice's frame. Nevertheless, the relation is true and we will rewrite it in a more convenient form below.

Now, what we want to do is in fact to derive a formula for the momentum of particle B. This formula should be the same whether or not the collision actually took place.

Thus, we should be able to forget entirely about particle A and rewrite the above expression purely in terms of things having to do with particle B. We can do this by a clever observation.

We originally set things up in a way that was symmetric with re-spect to particles A and B.

Thus, if we watched the collision from particle A's perspective, it would look just the same as if we watched it from particle B's perspective. In particular, we can see that the proper time $\Delta\tau$ between the collision and the event where particle A hits the north wall must be exactly the same as the proper time between the collision and the event where particle B hits the south wall.

Further, recall that we are interested in the formula above only in the limit of small vA. However, in this limit Alice's reference frame coincides with that of particle A.

As a result, the proper time $\Delta\tau$ is just the time Δt_A measured by Alice. Thus, we may replace Δt_A above with $\Delta\tau$.

$$\vec{p}_B = \frac{m_0(\Delta \vec{x}_B)}{\Delta\tau}.$$

The point is that $\Delta\tau$ (proper time) is a concept we undersand in any frame of reference. In particular, we understand it in the lab frame where the two particles (A and *B*) behave in a symmetric manner. Thus, $\Delta\tau$ is identical for both particles.

Note that since $\Delta\tau$ is independent of reference frame, this statement holds in any frame - in particular, it holds in Alice's frame. Thus, the important point about equation is that all of the quantities on the right hand side can be taken to refer only to particle *B*!

In particular, the expression no longer depends on particle A, so the limit is trivial. We have:

Since the motion of *B* is uniform after the collision, we can replace this ratio with a derivative:

$$|\vec{p}_B| = m_0 \frac{d\vec{x}_B}{d\tau} = m_0 \frac{1}{\sqrt{1-v^2/c^2}} \frac{d\vec{x}_B}{dt}$$

Thus, we have derived

$$|\vec{p}| = m_0 \frac{\vec{v}}{\sqrt{1-v^2/c^2}},$$

the relativistic formula for momentum.

Now, the form of equation is rather suggestive. It shows that the momentum forms the spatial components of a spacetime vector:

$$p = m_0 \frac{dx}{d\tau}$$

where x represents all of the spacetime coordinates (t, x, y, z). One is tempted to ask, "What about the time component $m_0 dt/d\tau$ of this vector?"

We have assumed that the momentum is conserved, and that this must therefore hold in every inertial frame. If 3 components of a spacetime vector are conserved in every inertial frame, then it follows that the fourth one does as well. So, this time component does represent some conserved quantity.

We can get an idea of what it is by expanding the associated formula in a Taylor series for small velocity:

$$m_0 \frac{dt}{d\tau} = m_0 \cosh\theta = m_0 \frac{1}{1-v^2/c^2}$$

$$= m_0 \left(1 + \frac{1}{2}\frac{v^2}{c^2} + \cdots\right) = c^{-2}\left(m_0 c^2 + \frac{1}{2} m_0 v^2 + \cdots\right)$$

In Newtonian physics, the first term is just the mass, which is conserved separately. The second term is the kinetic energy. So, we identify this time component of the spacetime momentum as the (c^{-2} times) the energy:

$$E = cp^t = \frac{m_0 c^2}{\sqrt{1 - v^2/c^2}}.$$

In relativity, mass and energy are not conserved separately. Mass and energy in some sense merge into a single concept 'mass-energy.'

Also, we have seen that energy and momentum fit together into a single spacetime vector just as space and time displacements fit together into a 'spacetime displacement' vector.

Thus, the concepts of momentum and energy also merge into a single 'energy-momentum vector.'

Chapter 7

Relativity and Gravitational Field

We finished the part of the course that is re- ferred to as 'Special Relativity'. Now, special relativity by itself was a real achievement. In addition to revolutionizing our conceptions of time and space, uncovering new phenomena, and dramatically changing our understanding of mass, energy, and momentum, Minkowskian geometry finally gave a good picture of how it can be that the speed of light (in a vacuum!) is the same in all frames of reference. However, in some sense there is still a large hole to be filled. We've talked about what happens when objects accelerate, but we have only begun to discuss why they accelerate, in terms of why and how various forces act on these objects.

We have the relation $E = \frac{m_0 c^2}{\sqrt{1 - v^2 / c^2}}$ so we know that, when we feed an object a certain amount of energy it will speed up, and when we take energy away it will slow down. We can even use this formula to calculate exactly how much the object will speed up or slow down. But what we haven't talked about are the basic mechanisms that add and subtract energy - the 'forces' themselves. Of course, physicists already had some understanding of these forces when Einstein broke onto the scene. The important question, of course, is whether this understanding fit well with relativity or whether relativity would force some major change the understanding of the forces themselves.

Physicists in Einstein's time knew about many kinds of forces:

- Electricity.
- Magnetism.
- Gravity.
- Friction.
- One object pushing another.
- Pressure.

and so on..... Now, the first two of these forces are described by Maxwell's equations. As we have discussed, Maxwell's equations fit well with (and even led to!) relativity. Unlike Newton's laws, Maxwell's equations are fully compatible with relativity and require no modifications at all. Thus, we may set these forces aside as 'complete' and move on to the others.

Let's skip ahead to the last three forces. These all have to do in the end with atoms pushing and pulling on each other. In Einstein's time, such things we believed1 to be governed by the electric forces between atoms. So, it was thought that this was also properly described by Maxwell's equations and would fit well with relativity.

You may have noticed that this leaves one force (gravity) as the odd one out. Einstein wondered: how hard can it be to make gravity consistent with relativity?

THE GRAVITATIONAL FIELD

Let's begin by revisiting the pre-relativistic understanding of gravity. Perhaps we will get lucky and find that it too requires no modification.

Newtonian Gravity vs. relativity

Newton's understanding of gravity was as follows:

Newton's Universal Law of Gravity Any two objects of masses m_1 and m_2 exert 'gravitational' forces on each other of magnitude

$$F = G\frac{m_1 m_2}{d^2},$$

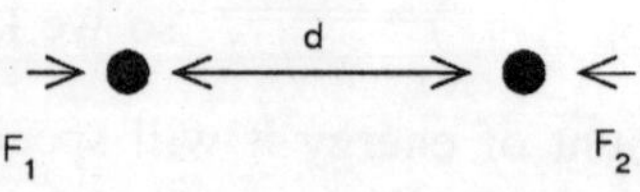

directed toward each other, where $G = 6.67 \times 10^{-11}$ Nm^2/kg^2 is called "Newton's Gravitational Constant." G is a kind of intrinsic measure of how strong the gravitational force is.

It turns out that this rule is not compatible with special relativity. In particular, having learned relativity we now believe that it should not be possible to send messages faster than the speed of light. However, Newton's rule above would allow us to do so using gravity. The point is that Newton said that the force depends on the separation between the objects at this instant.

Example: The earth is about eight light-minutes from the sun. This means that, at the speed of light, a message would take eight minutes to travel from the sun to the earth. However, suppose that, unbeknownst to us, some aliens are about to move the sun.

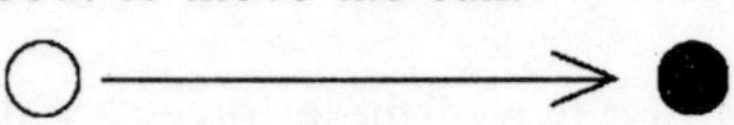

Then, based on our understanding of relativity, we would expect it to take eight minutes for us to find out! But Newton would have expected us to find out instantly because the force on the earth would shift (changing the tides and other things.....)

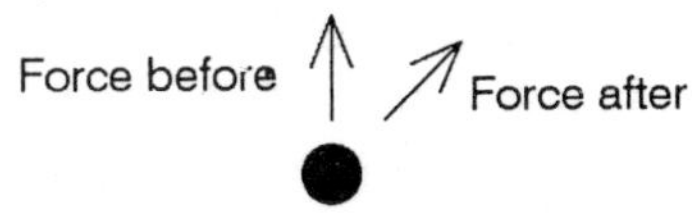

The Importance of the Field

Now, it is important to understand how Maxwell's equations get around this sort of problem. That is to say, what if the Sun were a positive electric charge, the earth were a big negative electric charge, and they were held together by an Electro-Magnetic field? We said that Maxwell's equations are consistent with relativity - so how what would they tell us happens when the aliens move the sun?

The point is that the positive charge does not act directly on the negative charge. Instead, the positive charge sets up an electric field which tells the negative charge how to move.

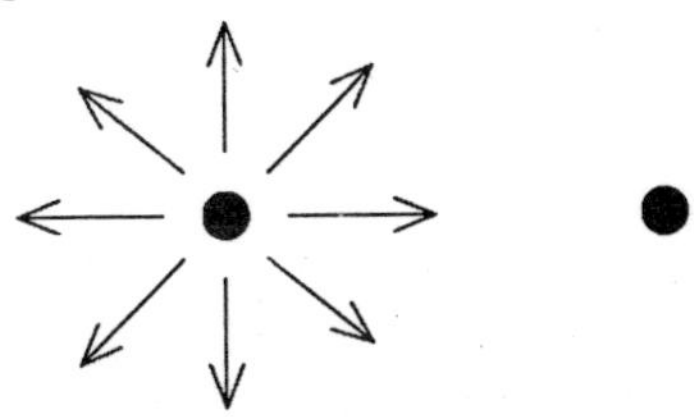

When the positive charge is moved, the electric field around it must change, but it turns out that the field does not change everywhere at the same time.

Instead, the movement of the charge modifies the field only where the charge actually is. This makes a 'ripple' in the field which then moves outward at the speed of light. In the figure below, the black circle is centered on the original position of the charge and is of a size ct, where *t* is the time since the movement began.

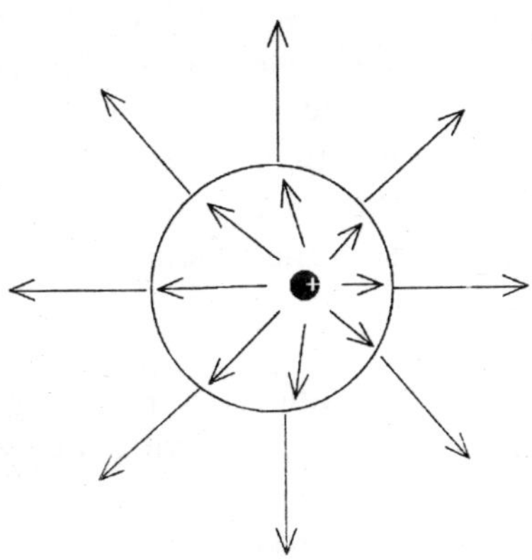

Thus, the basic way that Maxwell's equations get around the problem of instant reaction is by having a field that will carry the message to the other charge (or, say, to the planet) at a finite speed.

Oh, and remember that having a field that could carry momentum was also what allowed Maxwell's equations to fit with momentum conservation in relativity. What we see is that the field concept is the essential link that allows us to understand electric and magnetic forces in relativity.

Something like this must happen for gravity as well. Let's try to introduce a gravitational field by breaking Newton's law of gravity up into two parts.

The idea will again be than an object should produce a gravitational field (g) in the spacetime around it, and that this gravitational field should then tell the other objects how to move through spacetime. Any information about the object causing the gravity should not reach the other objects directly, but should only be communicated through the field.

$$\textit{Old: } F = \frac{m_1 m_2 G}{d^2}$$

$$\textit{New: } F_{\text{on}}\, m_1 = m_1\, g,$$

$$g = \frac{m_2 G}{d}.$$

SOME OBSERVATIONS

General Relativity from a somewhat different point of view than your readings do. The readings are simply stressing different aspects of the various thoughts that were rattling around inside Albert Einstein's head in the early 1900's. BTW, figuring out General Relativity was much harder than figuring out special relativity.

Einstein worked out special relativity is about a year (and he did many other things in that year). In contrast, the development of general relativity required more or less continuous work from 1905 to 1916.

For future reference, they are:

- Free fall and the gravitational field.
- The question of whether light is a ected by gravity.
- Further reflection on inertial frames.

Free Fall

Before going on to the other important ingredients, let's take a moment to make a few observations about gravitational fields and to introduce some terminology.

Notice an important property of the gravitational field. The gravitational force on an object of mass m is given by $F = mg$. But, in Newtonian physics, we also have $F = ma$. Thus, we have

$$a = \frac{mg}{m} = g \ .$$

The result is that all objects in a given gravitational field accelerate at the same rate (if no other forces act on them). The condition where gravity is the only influence on an object is known as "free fall." So, the gravitational field g has a direct meaning: it gives the acceleration of "freely falling" objects.

A particularly impressive example of this is called the 'quarter and feather experiment.' Imagine taking all of the air out of a cylinder (to remove air resistance which would be an extra force), and then releasing a quarter and a feather at the same time. The feather would then then "drops like a rock." In particular, the quarter and the feather fall together in exactly the same way.

Now, people over the years have wondered if it was really true that all objects fall at exactly the same rate in a gravitational field, or if this was only approximately true. If it is exactly correct, they wondered why it should be so. It is certainly a striking fact.

For example, we have seen that energy is related to mass through $E = mc^2$. So, sometimes in order to figure out the exact mass of an object (like a hot wall that a laser has been shining on....) you have to include some things (like heat) that we used to count separately as 'energy'.... Does this E/c^2 have the same effect on gravity as the more familiar notion of mass?

In order to be able to talk about all of this without getting too confused, people invented two distinct terms for the following two distinct concepts:

- Gravitational mass m_G. This is the kind of mass that interacts with the gravitational field. Thus, we have $F = m_G g$.
- Inertial mass m_I. This is the kind of mass that goes into Newton's second law. So, we have $F = m_I a$.

Now, we can ask the question we have been thinking of in the clean form: is it always true that gravitational mass and inertial mass are the same? That is, do we always have $m_G = m_I$?

The 2nd Ingredient: The Effects of Gravity on Light

Let's leave aside for the moment further thought about fields as such and turn to another favourite question: to what extent is light affected by gravity?

Now, first, why do we care? Well, we built up our entire discussion of special relativity using light rays and we assumed in the process that light always traveled at a constant speed in straight lines! So, what if it happens that gravity can pull on light? If so, we may have to modify our thinking.

Clearly, there are two possible arguments:

- No. Light has no mass ($m_{\text{light}} = 0$). So, gravity cannot exert a force on light and should not a ect it.
- Yes. After all, all things fall at the same rate in a gravitational field, even things with a very small mass. So, light should fall.

Well, we could go back and forth between these two points of view for quite awhile.... but let's proceed by introducing a third argument in order to break the tie. We'll do it by recalling that there is a certain equivalence between energy and mass.

In fact, in certain situations, "pure mass" can be converted into "pure energy" and vice versa. A nice example of this happens all the time in particle accelerators when an electron meets a positron (it's 'anti-particle').

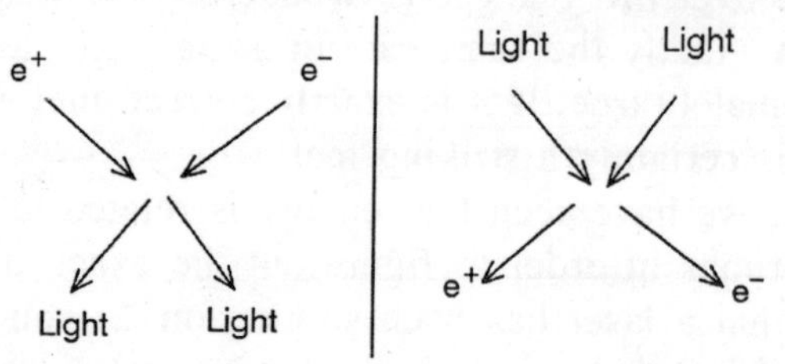

Let us suppose that gravity does not effect light and consider the following process:

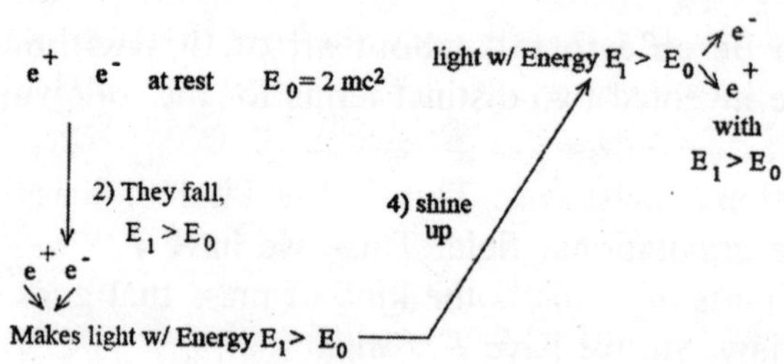

- First, we start with an electron (mass m) and a positron (also mass m) at rest. Thus, we have a total energy of $E_0 = 2mc^2$.
- Now, these particles fall a bit in a gravitational field. They speed up and gain energy.
  We have a new larger energy $E_1 > E_0$.
- Suppose that these two particles now interact and turn into some light. By conservation of energy, this light has the same energy $E_1 > E_0$.
- Let us take this light and shine it upwards, back to where the

particles started. (This is not hard to do - one simply puts enough mirrors around the region where the light is created.) Since we have assumed that gravity does not a ect the light, it must still have an energy $E_1 > E_0$.

- Finally, let us suppose that this light interacts with itself to make an electron and a positron again. By energy conservation, these particles must have an energy of $E_1 > E_0$.

Now, at the end of the process, nothing has changed except that we have more energy than when we started. And, we can keep repeating this to make more and more energy out of nothing. Just think about what this would do, for example, to ideas about energy conservation!

We have seen some hard to believe things turn out to be true, but such an infinite free source of energy seems especially hard to believe. This strongly suggests that light is in fact affeected by gravity in such a way that, when the light travels upwards though a gravitational field, it loses energy in much the same way as would a massive object.

Gravity, Light, Time, and all That

In the previous subsection we argued that light is in fact affected by gravity. In particular, when light travels upwards though a gravitational field, it looses just as much energy as would a massive object. Now, what happens to light when it looses energy? Well, it happens that light comes in little packages called 'photons.' This was only beginning to be understood when Einstein started thinking about gravity, but it is now well established and it will be a convenient crutch for us to use in assembling our own understanding of gravity. The amount of energy in a beam of light depends on how many photons are in the beam, and on how much energy each photon has separately. You can see that there are two ways for a beam of light to loose energy. It can either actually loose photons, or each photon separately can loose energy.

As the light travels up through the gravitational field, it should loose energy continuously. Loosing photons would not be a continuous, gradual process - it would happen in little discrete steps, one step each time a photon was lost. So, it is more likely that light looses energy in a gravitational field by each photon separately loosing energy.

How does this work? The energy of a single photon depends on something called the frequency of the light. The frequency is just a measure of how fast the wave oscillates. The energy E is in fact proportional to the frequency f, through something called "Plank's constant" (h). In other words, E = hf for a photon.

So, as it travels upwards in our gravitational field, this means that our light wave must loose energy by changing frequency and oscillating more slowly. It may please you to know that, long after this effect was suggested by Einstein, it was measured experimentally. The experiment was done by Pound and Rebke at Harvard in 1959.

Now, a light wave is a bunch of wave crests and wave troughs that chase each other around through spacetime. Let's draw a spacetime diagram showing the motion of, say, a bunch of the wave crests. Note that, if the wave oscillates more slowly at the top, then the wave crests must be farther apart at the top than at the bottom.

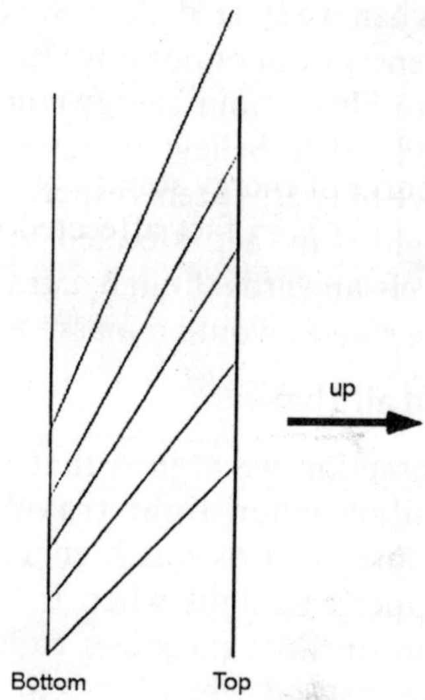

But, isn't each wave crest supposed to move at the same speed c in a vacuum?

It looks like the speed of light gets faster and faster as time passes! Perhaps we have done something wrong? By the way, do you remember any time before when we saw light doing weird stuff?

Nothing is really changing with time, so each crest should act the same as the one before and move at the same speed, at least when the wave is at the same place.

Let's choose to draw this speed as a 45° line as usual. In that case, our diagram must look like the one below.

However, we know both from our argument above and from Pound and Rebke's experiment that the time between the wave crests is larger at the top. So, what looks like the same separation must actually represent a greater proper time at the top.

This may seem very odd. Should we believe that time passes at a faster rate higher up? Note that we are really comparing time as measured by two different clocks, one far above the other. Also note that these clocks have no relative motion.

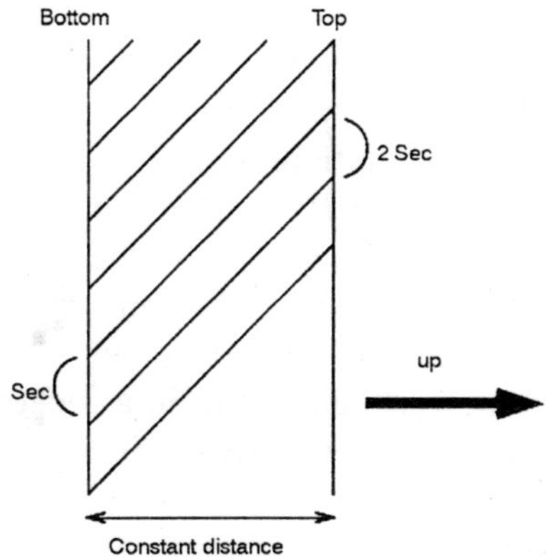

In fact, this does really occur! The Pound and Rebke experiment is an observation of this kind, but it direct experimental verification was made by precise atomic clocks maintained by the National Bureau of Standards in the 1960's. They kept one clock in Washington *D*.C. (essentially at sea level) and one clock in Denver (much higher up). The one in Denver measured more time to pass (albeit only by a very small amount, one part in 10^{15}!).

So, we have clocks with no relative motion that run at different rates. Is this absurd?

Well, no, and actually it should sound somewhat familiar. Do you recall seeing something like this before? (Hint: remember the accelerating rocket?)

Ah, yes. This sounds very much like the phenomenon in which clocks at the front and back of a uniformly rocket ship experienced no relative motion but had clocks that ran at different rates. If one works out the math based on our discussion of energy and frequency above one finds that, at least over small distances, a gravitational field is not just qualitatively, but also quantitatively like an accelerating rocket ship with a = g.

Gravity and Locality

But, what if we were not allowed to look at the Sun? What if we were only allowed to make measurements here in this room? [Such measurements are called local measurements.] What objects in this room are in inertial frames?

How do we know? Should we drop a sequence of rocks, as we would in a rocket ship?

If only local measurements are made, then it is the state of free-fall that is much like being in an inertial frame. In particular, a person in free-fall in a gravitational field feels just like an inertial observer!

Note how this fits with our observation about clocks higher up running faster than clocks lower down. We said that this exactly matches the results for an accelerating rocket with $a = g$. As a result, things that accelerate relative to the lab will behave like things that accelerate relative to the rocket.

In particular, it is the freely falling frame that accelerates downward at g relative to the lab, while it is the inertial frame that accelerates downward at g relative to the rocket! Thus, clocks in a freely falling frame act like those in an inertial frame, and it is in the freely falling frame that clocks with no relative motion in fact run at the same rate!!

Similarly, a lab on the earth and a lab in a rocket (with it's engine on, and, say, accelerating at $10 m/s^2$) are very similar. They have the following features in common:

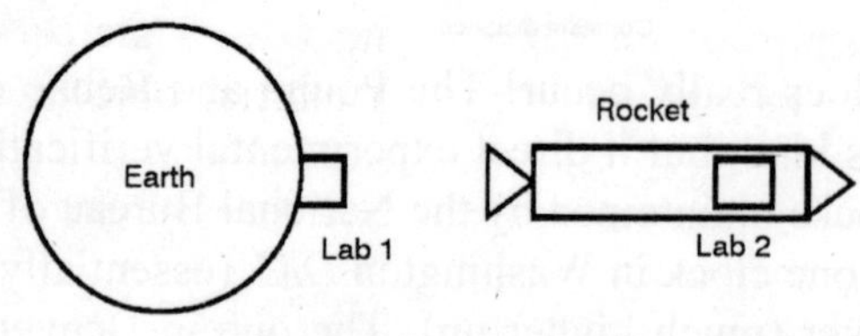

- Clocks farther up run faster in both cases, and by the same amount!
- If the non-gravitational force on an object is zero, the object "falls" relative to the lab at a certain acceleration that does not depend on what the object is!
- If you are standing in such a lab, you feel exactly the same in both cases.

Einstein's guess (insight?) was that, in fact:

Under local measurements, a gravitational field is completely equivalentto an acceleration.

This statement is known as The Equivalence Principle.

In particular, gravity has NO local effects in a freely falling reference frame. This ideas turns out to be useful even in answering non-relativistic problems. For example, what happens when we drop a hammer held horizontally? Does the heavy end hit first, or does the light end?

So then, what would be the best way to draw a spacetime diagram for a tower sitting on the earth? The answer of course is the frame that acts like an inertial frame. In this case, this is the freely falling reference frame. We have learned that, in such a reference frame, we can ignore gravity completely.

Now, how much sense does the above picture really make? Let's make this easy, and suppose that the earth were really big.... it turns out that, in this case, the earth's gravitational field would be nearly constant, and would weaken only very slowly as we go upward. Does this mesh with the diagram above?

Not really..... We said that the diagram above is effectively in an inertial frame. However, in this case we know that, if the distance between the bottom and top of the tower does not change, then the bottom must accelerate at a faster rate than the top does! But we just said that we want to consider a constant gravitational field.

Side note: No, it does not help to point out that the real earth's gravitational field is not constant.

How Local?

Well, we do have a way out of this: We realized before that the idea of freely falling frames being like inertial frames was not universally true. After all, freely falling objects on opposite side of the earth do accelerate towards each other. In contrast, any two inertial objects experience zero relative acceleration.

However, we did say that inertial and freely falling frames are the same 'locally.'

Let's take a minute to refine that statement.

How local is local? Well, this is much like the question of "when is a velocity small compared to the speed of light?" What we found before was that Newtonian physics held true in the limit of small velocities. In the same way, our statement that inertial frames and freely-falling frames are similar is supposed to be true in the sense of a limit. This comparison becomes more and more valid the smaller a region of spacetime we use to compare the two.

Nevertheless, it is still meaningful to ask how accurate this comparison is. In other words, we will need to know exactly which things agree in the above limit.

To understand Einstein's answer, let's consider a tiny box of spacetime from our diagram above.

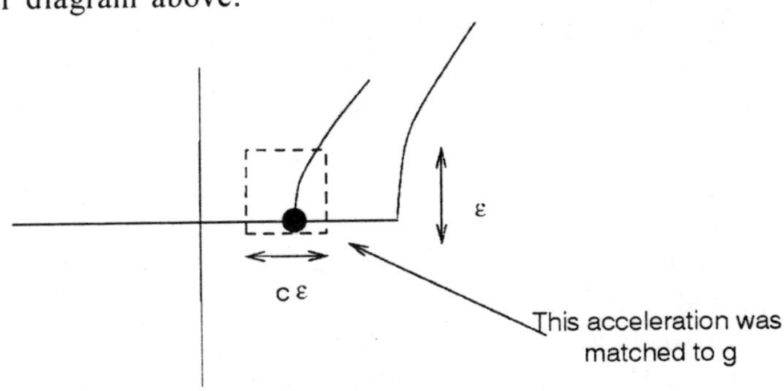

For simplicity, consider a 'square' box of height ε and width $c\varepsilon$. This square should contain the event at which we matched the "gravitational field g" to the acceleration of the rocket.

In this context, Einstein's proposal was that

Errors in dimensionless quantities like angles, v/c, and boost parameters should be proportional to ε^2.

Let us motivate this proposal through the idea that the equivalence principle should work "as well as it possibly can." Suppose for example that the gravitational field is really constant, meaning that static observers at any position measure the same gravitational field g.

We then have the following issue: when we match this gravitational field to an accelerating rocket in flat spacetime, do we choose a rocket with $\alpha_{top} = g$ or one with $\alpha_{bottom} = g$. Any rigid rocket will have a different acceleration at the top than it does at the bottom. So, what we mean by saying that the equivalence principle should work 'as well as it possibly can' is that it should predict any quantity that does not depend on whether we match $\alpha = g$ at the top or at the bottom, but it will not directly predict any quantity that would depend on this choice.

To see how this translates to the ε^2 criterion above, let us consider a slightly simpler setting where we have only two freely falling observers. Again, we will study such observers inside a small box of spacetime of dimensions $\delta t = \varepsilon$. Let's assume that they are located on opposite sides of the box, separated by a distance δx.

In general, we have seen that two freely falling observers will accelerate relative to each other. Let us write a Taylor's series expansion for this relative acceleration a as a function of the separation δx. In general, we have

$$a(\delta x) = a_0 + a_1 \delta_x + O(\delta x^2).$$

But, we know that this acceleration vanishes in the limit $\delta x \to 0$ where the two observers have zero separation. As a result, $a_0 = 0$ and for small δx we have the approximation $a \approx a_1 \delta x$.

Now, if this were empty space with no gravitational field, everything would be in a single inertial frame. As a result there would be no relative acceleration and, if we start the observers at rest relative to each other, their relative velocity would always remain zero. This is an example of an error we would make if we tried to use the equivalence principle in too strong of a fashion. What is the correct answer? Well, the relative acceleration is $a_1 \delta x = a_1 c\varepsilon$ and they accelerate away from each other for a time (within our box) $\delta t = \varepsilon$. As a result, they attain a relative velocity of $v = a_1 c\varepsilon^2$. But since our inertial frame model would have predicted $v = 0$, the error in v/c is also $a_1\varepsilon^2$. examples turn out to work in much the same way, and this is why Einstein made the proposal above.

- ***To summarize:*** what we have found is that locally a freely falling reference frame is almost the same as an inertial frame. If we think about a freely falling reference frame as being exactly like an inertial frame, then we make a small error in computing things. The fractional error is proportional to ε^2, where ε is the size of the spacetime region needed to make the measurement.

The factor of proportionality is called R after the mathematician Riemann. Note that R is not a radius. Since the error in an angle θ is $R\varepsilon^2$, R has dimensions of $(\text{length})^{-2}$.

GOING BEYOND LOCALITY

The fact that locally a freely falling frame in a gravitational field acts like an inertial frame does in the absence of gravity. However, we saw that freely falling frames and inertial frames are not exactly the same if they are compared over any bit of spacetime of finite size. No matter how small of a region of spacetime we consider, we always make some error if we interpret a freely falling frame as an inertial frame. So, since any real experiment requires a finite piece of spacetime, how can our local principle be useful in practice?

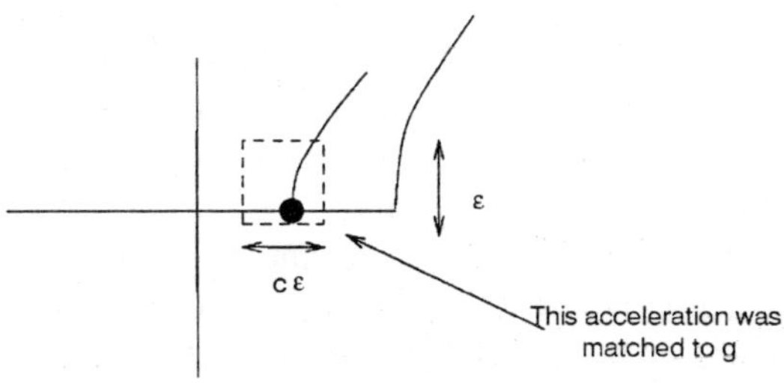

The answer lies in the fact that we were able to quantify the error that we make by pretending that a freely falling frame is an inertial frame. We found that if we consider a bit of spacetime of size ε, then the error in dimensionless quantities like angle or velocity measurements is ε^2. Note that ratios of lengths (L_1/L_2) or times (T_1/T_2) are also dimensionless. In fact, an angle is nothing but a ratio of an arc length to a radius ($\theta = s/r$)! So, this principle should also apply to ratios of lengths and/or times:

$$\delta\frac{T_1}{T_2}\propto\varepsilon^2,$$

where δ denotes the error.

Let me pause here to say that the conceptual setup with which we have surrounded equation is much like what we find in calculus. In calculus, we learned that locally any curve was essentially the same as a straight line. Over a region of finite size, curves are generally not straight lines. However, the error we make by pretending a curve is straight over a small finite region is small. Calculus is the art of carefully controlling this error to build up curves out of lots of tiny pieces of straight lines. Similarly, the main idea of general relativity is to build up a gravitational field out of lots of tiny pieces of inertial frames. Suppose, for example, that we wish to compare clocks at the top and bottom of a tall tower. We begin by breaking up this tower into a larger number of short towers, each of size Δl.

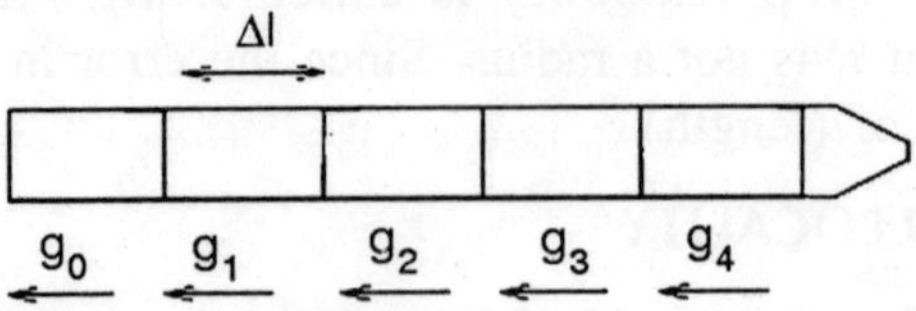

If the tower is tall enough, the gravitational field may not be the same at the top and bottom - the top might be enough higher up that the gravitational field is measurably weaker.

So, in general each little tower (0,1,2...) will have a different value of the gravitational field g (g_0, g_1, g_2...). If l is the distance of any given tower from the bottom, we might describe this by a function $g(l)$.

A Tiny Tower

Let's compare the rates at which clocks run at the top and bottom of one of these tiny towers. We will try to do this by using the fact that a freely falling frame is much like an inertial frame. Of course, we will have to keep track of the error we make by doing this.

In any accelerating rocket, the front and back actually do agree about simultaneity. As a result, all of our clocks in the towers will also agree about simultaneity. Thus, we can summarize all of the interesting information in a 'rate function' $\rho(l)$ which tells us how fast the clock at position l runs compared to the clock at position zero:

$$\rho(l) = \frac{\Delta\tau_l}{\Delta\tau_0}.$$

We wish to consider a gravitational field that does not change with time, so that ρ is indeed a function only of l and not of t. So, let us model our tiny tower as a rigid rocket accelerating through an inertial frame. A spacetime diagram drawn in the inertial frame is shown below.

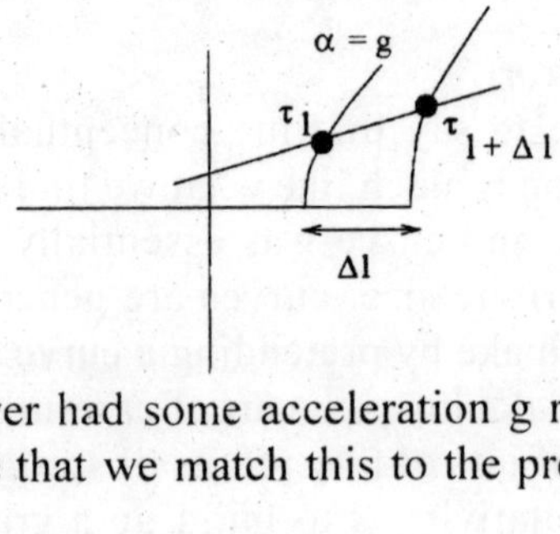

Now, the tiny tower had some acceleration g relative to freely falling frames. Let us suppose that we match this to the proper acceleration of the back of the rocket.

In this case, the back of the rocket will follow a worldline that remains a constant proper distance $d = c^2/\alpha$ from some fixed event.

Note that the top of the rocket remains a constant distance $d + \Delta l$ from this event. As a result, the top of the rocket has a proper acceleration $\alpha_{top} = \frac{c^2}{d+\Delta l}$. As we have learned, this means that the clocks at the top and bottom run at different rates:

$$\frac{\Delta\tau_{top}}{\Delta\tau_{bottom}} = \frac{\alpha_{bottom}}{\alpha_{top}} = \frac{d=\Delta l}{d} = 1+\frac{\Delta l}{d}.$$

In terms of our rate function, this is just

$$\frac{\rho(l+\Delta l)}{\rho(l)} = \frac{\rho(l)+\Delta\rho}{\rho(l)} = 1+\frac{\Delta\rho}{\rho(l)}.$$

Thus, we have

$$\frac{\Delta\rho}{\rho(l)} = \frac{\Delta l}{d} = \frac{\alpha\Delta l}{c^2}.$$

Now, how much of an error would we make if we use this expression for our tiny tower in the gravitational field? Well, the above is in fact a fractional change in a time measurement. So, the error must be of size Δl^2. So, for our tower case, we have

$$\frac{\Delta\rho}{\rho(l)} = \frac{g(l)+\Delta l}{c^2} + k(\Delta l)^2$$

for some number k. Here, we have replaced α with g, since we matched the acceleration $g(l)$ of our tower (relative to freely falling frames) to the proper acceleration α.

Actually, we might have figured out the error directly from expression above. The error can be seen in the fact that the acceleration does not change in the same way from the bottom of the tower to the top of the tower as it did from the bottom of the rocket to the top of the rocket. The equivalence principle directly predicts only quantities that are independent of such matching details. So what would is the difference between these two options? Well, the difference in the accelerations is just $\Delta\alpha \approx \frac{d\alpha}{dl}\Delta l$ Note that α is already multiplied by Δl in the expression above. This means that, if we were to change the value if Δl by $\frac{d\alpha}{dl}\Delta l$, we would indeed create a term of the form $k(\Delta l)^2$! So, we see again that this term really does capture well all of the errors we might possibly make in matching a freely falling frame to an inertial frame.

Let us write our relation above as

$$\frac{\Delta\rho}{\Delta l} = \left(\frac{g}{c^2} + k\Delta l\right)\rho(l).$$

As in calculus, we wish to consider the limit as $\Delta l \to 0$. In this case, the left $\frac{d\rho}{dl}$. On the right hand side, the first term does not depend on Δl at all, while the second term vanishes in this limit. Thus, we obtain

$$\frac{d\rho}{dl} = \frac{g(l)\,\rho(l)}{c^2}.$$

Note that the term containing k (which encodes our error) has disappeared entirely. We have managed to use our local matching of freely falling and inertial frames to make an exact statement not directly, but about the derivative $\frac{d\rho}{dl}$.

The Tall Tower

Of course, we have still not answered the question about how the clocks actually run at different heights in the tower. To do so, we need to solve the equation for $\rho(l)$. We can do this by multiplying both sides by dl and integrating:

$$\int_{\rho(0)}^{\rho} \frac{d\rho}{\rho} = \int_0^l \frac{g(l)}{c^2}\,dl.$$

Now, looking at our definition above we find that $\rho_0 = 1$. Thus, we have

$$\int_{\rho(0)}^{\rho} \frac{d\rho}{\rho} = \ln \rho - \ln 1 = \ln \rho$$

and so

$$\ln \rho = \int_0^l \frac{g(l)}{c^2}\,dl,$$

or,

$$\frac{\Delta\tau_l}{\Delta\tau_0}\ \rho(1) = \exp\left(\int_0^l \frac{g(l)}{c^2}\,dl\right)$$

Expression is the exact relation relating clocks at different heights l in a gravitational field. One important property of this formula is that the factor inside the exponential is always positive. As a result, we find that clocks higher up in a gravitational field always run faster, regardless of whether the gravitational field is weaker or stronger higher up!

Note that, due to the properties of exponential functions, we can also write this as:

$$\frac{\Delta\tau_b}{\Delta\tau_a} = \rho(l) = \exp\left(\int_a^b \frac{g(l)}{c^2}\,dl\right)$$

Gravitational time Dilation Near the Earth

The effect described in equation is known as gravitational time dilation. There are a couple of interesting special cases of this effect that are worth investigating in detail. The first is a uniform gravitational field in which $g(l)$ is constant.

This is not in fact the same as a rigid rocket accelerating through an inertial frame, as the acceleration is actually different at the top of the rigid rocket than at the bottom.

Still, in a uniform gravitational field with $g(l) = g$ the integral is easy to do and we get just:

$$\frac{\Delta\tau_l}{\Delta\tau_0} = e^{gl/c^2}.$$

In this case, the difference in clock rates grows exponentially with distance. The other interesting case to consider is something that describes (to a good approximation) the gravitational field near the earth. We have seen that Newton's law of gravity is a pretty good description of gravity near the earth, so we should be able to use the Newtonian form of the gravitational field:

$$g = \frac{m_E G}{r^2},$$

where r is the distance from the centre of the earth. Let us refer to the radius of the earth as r0. For this case, it is convenient to compare the rate at which some clock runs at radius r to the rate at which a clock runs on the earth's surface (i.e., at $r = r_0$).

Since $\int_{r_1}^{r_2} r^{-2}dr = r_1^{-1} - r_2^{-1}$, we have $\frac{\Delta\tau(r)}{\Delta\tau(r_0)}$

$$= \exp\left(\int_{r_0}^{r} \frac{m_E G}{c^2 r^2}\,dr\right) = \left[\frac{m_E G}{c^2}\left(\frac{1}{r_0} - \frac{1}{r}\right)\right].$$

Here, it is interesting to note that the r dependence drops out as $r \to \infty$, so that the gravitational time dilation factor between the earth's surface (at r0) and infinity is actually finite. The result is

$$\frac{\Delta\tau(\infty)}{\Delta\tau(r_0)} = e^{\frac{m_E G}{r_0 c^2}}.$$

So, time is passing more slowly for us here on earth than it would be if we were living far out in space..... By how much? Well, we just need to put in some numbers for the earth. We have

$$m_E = 6 \times 10^{24}\ \text{kg},$$
$$G = 6 \times 10^{-11}\ \text{Nm}^2/\text{kg}^2,$$
$$r_0 = 6 \times 10^6\text{m}.$$

Putting all of this into the above formula gives a factor of about $e^{\frac{2}{3}\times 10^{-10}}$. Now, how big is this? Well, here it is useful to use the Taylor series expansion $e^x = 1 + x$ + small corrections for small x. We then have

$$\frac{\Delta\tau(\infty)}{\Delta\tau(r_0)} \approx 1+\frac{2}{3}\times 10^{-10}$$

This means that time passes more slowly for us than it does far away by roughly one part in 1010, or, one part in ten billion! This is an incredibly small amount - one that can easily go unnoticed. However, as mentioned earlier, the national bureau of standards was in fact able to measure this back in the 1960's, by comparing very accurate clocks in Washington, *D*.C. with very accurate clocks in Denver! Their results were of just the right size to verify the prediction above.

In fact, there is an even more precise version of this experiment that is going on right now - constantly verifying Einstein's prediction every day! It is called the "Global Positioning System" (GPS). Perhaps you have heard of it?

The Global Positioning System

The Global Positioning System is a setup that allows anyone, with the aid of a small device, to tell exactly where they are on the earth's surface. It is made up of a number of satellites in precise, well-known orbits around the earth. Each of these satellites contains a very precise clock and a microwave transmitter.

Each time the clock 'ticks' (millions of times every second!) it sends out a microwave pulse which is 'stamped' with the time and the ID of that particular satellite.

A hand-held GPS locator then receives these pulses. Because it is closer to some satellites than to others, the pulses it receives take less time to reach it from some satellites than from others. The result is that the pulses it receives at a given instant are not all stamped with the same time.

The locator then uses the differences in these time-stamps to figure out which satellites it is closest to, and by how much. Since it knows the orbits of the satellites very precisely, this tells the device exactly where it itself it located. This technology allows the device to pinpoint its location on the earth's surface to within a one meter circle.

To achieve this accuracy, the clocks in the satellites must be very precise, and the time stamps must be very accurate. In particular, they must be much more accurate than one part in ten billion.

If they were 0 by that much, then every second the time stamps would become 0 by 10^{-10} seconds. But, in this time, microwaves (or light) travel a distance $(3 \times 10^8$ m/s$)$ $(10^{-10}$ sec$) = 3 \times 10^{-2}$ m = 3cm and the GPS locator would think it was 'drifting away' at 3cm/sec. While this is not very fast, it would add up over time.

This drift rate is 72 m/hr, which would already spoil the accuracy of the GPS system.

Over long times, the distance becomes even greater. The drift rate can also be expressed as 1.5km/day or 500km/year. So, after one year, a GPS device in Syracuse, NY might think that it is in Philadelphia!

By the way, since the GPS requires this incredible precision, you might ask if it can measure the effects of regular speed-dependent special relativity time dilation as well (since the satellites are in orbit and are therefore 'moving.') The answer is that it can. In fact, for the particular satellites used in the GPS system, these speed-dependent effects turn out to be of a comparable size to the gravitational time dilation effect.

Note that these effects actually go in opposite directions: the gravity effect makes the higher (satellite) clock run fast while the special relativity effect makes the faster (satellite) clock run slow.

Which effect is larger turns out to depend on the particular orbit. Low orbits (like that of the space shuttle) are higher speed, so in this case the special relativity effect dominates and the orbiting clocks run more slowly than on the earth's surface.

High orbits (like that of the GPS satellites) are lower speed, so the gravity effect wins and their clocks run faster than clocks on the earth's surface. For the case of GPS clocks, the special relativity effect means that the amount of the actual time dilation is less than the purely gravitational effect by about a factor of two.

THE MORAL OF THE STORY

We were able to figure out how clocks run at different heights in a gravitational field. We have also seen how important this is for the running of things like GPS. But, what does all of this mean? And, why is this often considered a new subject (called 'General Relativity'), different from our old friend Special Relativity?

Local Frames vs. Global frames

Let us briefly retrace our logic. While thinking about various frames of reference in a gravitational field, we discovered that freely falling reference frames are useful. In fact, they are really the most useful frames of reference, as they are similar to inertial frames. This fact is summarized by the equivalence principle which says "freely falling frames are locally equivalent to inertial frames."

The concept of these things being locally equivalent is a subtle one, so let me remind you what it means. The idea is that freely falling reference frames are indistinguishable from inertial reference frames so long as we are only allowed to perform experiments in a tiny region of spacetime.

More technically, suppose that we make then mistake of pretending that a freely falling frame actually is an inertial frame of special relativity, but that we limit ourselves to measurements within a region of spacetime of size $\in$.

When we then go and predict the results of experiments, we will make small errors in, say, the position of objects. However, these errors will be very small when? is small; in fact, the per cent error will go to zero like $\in^2$.

The same sort of thing happens in calculus. There, the corresponding statement is that a curved line is locally equivalent to a straight line.

Anyway, the important point is that we would make an error by pretending that freely falling frames are exactly the same as inertial frames. Physicists say that the two are locally equivalent, but are not "globally" equivalent. The term 'global' (from globe, whole, etc.) is the opposite of local and refers to the frame everywhere (as opposed to just in a small region).

So, if freely falling frames are not globally inertial frames, then where are the inertial frames? They cannot be the frames of reference that are attached to the earth's surface. After all, if a frame is globally like an inertial frame then it must also be like an inertial frame locally. However, frames tied to the surface of the earth are locally like uniformly accelerated frames, not inertial frames.

But, there are really not any other frames left to consider. To match an inertial frame locally requires free fall, but that will not let us match globally. We are left with the conclusion that:

In a generic gravitational field, there is no such thing as a global inertial frame.

One can take various perspectives on this, but the bottom line is that we (following Einstein) merely assumed that the speed of light was constant in all (globally) inertial frames of reference. However, no such reference frame will exist in a generic gravitational field.

And what if we retreat to Newton's first law, asking about the behaviour of objects on which no forces act? The trouble is that, as we have discussed, to identify an inertial frame in this way we would need to first identify an object on which no forces act.

But, which object is this? Any freely falling object seems to pass the 'no forces' tests as well (or better than!) an object sitting on the earth! However, if freely falling objects are indeed free of force, then Newton's first law tells us that they do not accelerate relative to each other..... in gross contradiction with experiment.

This strongly suggests that global inertial frames do not exist and that we should therefore abandon the concept and move on. In its place, we will now make use of local inertial frames, a.k.a. freely falling frames. It is just this change that marks the transition from 'special' to 'general' relativity. Special relativity is just the special case in which global inertial frames exist. Actually, there is another reason why the study of gravity is known as "General Relativity."

The point is that in special relativity (actually, even before) we noticed that the concept of velocity is intrinsically a relative one. That is to say, it does not make sense to talk about whether an object is moving or at rest, but only whether it is moving or at rest relative to some other object. However, we did have an absolute notion of acceleration: an object could be said to be accelerating without stating explicitly what frame was being used to make this statement. The result would be the same no matter what inertial frame was used.

However, now even the concept of acceleration becomes relative in a certain sense. Suppose that you are in a rocket in deep space and that you cannot look outside to see if the rockets are turned on. You drop an object and it falls.

Are you accelerating, or are you in some monster gravitational field? There is no right answer to this question as the two are identical. In this sense, the concept of acceleration is now relative as well - it is equivalent to being in a gravitational field.

While this point is related to why the study of gravity historically acquired the name "General Relativity," it is not clear that this is an especially useful way to think about things.One can still measure one's proper acceleration as the acceleration relative to a nearby (i.e., local!) freely falling frame.

Thus, there is an absolute distinction between freely falling and not freely falling. Whether you wish to identify these terms with non-accelerating and accelerating is just a question of semantics - though most modern relativists find it convenient to do so. A language in which acceleration is not a relative concept but in which it implicitly means "acceleration measured locally with respect to freely falling frames."

And what About the Speed of Light?

There is a question that you probably wanted to ask a few paragraphs back, but then A general gravitational field there are no frames of reference in which light rays always travel in straight lines at constant speed. So, after all of our struggles, have we finally thrown out the constancy of the speed of light? No, not completely.

There is one very important statement left. Suppose that we measure the speed of light at some event (E) in a frame of reference that falls freely at event E. Then, near event E things in this frame work just like they do in inertial frames - so, light moves at speed c and in a straight line. Said in our new language:

As measured locally in a freely falling frame, light always moves in straight lines at speed c.

Chapter 8

Relativity and Curved Spacetime

The equivalence principle to calculate the effects of a gravitational field over a finite distance by carefully patching together local inertial frames. If we are very, very careful, we can calculate the effects of any gravitational field in this way.

However, this approach turns out to be a real mess. Consider for example the case where the gravitational field changes with time. Then, it is not enough just to patch together local inertial frames at different positions. One must make a quilt of them at different places as well as at different times!

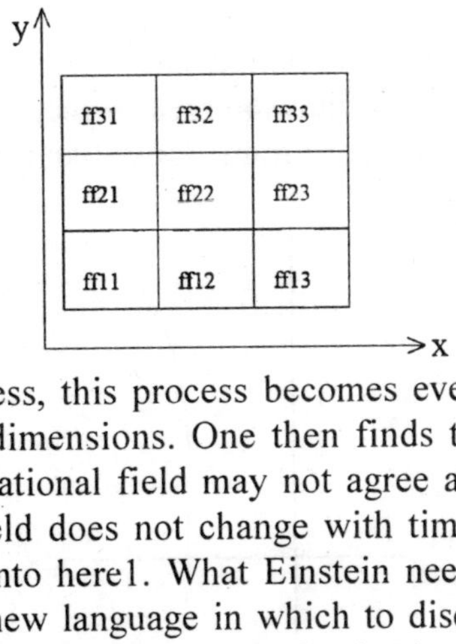

As you might guess, this process becomes even more complicated if we consider all 3+1 dimensions. One then finds that clocks at different locations in the gravitational field may not agree about simultaneity even if the gravitational field does not change with time.... but that is a story that we need not go into here[1]. What Einstein needed was a new way of looking at things - a new language in which to discuss gravity that would organize all of this into something relatively simple.

Another way to say this is that he needed a better conception of what a gravitational field actually is. This next step was very hard for Albert. It took him several years to learn the appropriate mathematics and to make that mathematics into useful physics. Instead of going through all of the twists and turns in the development of the subject.

A RETURN TO GEOMETRY

You see, Einstein kept coming back to the idea that freely falling observers are like inertial observers - or at least as close as we can get. In the presence of a general gravitational field, there really are no global inertial frames.

When we talked about our 'error' in thinking of a freely falling frame as inertial, it is not the case that there is a better frame which is more inertial than is a freely falling frame.

Instead, when gravity is present there are simply no frames of reference that act precisely in the way that global inertial frames act. Anyway, Einstein focussed on the fact that freely falling frames are locally the same as inertial frames.

However, he knew that things were tricky for measurements across a finite distance. Consider, for example, the reference frame of a freely falling person. Suppose that this person holds out a rock and releases it. The rock is then also a freely falling object, and the rock is initially at rest with respect to the person.

However, the rock need not remain exactly at rest with respect to the person. Suppose, for example, that the rock is released from slightly higher up in the gravitational field.

Then, Newton would have said that the gravitational field was weaker higher up, so that the person should accelerate toward the earth faster than does the rock.

This means that there is a relative acceleration between the person and the rock, and that the person finds the rock to accelerate away! A spacetime diagram in the person's reference frame looks like this:

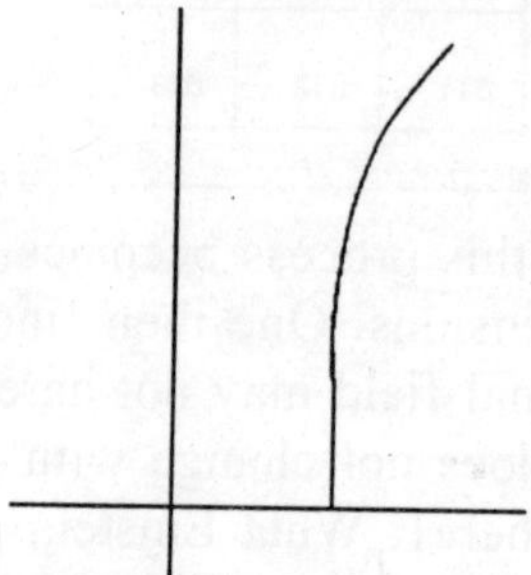

Suppose, on the other hand, that the rock is released to the person's side. Then, Newton would say that both person and rock accelerate toward the centre of the earth.

However, this is not in quite the same direction for the person as for the rock:

So, again there is a relative acceleration. This time, however, the person finds the rock to accelerate toward her. So, she would draw a spacetime diagram for this experiment as follows:

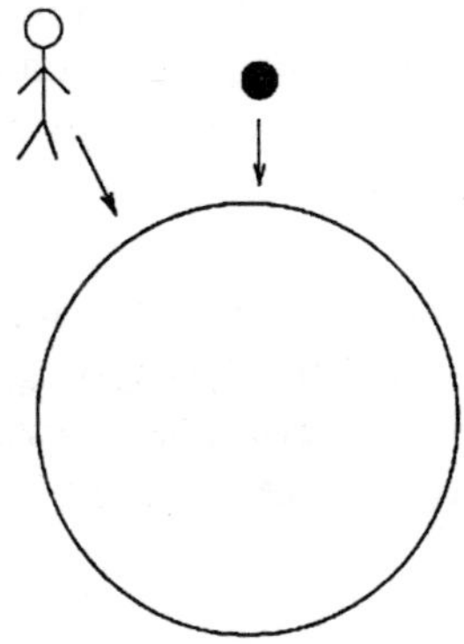

The issue is that we would like to think of the freely falling worldlines as inertial worldlines.

That is, we would like to think of them as being 'straight lines in spacetime.' However, we see that we are forced to draw them on a spacetime diagram as curved.

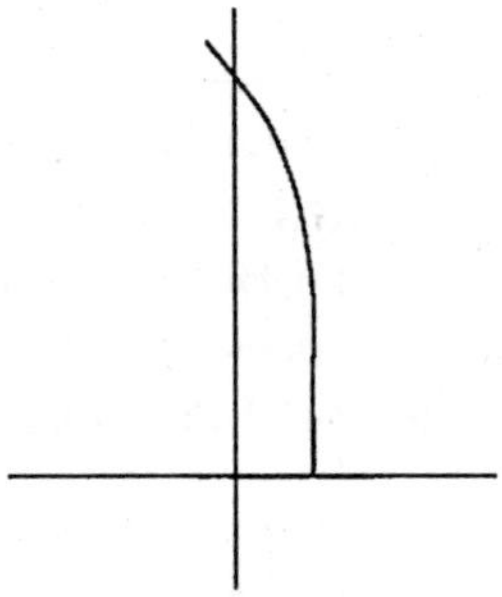

Now, we can straighten out any one of them by using the reference frame of an observer moving along that worldline. However, this makes the other freely falling worldlines appear curved. How are we to understand this?

Straight Lines in Curved Space

Eventually Einstein found a useful analogy with something that at first sight appears quite different - a curved surface. The idea is captured by the question "What is a straight line on a curved surface?"

To avoid language games, mathematicians made up a new word for this idea: "geodesic." A geodesic can be thought of as the "straightest possible line on a curved surface."

More precisely, we can define a geodesic as a line of minimal distance - the shortest line between two points2. The idea is that we can define a straight line to be the shortest line between two points. Actually, there is another definition of geodesic that is even better, but requires more mathematical machinery to state precisely. Intuitively, it just captures the idea that the geodesic is 'straight.' It tells us that a geodesic is the path on a curved surface that would be traveled, for example, by an ant (or a person) walking on the surface who always walks straight ahead and does not turn to the right or left.

As an example, suppose you stand on the equator of the earth, face north, and then walk forward. Where do you go? If you walk far enough (over the ocean, etc.) you will eventually arrive at the north pole. The path that you have followed is a geodesic on the sphere.

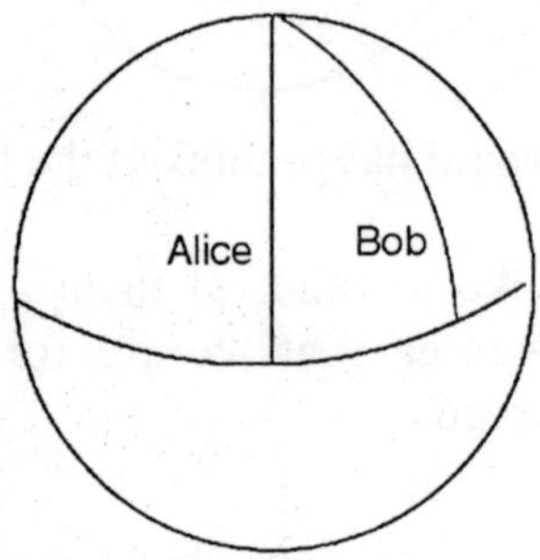

Note that this is true no matter where you start on the equator. So, suppose there are in fact two people walking from the equator to the north pole, Alice and Bob. As you can see, Alice and Bob end up moving toward each other. So, if we drew a diagram of this process using Alice's frame of reference (so that her own path is straight), it would look like this:

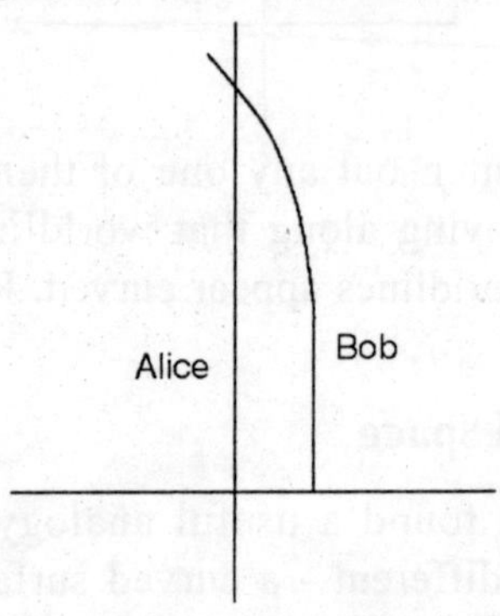

By the way, the above picture is not supposed to be a spacetime diagram. It is simply supposed to be a map of part of the (two dimensional) earth's surface, on which both paths have been drawn.

This particular map is drawn in such a way that Alice's path appears as a straight line. As you probably know from looking at maps of the earth's surface, no flat map will be an accurate description globally, over the whole earth. There will always be some distortion somewhere. However, a flat map is perfectly fine locally, say in a region the size of the city of Syracuse (if we ignore the hills). Now, does this look or sound at all familiar? What if we think about a similar experiment involving Alice and Bob walking on a funnel-shaped surface:

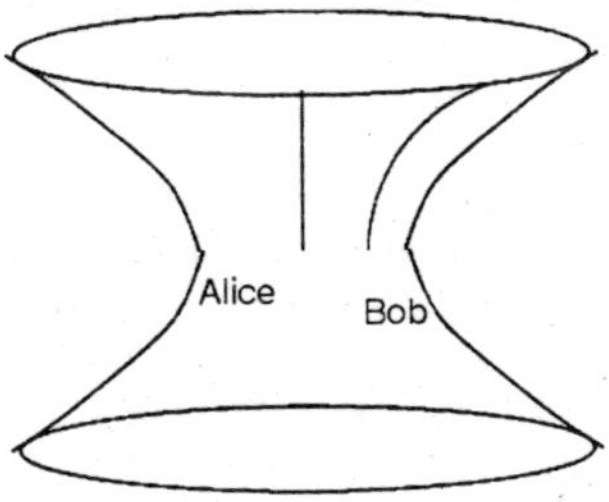

In this case they begin to drift apart as they walk so that Alice's map would look like this:

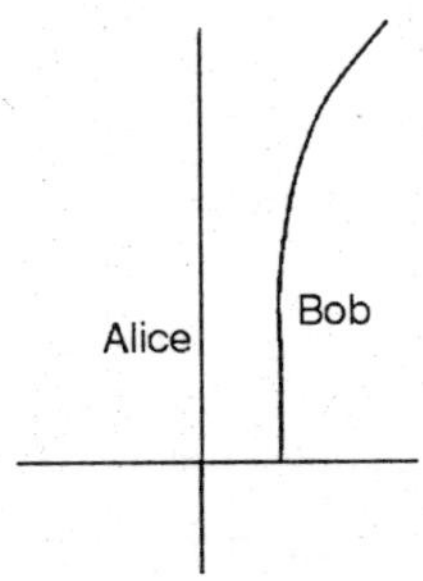

So, we see that straight lines (geodesics) on a curved surface act much like freely falling worldlines in a gravitational field. It is useful to think through this analogy at one more level: Consider two people standing on the surface of the earth. We know that these two people remain the same distance apart as time passes.

Why do they do so? Because the earth itself holds them apart and prevents gravity from bringing them together. The earth exerts a force on each person, keeping them from falling freely.

Now, what is the analogy in terms of Alice and Bob's walk across the sphere or the funnel? Suppose that Alice and Bob do not simply walk independently, but that they are actually connected by a sti bar. This bar will force them to always remain the same distance apart as they walk toward the north pole. The point is that, in doing so, Alice and Bob will be unable to follow their natural (geodesic) paths. As a result, Alice and Bob will each feel some push or pull from the bar that keeps them a constant distance apart. This is much like our two people standing on the earth who each feel the earth pushing on their feet to hold them in place.

Curved Surfaces are Locally Flat

Note that straight lines (geodesics) on a curved surface act much like freely falling worldlines in a gravitational field. In particular, exactly the same problems arise in trying to draw a flat map of a curved surface as in trying to represent a freely falling frame as an inertial frame.

A quick overview of the errors made in trying to draw a flat map of a curved surface are shown below:

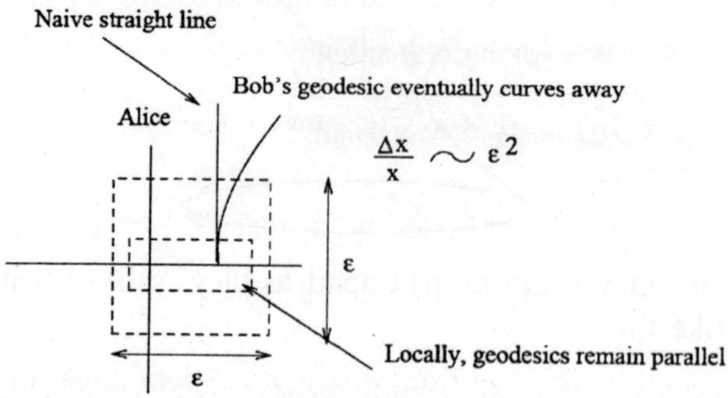

We see that something like the equivalence principle holds for curved surfaces: flat maps are very accurate in small regions, but not over large ones.

In fact, we know that we can in fact build up a curved surface from a bunch of flat ones. One example of this happens in an atlas. An atlas of the earth contains many flat maps of small areas of the earth's surface (the size of states, say). Each map is quite accurate and together they describe the round earth, even though a single flat map could not possibly describe the earth accurately.

Computer graphics people do much the same thing all of the time. They draw little flat surfaces and stick them together to make a curved surface.

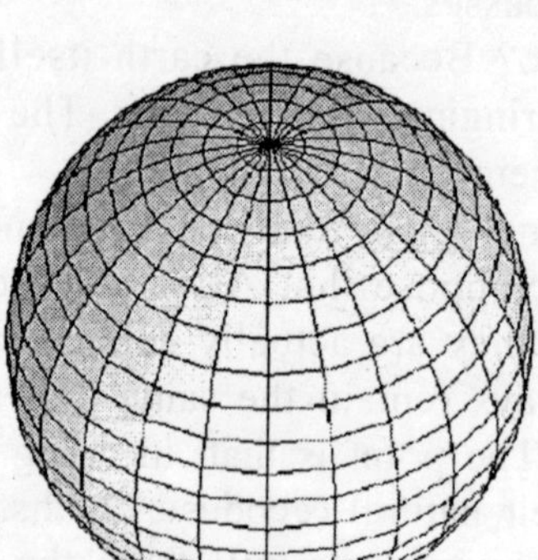

This is much like the usual calculus trick of building up a curved line from little pieces of straight lines. In the present context with more than one dimension, this process has the technical name of "differential geometry."

From Curved Space to Curved Spacetime

The point is that this process of building a curved surface from flat ones is just exactly what we want to do with gravity! We want to build up the gravitational field out of little pieces of "flat" inertial frames. Thus, we might say that gravity is the curvature of spacetime. This gives us the new language that Einstein was looking for:

- (Global) Inertial Frames ⇔ Minkowskian Geometry ⇔ Flat Spacetime: We can draw it on our flat paper or chalk board and geodesics behave like straight lines.
- Worldlines of Freely Falling Observers ⇔ Straight lines in Spacetime
- Gravity ⇔ The Curvature of Spacetime

Similarly, we might refer to the relation between a worldline and a line of simultaneity as the two lines being at a "right angle in spacetime." It is often nice to use the more technical term "orthogonal" for this relationship.

By the way, the examples (spheres, funnels, etc.) that we have discussed so far are all curved spaces. A curved spacetime is much the same concept. However, we can't really put a curved spacetime in our 3-D Euclidean space. This is because the geometry of spacetime is fundamentally Minkowskian, and not Euclidean.

Remember the minus sign in the interval? Anyway, what we can do is to once again think about a spacetime diagram for 2+1 Minkowski space - time will run straight up, and the two space directions (x and y) will run to the sides.

Light rays will move at 45 degree angles to the (vertical) t-axis as usual. With this understanding, we can draw a (1+1) curved spacetime inside this 2+1 spacetime diagram. An example is shown below:

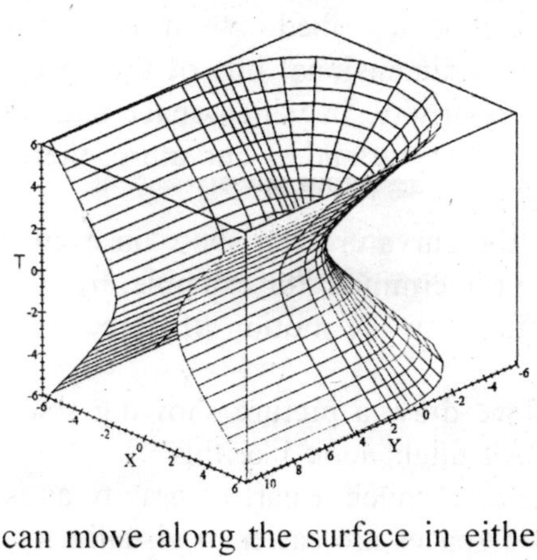

Note that one can move along the surface in either a timelike manner (going up the surface) or a spacelike manner (going across the surface), so that this surface does indeed represent a (1+1) spacetime. The picture above turns out to represent a particular kind of gravitational field that we will be discussing more in a few weeks.

To see the similarity to the gravitational field around the earth, think about two freely falling worldlines (a.k.a. "geodesics," the straightest possible lines) that begin near the middle of the diagram and start out moving straight upward. Suppose for simplicity that one geodesic is on one side of the fold while the second is on the other side. You will see that the two worldlines separate, just as two freely falling objects do at different heights in the earth's gravitational field.

Thus, if we drew a two-dimensional map of this curved spacetime using the reference frame of one of these observers, the results would be just like the spacetime diagram we drew for freely falling stones at different heights! This is a concrete picture of what it means to say that gravity is the curvature of spacetime.

Well, there is one more subtlety that we should mention...... it is important to realise that the extra dimension we used to draw the picture above was just a crutch that we needed because we think best in flat spaces. One can in fact talk about curved spacetimes without thinking about a "bigger space" that contains points "outside the spacetime." This minimalist view is generally a good idea.

MORE ON CURVED SPACE

Let us remember that the spacetime in which we live is fundamentally four (=3+1) dimensional and ask if this will cause any new wrinkles in our story. It turns out to create only a few. The point is that curvature is fundamentally associated with two-dimensional surfaces.

Roughly speaking, the curvature of a four-dimensional spacetime (labelled by x, y, z, t) can be described in terms of xt curvature, yt curvature, etc. associated with two-dimensional bits of the spacetime.

However, this is relativity, in which space and time act pretty much the same. So, if there is xt, yt, and zt curvature, there should also be xy, yz, and xz curvature!

This means that the curvature can show up even if we consider only straight lines in space (determined, for example, by stretching out a string) in addition to the effects on the motion of objects that we have already discussed.

For example, if we draw a picture showing spacelike straight lines (spacelike geodesics), it might look like this:

So, curved space is as much a part of gravity as is curved spacetime. This is nice, as curved spaces are easier to visualize.

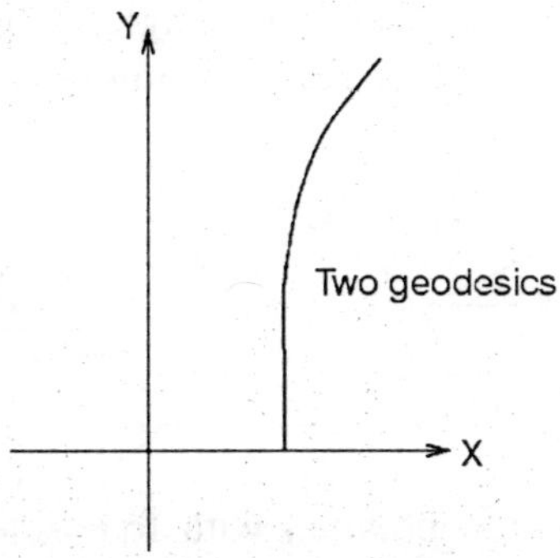

Let us now take a moment to explore these in more depth and build some intuition about curvature in general. Curved spaces have a number of fun properties. Some of my favorites are:

$C \neq 2\pi R$: The circumference of a circle is typically not 2π times its radius. Letus take an example: the equator is a circle on a sphere. What is it's centre? We are only supposed to consider the two-dimensional surface of the sphere itself as the third dimension was just a crutch to let us visualize the curved two-dimensional surface. So this question is really 'what point on the sphere is equidistant from all points on the equator?' In fact, there are two answers: the north pole and the south pole. Either may be called the centre of the sphere. Now, how does the distance around the equator compare to the distance (measured along the sphere) from the north pole to the equator? The arc running from the north pole to the equator goes 1/4 of the way around the sphere. This is the radius of the equator in the relevant sense. Of course, the equator goes once around the sphere. Thus, its circumference is exactly four times its radius.

$A \neq \pi R^2$: The area of a circle is typically not π times the square of its radius. Again, the equator on the sphere makes a good example. With the radius defined as above, the area of this circle is much less than πR^2.

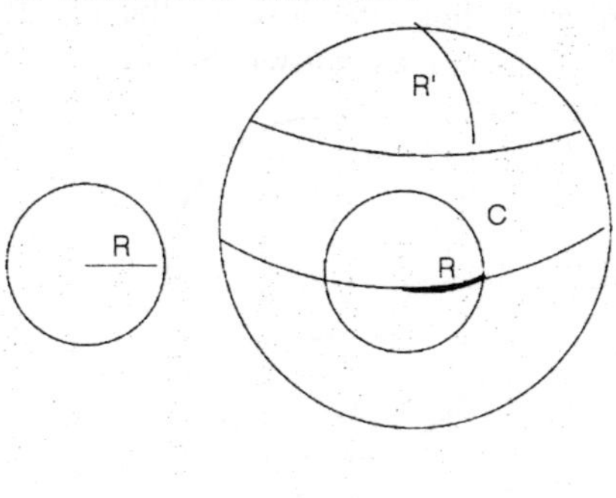

Σ (angles) $\neq 180°$: The angles in a triangle do not in general add up to 180°. An example on a sphere is shown below.

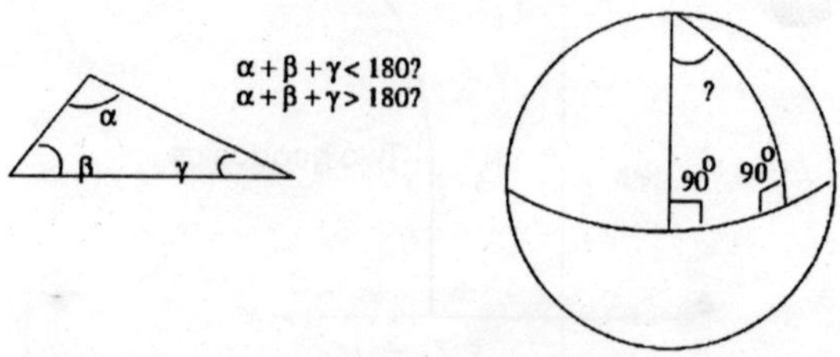

Squares do not close: A polygon with four sides of equal length and four right angles (a.k.a., a square) in general does not close.

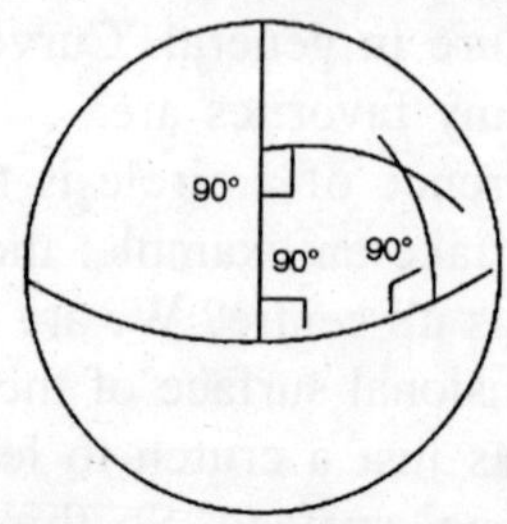

Vectors (arrows) "parallel transported" around closed curves are rotated: This one is a bit more complicated to explain. Unfortunately, to describe this property as precisely as the ones above would require the introduction of more complicated mathematics. Nevertheless, the discussion below should provide you with both the flavour of the idea and an operational way to go about checking this property.

In a flat space (like the 3-D space that most people think we live in until they learn about relativity....), we know what it means to draw an arrow, and then to pick up this arrow and carry it around without turning it. The arrow can be carried around so that it always remains parallel to its original direction.

Now, on a curved surface, this is not possible. Suppose, for example, that we want to try to carry an arrow around a triangular path on the sphere much like the one that we discussed a few examples back. For concreteness, let's suppose that we start on the equator, with the arrow also pointing along the equator as shown below:

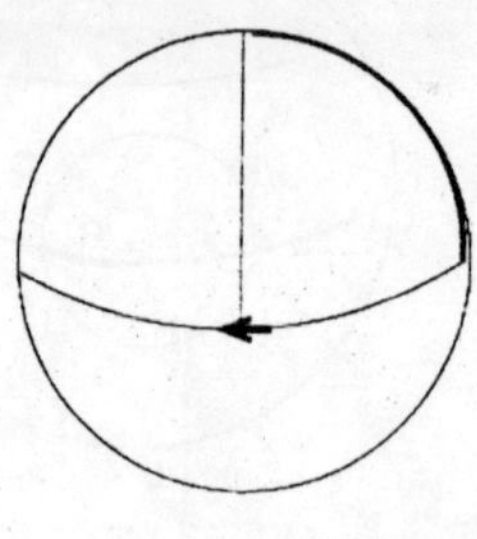

We now wish to carry this vector to the north pole, keeping it always pointing in the same direction as much as we can. Well, if we walk along the path shown, we are going in a straight line and never turning. So, since we start with the arrow pointing to our left, we should keep the arrow pointing to our left at all times. This is certainly what we would do in a flat space. When we get to the north pole, the arrow looks like this:

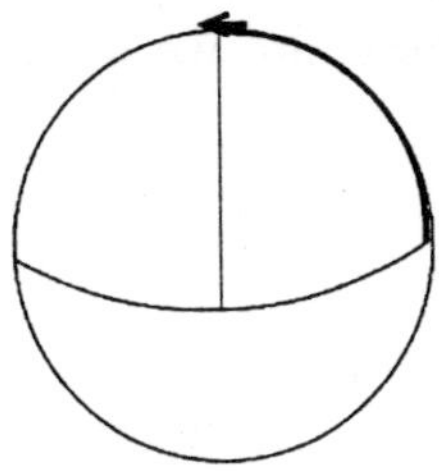

Now we want to turn and walk toward the equator along a different side of the triangle. We turn (say, to the right), but we are trying to keep the arrow always pointing in the same direction. So the arrow should not turn with us. As a result, it points straight behind us. We carry it down to the equator so that it points straight behind us at every step:

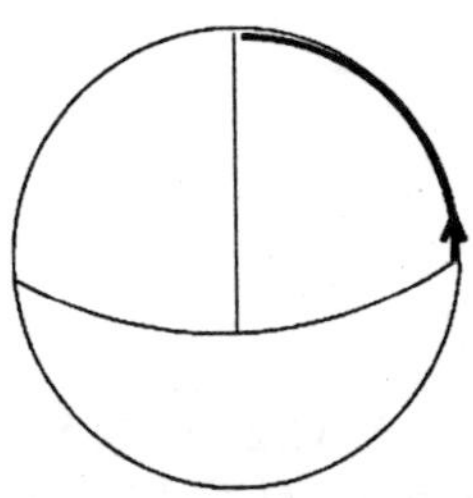

Finally, we wish to bring the arrow back to where it started. We see that the arrow has rotated 90o relative to the original direction:

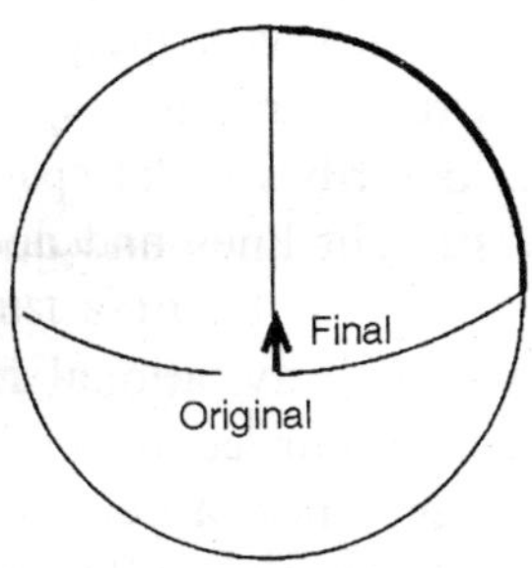

All of these features will be present in any space (say, a surface of simultaneity) in a curved spacetime. Now, since we identify the gravitational field with the curvature of spacetime then the above features must also be encoded in the gravitational field. But there is a lot of information in these features. In particular there are independent curvatures in the xy, yz, and xz planes that control, say, the ratio of circumference to radius of circles in these various planes.

But wait! Doesn't this seem to mean that the full spacetime curvature (gravitational field) contains a lot more information than just specifying an acceleration g at each point?

After all, acceleration is related to how thing behave in time, but we have just realized that at least parts of the spacetime curvature are associated only with space. How are we to deal with this?

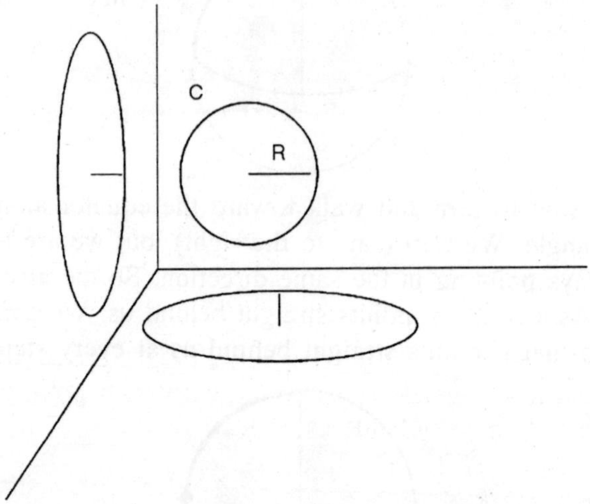

GRAVITY AND THE METRIC

Let's recall where we are. A while back we discovered the equivalence principle: that locally a gravitational field is equivalent to an acceleration in special relativity. Another way of stating this is to say that, locally, a freely falling frame is equivalent to an inertial frame in special relativity. We noticed the parallel between this principle and the underlying ideas being calculus: that locally every curve is a straight line.

A global inertial frame describes a flat spacetime - one in which, for example, geodesics follow straight lines and do not accelerate relative to one another.

A general spacetime with a gravitational field can be thought of as being curved. Just as a general curved line can be thought of as being made up of tiny bits of straight lines, a general curved spacetime can be thought of as being made of of tiny bits of flat spacetime - the local inertial frames of the equivalence principle.

This gives a powerful geometric picture of a gravitational field. It is nothing else than a curvature of spacetime itself. Now, there are several ways to discuss curvature. We are used to looking at curved spaces inside of some larger (flat) space. Einstein's idea was that the only relevant things are those that can be measured in terms of the curved surface itself and which have nothing to do with it (perhaps) being part of some larger flat space. As a result, one would gain nothing by assuming that there is such a larger flat space. In Einstein's theory, there is no reason to suppose that one exists.

For example, we noticed above that this new understanding of gravity means that the gravitational field contains more information than just giving an accelera- tion at various points in spacetime. The acceleration is related to curvature in spacetime associated with a time direction (say, in the xt plane), but there are also parts of the gravitational field associated with the (purely spatial) xy, xz, and yz planes.

Let's begin by thinking back to the flat spacetime case (special relativity). What was the object which encoded the flat Minkowskian geometry? It was the interval: (interval)^2 = $-c^2 \, \Delta t^2 + \Delta x^2$.

Building Intuition in Flat Space

To understand fully what information is contained in the interval, it is perhaps even better to think first about flat space, for which the analogous quantity is the distance Δs between two points: $\Delta s^2 = \Delta x^2 + \Delta y^2$. Much of the important information in geometry is not the distance between two points per se, but the closely related concept of length. For example, one of the properties of flat space is that the length of the circumference of a circle is equal to 2π times the length of its radius.

Now, in flat space, distance is most directly related to length for straight lines: the distance between two points is the length of the straight line connecting them. To link this to the length of a curve, we need only recall that locally every curve is a straight line.

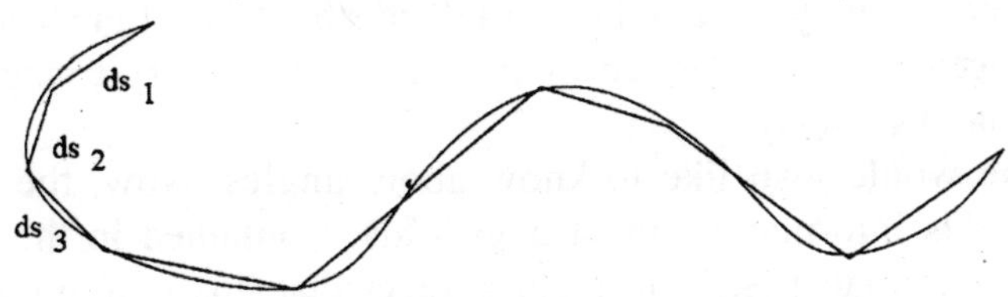

In particular, what we need to do is to approximate any curve by a set of tiny (infinitesimal) straight lines. Because we wish to consider the limit in which these straight lines are of zero size, let us denote the length of one such line by ds.

The relation of Pythagoras then tells us that $ds^2 = dx^2 + dy^2$ for that straight line, where dx and dy are the infinitesimal changes in the x and y coordinates between the two ends of the infinitesimal line segment. To find the length of a curve, we need only add up these lengths over all of the straight line segments. In the language of calculus, we need only perform the integral:

$$\text{Length} = \int_{\text{curve}} ds = \int_{\text{curve}} \sqrt{dx^2 + dy^2}$$

You may not be used to seeing integrals written in a form like the one above. Let me just pause for a moment to note that this can be written in a more familiar form by, say, taking out a factor of dx from the square root. We have

$$\text{Length} = \int_{\text{curve}} dx\sqrt{1+\left(\frac{dy}{dx}\right)^2} .$$

So, if the curve is given as a function $y = y(x)$, the above formula does indeed allow you to calculate the length of the curve.

Now, what does this all really mean? What is the 'take home' lesson from this discussion? The point is that the length of every curve is governed by the formula

$$ds^2 = dx^2 + dy^2.$$

Thus, this formula encodes lots of geometric information, such that the fact that the circumference of a circle is 2π times its radius. As a result, Above equation will be false on a curved surface like a sphere.

A formula of the form ds^2 = stuff is known as a metric, as it tells us how to measure things (in particular, it tells us how to measure lengths). What we are saying is that this formula will take a different form on a curved surface and will not match with equation.

On to Angles

What other geometric information is there aside from lengths? Here, you might consider the examples we talked about last time during class: that flat spaces are characterized by having 180o in every triangle, and by squares behaving nicely.

So, one would also like to know about angles. Now, the important question is: "Is information about angles also contained in the metric?"

It turns out that it is. You might suspect that this is true on the basis of trigonometry, which relates angles to (ratios of) distances. Of course, trigonom- etry is based on flat space, but recall that any space is locally flat, and notice that an angle is something that happens at a point (and so is intrinsically a local notion).

To see just how angular information is encoded in the metric, let's look at an example.

The standard (Cartesian) metric on flat space $ds^2 = dx^2 + dy^2$ is based on an 'orthogonal' coordinate system - one in which the constant x lines intersect the constant y lines at right angles. What if we wish to express the metric in terms of x and, say, some other coordinate z which is not orthogonal to x?

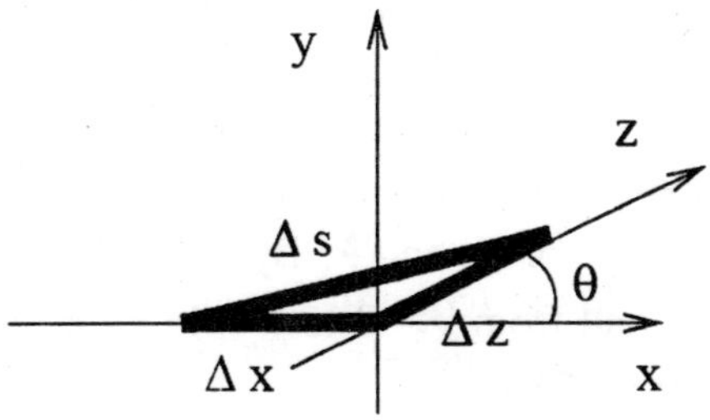

In this case, the distances Δx, Δz, and Δs are related in a slightly more complicated way. If you have studied much vector mathematics, you will have seen the relation:

$$\Delta s^2 = \Delta x^2 + \Delta z^2 + 2\Delta x\Delta z \cos \theta.$$

In vector notation, this is just $|\vec{x}+\vec{z}|^2 = |\vec{x}|^2 + |\vec{z}| + 2\vec{x}\cdot\vec{z}$.

Even if you have not seen this relation before, it should make some sense to you.

Note, for example, that if $\theta = 0$ we get $\Delta s = 2\Delta x$ (since x and z are parallel and our 'triangle' is just a long straight line), while for $\theta = 180°$ we get $\Delta s = 0$ (since x and z now point in opposite directions and, in walking along the two sides of our triangle, we cover the same path twice in opposite directions, returning to our starting point.).

For an infinitesimal triangle, we would write this as:

$$ds^2 = dx^2 + dz^2 - 2dxdz \cos \theta.$$

So, the angular information lies in the "cross term" with a $dxdz$. The coefficient of this term tells us the angle between the x and z directions.

Metrics on Curved Space

This gives us an idea of what a metric on a general curved space should look like. After all, locally (i.e., infinitesimally) it should looks like one of the flat cases above! Thus, a general metric should have a part proportional to dx^2, a part proportional to dy^2, and a part proportional to dxdy. In general, we write this as:

$$ds^2 = g_{xx}dx^2 + 2g_{xy}dxdy + g_{yy}dy^2.$$

What makes this metric different from the ones above (and therefore not necessarily flat) is that g_{xx}, g_{xy}, and g_{yy} are in general functions of the coordinates x, y. In contrast, the functions were constants for the flat metrics above.

Note that this fits with our idea that curved spaces are locally flat since, close to any particular point (x, y) the functions g_{xx}, g_{xy}, g_{yy} will not deviate too much from the values at that point. In other words, any smooth function is locally constant.

Now, why is there a 2 with the *dxdy* term? Note that since $dxdy = dydx$, there is no need to have a separate gyx term. The metric is always symmetric, with $g_{yx} = g_{xy}$. So, $g_{xy}dxdy + g_{yx}dydx = 2g_{xy}dxdy$.

If you are familiar with vectors, then a bit more about how lengths and angles are encoded.

Consider the 'unit' vectors $\hat{x}$ and $\hat{y}$. By 'unit' vectors, the vectors that go from $x = 0$ to $x = 1$ and from $y = 0$ to $y = 1$. As a result, their length is one in terms of the coordinates.

This may or may not be the physical length of the vectors. For example, to use coordinates with a tiny spacing (so that $\hat{x}$ is very short) or coordinates with a huge spacing (so that $\hat{x}$ is large). What the metric tells us directly are the dot products of these vectors:

$$\hat{x}\cdot\hat{x} = g_{xx},$$
$$\hat{x}\cdot\hat{y} = g_{xy},$$
$$\hat{y}\cdot\hat{y} = g_{yy}.$$

Anyway, this object ($g_{\alpha\beta}$) is called the metric (or, the metric tensor) for the space. It tells us how to measure all lengths and angles. The corresponding object for a spacetime will tell us how to measure all proper lengths, proper times, angles, etc. It will be much the same except that it will have a time part with gtt negative4 instead of positive, as did the flat Minkowski space.

Rather than write out the entire expression all of the time (especially when working in, say, four dimensions rather than just two) physicists use a condensed notation called the 'Einstein summation convention'.

To see how this works, let us first relabel our coordinates. Instead of using x and y, let's use x_1, x_2 with $x_1 = x$ and $x_2 = y$. Then we have:

$$ds^2 = \sum_{\alpha=1}^{2}\sum_{\beta=1}^{2} g_{\alpha\beta}dx^{\alpha}dx^{\beta} = g_{\alpha\beta}dx^{\alpha}dx^{\beta}.$$

It is in the last equality that we have used the Einstein summation convention instead of writing out the summation signs, the convention is that we implicitly sum over any repeated index.

A First Example

To get a better feel for how the metric works, let's look at the metric for a flat plane in polar coordinates (r, θ). It is useful to think about this in terms of the unit vectors $\hat{r}, \hat{\theta}$.

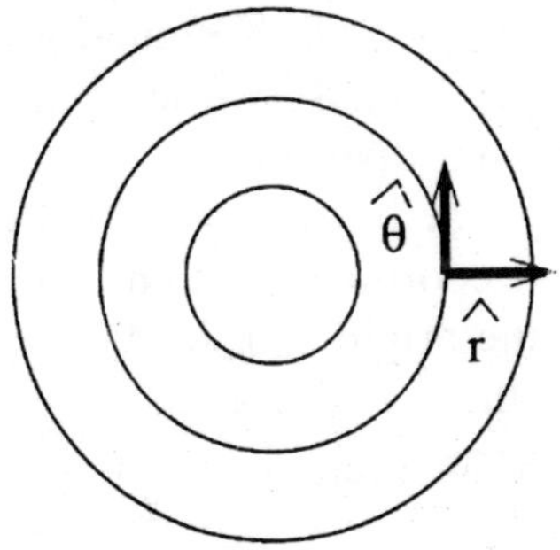

From the picture above, we see that these two vectors are perpendicular: $\hat{r} \cdot \hat{\theta} = 0$. Normally, we measure the radius in terms of length, so that $\hat{r}$ has length one and $\hat{r} \cdot \hat{r} = 1$.

The same is not true for θ: one radian of angle at large r corresponds to a much longer arc than does one radian of angle at small *r*. In fact, one radian of angle corresponds to an arc of length r. The result is that $\hat{\theta}$ has length *r* and $\hat{\theta} \cdot \hat{\theta} = r^2$. So, for theta measured in radians and running from 0 to 2π, the metric turns out to be:

$$ds^2 = dr^2 + r^2\, d\theta^2.$$

Now, let's look at a circle located at some constant value of *r*.

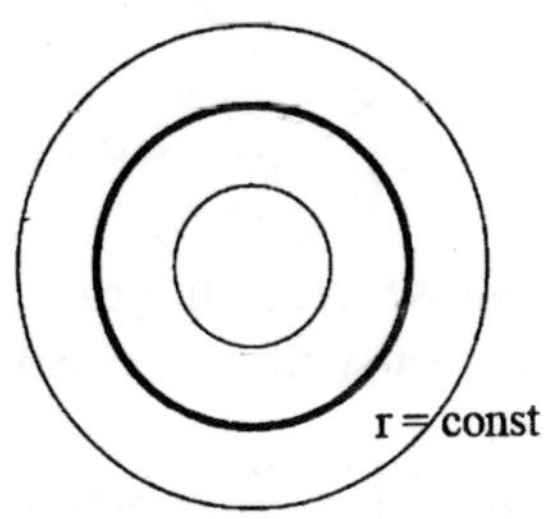

To find the circumference of the circle, we need to compute the length of a curve along the circle. Now, along the circle, r does not change, so we have dr = 0. So, we have $ds = r d\theta$. Thus, the length is:

$$C = \int_0^{2\pi} ds = \int_0^{2\pi} r d\theta = 2\pi r.$$

Let's check something that may seem trivial: What is the radius of this circle? The radius (R) is the length of the curve that runs from the origin out to the circle along a line of constant θ. Along this line, we have $d\theta = 0$. So, along this curve, we have $ds = dr$. The line runs from $r = 0$ to $r = r$, so we have

$$R = \int_0^r dr = r.$$

So, we do indeed have $C = 2\pi R$. Note that while the result $R = r$ may seem obvious it is true only because we used an r coordinate which was marked off in terms of radial distance.

In general, this may not be the case. There are times when it is convenient to use a radial coordinate which directly measures something other than distance from the origin and, in such cases, it is very important to remember to calculate the actual 'Radius' (the distance from the origin to the circle) using the metric.

A Aecond Example

Now let's look at a less trivial example. Suppose the metric of some surface is given by:

$$ds^2 = \frac{dr^2 + r^2 d\theta^2}{(1+r^2)}.$$

Is this space flat? Well, let's compare the circumference (C) of a circle at constant r to the radius (R) of that circle.

Again, the circumference is a line of constant r, so we have $dr = 0$ for this line and $ds = \frac{r}{\sqrt{1+r^2}} d\theta$. The circle as usual runs from $\theta = 0$ to $\theta = 2\pi$. So, we have

$$C = \int_0^{2\pi} \frac{r}{\sqrt{1+r^2}} d\theta = \frac{2\pi r}{\sqrt{1+r^2}}.$$

Now, how about the radius? Well, the radius R is the length of a line at, say, $\theta = 0$ that connects $r = 0$ with $r = r$. So, we have

$$R = \int_0^{\pi} \frac{dr}{\sqrt{1+r^2}} = \sinh^{-1} r$$

(This is yet another neat use of hyperbolic trigonometry.... it allows us to explicitly evaluate certain integrals that would otherwise be a real mess.) Clearly, $C \neq 2\pi R$. In fact, studying the large r limit of the circumference shows that the circumference becomes constant at large r. This is certainly not true of the radius: $R \to \infty$ as $r \to \infty$.

Thus, C is much less than 2πR for large R.

Some Parting Comments on Metrics

This is perhaps the right place to make a point: We often think about curved spaces as being curved inside some larger space. For example, the two-dimensional surface of a globe can be thought of as a curved surface that sits inside some larger (flat) three-dimensional space.

However, there is a notion of curvature (associated with the geometry of the surface - the measurements of circles, triangles, and rectangles drawn in that surface - and encoded by the metric) that does not refer in any way to anything outside the surface itself. So, in order for a four dimensional spacetime to be curved, there does not need to be any 'fifth dimension' for the universe to be 'curved into.'

The point is that what physicists mean by saying that spacetime is curved is not that it is 'bent' in some new dimension, but rather they mean that the geometry on the spacetime is more complicated than that of Minkowski space. For example, they mean that not every circle has circumference $2\pi R$.

Another comment that should be made involves the relationship between the metric and the geometry. We have seen that the metric determines the geometry: it allows us to compute, for example, the ratio of the circumference of a circle to its radius.

One might ask if the converse is true: Does the geometry determine the metric? The answer is a resounding "no." We have, in fact, already seen three metrics for flat space: We had one metric in (orthogonal) Cartesian coordinates, one in 'tilted' Cartesian coordinates where the axes were at some arbitrary angle θ , and one in polar coordinates. Actually, we have seen infinitely many different metrics since the metric was different for each value of the tilt angle θ for the tilted Cartesian coordinates. So, the metric carries information not only about the geometry itself, but also about the coordinates you happen to be using to describe it.

The idea in general relativity is that the real physical effects depend only on the geometry and not upon the choice of coordinates5. After all, the circumference of a circle does not depend on whether you calculate it in polar or in Cartesian coordinates.

As a result, one must be careful in using the metric to make physical predictions - some of the information in the metric is directly physical, but some is an artifact of the coordinate system and disentangling the two can sometimes be subtle.

The choice of coordinates is much like the choice of a reference frame. We saw this to some extent in special relativity. For a given observer (say, Alice) in a given reference frame, we would introduce a notion of position (xAlice) as measured by Alice, and we would introduce a notion of time (tAlice) as measured by Alice.

In a different reference frame (say, Bob's) we would use different coordinates (xBob and tBob). Coordinates describing inertial reference frames were related in a relatively simple way, while coordinates describing an accelerated reference frame were related to inertial coordinates in a more complicated way.

However, whatever reference frame we used and whatever coordinate system we chose, the physical events are always the same. Either a given clock ticks 2 at the event where two light rays cross or it does not. Either a blue paintbrush leaves a mark on a meter stick or it does not. Either an observer writes "I saw the light!" on a piece of paper or she does not. The true physical predictions do not depend on the choice of reference frame or coordinate system at all.

So long as we understand how to deal with physics in funny (say, accelerating) coordinate systems, such coordinate systems will still lead to the correct physical results.

The idea that physics should not depend on the choice of coordinates is called General Coordinate Invariance. Invariance is a term that captures the idea that the physics itself does not change when we change coordinates. This turns out to be an important principle for the mathematical formulation of General Relativity.

THE METRIC OF SPACETIME

We have now come to understand that the gravitational field is encoded in the metric. Once a metric has been given to us, we have also learned how to use it to compute various objects of interest. In particular, we have learned how to test a space to see if it is flat by computing the ratio of circumference to radius for a circle.

However, all of this still leaves open what is perhaps the most important question: just which metric is it that describes the spacetime in which we live?

First, let's again recall that there really is no one 'right' metric, since the metric will depend on the choice of coordinates and there is no one 'right' choice of coordinates.

But there is a certain part of the metric that is in fact independent of the choice of coordinates.

That part is called the 'geometry' of the spacetime. It is mathematically very complicated to write this part down by itself. So, in practice, physicists work with the metric and then make sure that the things they calculate do not depend on the choice of coordinates.

What determines the right geometry? The geometry is nothing other than the gravitational field. So, we expect that the geometry should in some way be tied to the matter in the universe: the mass, energy, and so on should control the geometry.

Figuring out the exact form of this relationship is a difficult task, and Einstein worked on it for a long time. We will not reproduce his thoughts in any detail here.

However, in the end he realized that there were actually not many possible choices for how the geometry and the mass, energy, etc. should be related.

The Einstein Equations

It turns out that, if we make five assumptions, then there is really just one family of possible relationships. These assumptions are:

- Gravity is spacetime curvature, and so can be encoded in a metric.
- General Coordinate Invariance: Real physics is independent of the choice of coordinates used to describe it.
- The basic equations of general relativity should give the dynamics of the metric, telling how the metric changes in time.
- Energy (including the energy in the gravitational field) is conserved.
- The (local) equivalence principle.

Making these five assumptions, one is led to a relation between an object Gab (called the Einstein Tensor) which encodes part of the spacetime curvature and an object Tab (called the Stress-Energy or Energy-Momentum Tensor) which encodes all of the energy, momentum, and stresses in everything else ("matter," electric and magnetic fields, etc.). Here, α, β run over the various coordinates (t, x, y, z). What is a stress? One example of a stress is pressure. It turns out that, in general relativity, pressure contributes to the gravitational field directly6, as do mass, energy, and momentum. The relationship can be written:

$$G_{\alpha\beta} = kT_{\alpha\beta} + Lg_{\alpha\beta}$$

The $g_{\alpha\beta}$ in the equation above is just the metric itself. This relation is known as The Einstein equations.

We see that κ controls just the overall strength of gravity. Making κ larger is the same thing as making $T_{\alpha\beta}$ bigger, which is the same as adding more mass and energy. On the other hand, Λ is something different. Note that it controls a term which relates just to the geometry and not to the energy and mass of the matter. However, this term is added to the side of the equation that contains the energy-momentum tensor. As a result, Λ can be said to control the amount of energy that is present in spacetime that has no matter in it at all. Partially for this reason, Λ is known as the Cosmological constant.

The Newtonian Approximation

As we said (but did not explicitly derive), equation can be deduced from the five above assumptions on purely mathematical grounds. It is not necessary to use Isaac Newton's theory of gravity here as even partial input. So, what is the connection to Newton's ideas about gravity?

Newton's law of gravity can only be correct when the objects are slowly moving - otherwise special relativity would be rele-vant and all sorts of things would go wrong. There is in fact another restriction on when Newtonian gravity is valid. The point is that, in Newtonian gravity, mass creates a gravitational field. But, we know now that energy and mass are very closely related. So, all energy should create some kind of gravity. How-ever, we have also seen that a field (like the gravitational field itself) can carry energy.

As a result, the gravitational field must also act as a source of further gravity. That is, once relativity is taken into account, gravitational fields should be stronger than Newton would have expected. For a very weak field (where the field itself would store little energy), this effect should be small. But, for a strong gravitational field, this effect should be large. So, Newton's law of grav-ity should only be correct for slowly moving objects in fairly weak gravitational fields.

If one does study the Einstein equations for the case of slowly moving objects and weak gravitational fields, one indeed obtains the Newtonian law of gravity for the case $\Lambda = 0$, $\kappa = 8\pi G$, where $G = 6.67 \times 10^{-11} Nm^2/kg^2$ is Newton's gravitational constant. So, to the extent that these numbers are determined by experimental data, they must be the correct values.

In summary, given a lot of thought, Einstein came up with the above five assumptions about the nature of gravity.

Then, by mathematics alone he was able to show that these assumptions lead to equation. For weak gravitational fields and slowly moving objects something like Newton's law of gravity also follows, but with two arbitrary parameters κ and Λ.

One of these (κ) is just 8? times Newton's own arbitrary parameter G. As a result, except for one constant (Λ) Newton's law of gravity has also followed from the five assumptions using only mathematics. Finally, by making use of experimental data (the same data that Newton used originally!) Einstein was able to determine the values of κ and Λ. The Einstein equations then take on the pleasing form:

$$G_{\alpha\beta} = 8\pi G T_{\alpha\beta}.$$

The Schwarzschild Metric

Of course, the natural (and interesting!) question to ask is "What happens when the gravitational field is strong and Newton's law of gravity does NOT hold?" We're not actually going to solve the Einstein equations ourselves they're pretty complicated even for the simplest of cases.

When an object is perfectly round (spherical), the high symmetry of the situation simplifies the mathematics.

The point is that if the object is round, and if the object completely determines the gravitational field, then the gravitational field must be round as well. So, the first simplification we will perform is to assume that our gravitational field (i.e., our spacetime) is spherically symmetric.

The second simplification we will impose is to assume that there is no matter (just empty spacetime), at least in the region of spacetime that we are studying.

In particular, the energy, momentum, etc., of matter are equal to zero in this region. As a result, we will be describing the gravitational field of an object (the earth, a star, etc.) only in the region outside of the object. This would describe the gravitational field well above the earth's surface, but not down in the interior.

For this case, the Einstein equations were solved by a young German mathematician named Schwarzschild. There is an interesting story here, as Schwarzschild solved these equations during his spare time while he was in the trenches fighting (on the German side) in World War I. The story is that Schwarzschild got his calculations published but, by the time this happened, he had been killed in the war.

Because of the spherical symmetry, it was simplest for Schwarzschild to use what are called spherical coordinates (r, θ, φ) as opposed to Cartesian Coordinates (x, y, z). Here, r tells us how far out we are, and θ, φ are latitude and longitude coordinates on the sphere at any value of r.

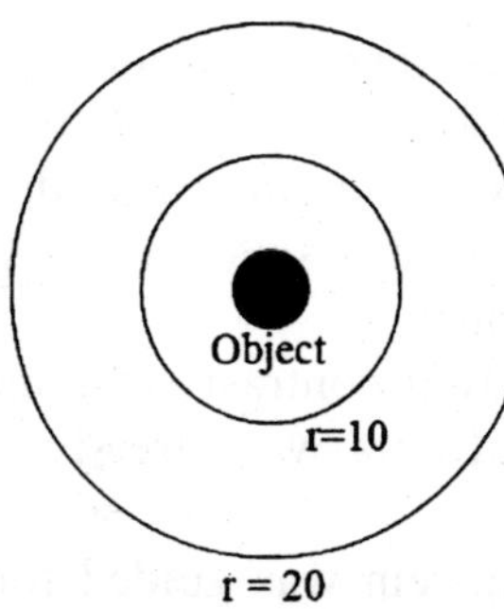

Schwarzschild found that, for any spherically symmetric spacetime and outside of the matter, the metric takes the form:

$$ds^2 = \left(1 - \frac{R_s}{r}\right)dt^2 + \frac{dr^2}{1 - \frac{R_s}{r}} + r^2(d\theta^2 + \sin^2\theta d\phi^2)$$

Here, the parameter R_s depends on the total mass of the matter inside. In particular, it turns out that $R_s = 2MG/c^2$.

The last part of the metric, $r^2(d\theta^2 + \sin^2\theta d\phi^2)$, is just the metric on a standard sphere of radius r.

This part follows just from the spherical symmetry itself. θ is a latitude coordinate and ϕ is a longitude coordinate. The factor of $\sin^2 \theta$ encodes the fact that circles at constant θ (i.e., with $d\theta = 0$) are smaller near the poles ($\theta = 0, \pi$) than at the equator ($\theta = \pi/2$).

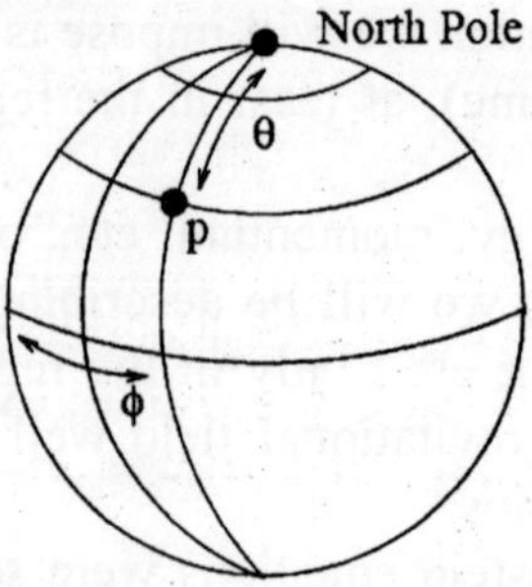

THE EXPERIMENTAL VERIFICATION OF GENERAL RELATIVITY

Now that the Schwarzschild metric is in hand, we know what is the spacetime geometry around any round object. Now, what can we do with it? Well, in principle, one can do just about anything. The metric encodes all of the information about the geometry, and thus all of the information about geodesics. Any freely falling worldline (like, say, that of an orbiting planet) is a geodesic. So, one thing that can be done is to compute the orbits of the planets. Another would be to compute various gravitational time dilation effects.

Having arrived at the Schwarzschild solution, we are finally at the point where Einstein's ideas have a lot of power. They now predict the curvature around any massive object (the sun, the earth, the moon, etc.). So, Einstein started looking for predictions that could be directly tested by experiment to check that he was actually right.

This makes an interesting contrast with special relativity, in which quite a bit of experimental data was already available before Einstein constructed the theory.

In the case of GR, Einstein was guided for a long time by a lot of intuition (i.e., guesswork) and, for the most part, after he had constructed the theory.

A few pieces of experimental evidence already (such as the Pound-Rebke and GPS experiments) these occurred only in 1959 and in the 1990's! Einstein finished developing General Relativity in 1916 and certainly wanted to find an experiment that could be done soon after.

The Planet Mercury

We have seen that Einstein's theory of gravity agrees with Newton's when the gravitational fields are weak (i.e., far away from any massive object). But, the discrepancy increases as the field gets stronger. So, the best place (around here) to look for new effects is close to the sun. One might therefore start by considering the orbit of Mercury. Actually, there is an interesting story about Mercury and its orbit. Astronomers had been tracking the motion of the planets for hundreds of years. Ever since Newton, they had been comparing these motions to what Newton's law of gravity predicted.

The agreement was incredible. In the early 1800's, they had found small dis-crepancies (30 seconds of arc in 10 years) in the motion of Uranus. For awhile people thought that Newton's law of gravity might not be exactly right. However, someone then had the idea that maybe there were other objects out there whose gravity a ected Uranus. They used Newton's law of gravity to predict the existence of new planets: Neptune, and later Pluto. They could even tell astronomers where to look for Neptune within about a degree of angle on the sky.

However, there was one discrepancy with Newton's laws that the astronomers could not explain. This was the 'precession' of Mercury's orbit. The point is that, if there were nothing else around, Newton's law of gravity would say that Mercury would move in a perfect ellipse around the sun, retracing its path over, and over, and over...

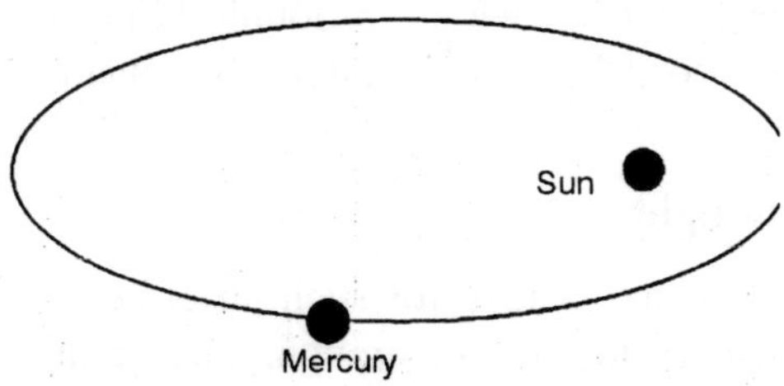

Of course, there are small tugs on Mercury by the other planets that modify this behaviour. However, the astronomers knew how to account for these effects. Their results seemed to say that, even if the other planets and such were not around, Mercury would do a sort of spiral dance around the sun, following a path that looks more like this:

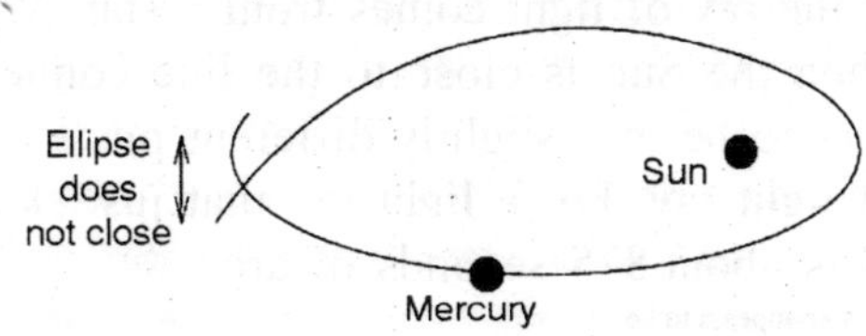

Here, the ellipse itself as rotating (a.k.a. 'precessing') about the sun. After all known effects had been taken into account, astronomers found that Mercury's orbit precessed by an extra 43 seconds of arc per century. This is certainly not very much, but the astronomers already understood all of the other planets to a much higher accuracy. So, what was going wrong with Mercury? Most astronomers thought that it must be due to some sort of gas or dust surrounding the Sun (a big 'solar atmosphere') that was somehow a ecting Mercury's orbit.

However, Einstein knew that his new theory of gravity would predict a preces-sion of Mercury's orbit for two reasons. First, he predicted a slightly stronger gravitational field (since the energy in the gravitational field itself acts as a source of gravity). Second, in Einstein's theory, space itself is curved and this effect will also make the ellipse precess (though, since the velocity of mercury is small, this effect turns out to be much smaller than the one due to the stronger gravitational field).

The number that Einstein calculated from his theory was 43 seconds of arc per century. That is, his prediction agreed with the experimental data to better than 1%! Clearly, Einstein was thrilled.

This was big news. However, it would have been even bigger news if Einstein had predicted this result before it had been measured. Physicists are always skeptical of just explaining known effects. After all, maybe the scientist (intentionally or not) fudged the numbers or the theory to get the desired result? So, physicists tend not to really believe a theory until it predicts something new that is then verified by experiments.

This is the same sort of idea as in double blind medical trials, where even the researchers don't know what effect they want a given pill to have on a patient!

The Bending of Starlight

Luckily, Einstein had an idea for such an effect and now had enough confidence in his theory to push it through. The point is that, as we have discussed, light will fall in a gravitational field. For example, a laser beam fired horizontally across the classroom will be closer to the ground on the side where it hits the far wall it was when it left the laser.

Similarly, a ray of light that goes skimming past a massive object (like the sun) will fall a bit toward the sun. The net effect is that this light ray is bent. Suppose that the ray of light comes from a star. What this means in the end is that, when the Sun is close to the line connecting us with the star, the star appears to be in a slightly different place than when the Sun is not close to that light ray. For a light ray that just skims the surface of the Sun, the effect is about.875 seconds of arc.

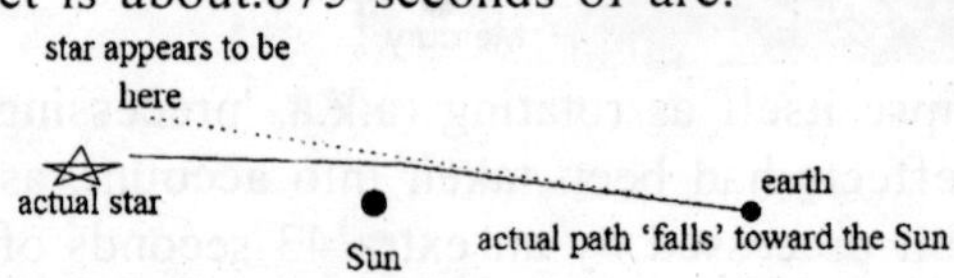

However, this is not the end of the story. It turns out that there is also another effect which causes the ray to bend. This is due to the effect of the curvature of space on the light ray. This effect turns out to be exactly the same size as the first effect, and with the same sign. As a result, Einstein predicted a total bending angle of 1.75 seconds of arc - twice what would come just from the observation that light falls in a gravitational field.

This is a tricky experiment to perform, because the Sun is bright enough that any star that close to the sun is very hard to see. One solution is to wait for a solar eclipse (when the moon pretty much blocks out the light from the sun itself) and then one can look at the stars nearby.

Just such an observation was performed by the British physicist Sir Arthur Eddington in 1919. The result indicated a bending angle of right around 2 seconds of arc.

More recently, much more accurate versions of this experiment have been performed which verify Einstein's theory to high precision. See Theory and experiment in gravitational physics by Cli ord *M.* Will, (Cambridge University Press, New York, 1993) QC178.W47 1993 for a modern discussion of these issues.

Other Experiments: Radar Time Delay

The bending of starlight was the really big victory for Einstein's theory. However, there are two other classic experimental tests of general relativity that should be mentioned. One of these is just the effect of gravity on the frequency of light that we have already discussed. As we said before, this had to wait quite a long time (until 1959) before technology progressed to the stage where it could be performed.

The last major class of experiments is called 'Radar Time delay.' These turn out to be the most accurate tests of Einstein's theory, but they had to wait until even more modern times. The point is that the gravitational field effects not only the path through space taken by a light ray, but that it also effects the time that the light ray takes to trace out that path. As we have discussed once or twice before, time measurements can be made extremely accurately. So, these experiments can be done to very high precision.

The idea behind these experiments is that you then send a microwave (a.k.a. radar) signal (which is basically a long wavelength light wave) over to the other side of the sun and back.

You can either bounce it o a planet (say, Venus) or a space probe that you have sent over there for just this purpose. If you measure the time it takes for the signal to go over and then return, this time is always longer than it would have been in flat spacetime. In this way, you can carefully test Einstein's theory.

Chapter 9

Black Holes

INVESTIGATING THE SCHWARZSCHILD METRIC

We computed the gravitational effect on time dilation back. However, in this computation we needed to know the gravitational acceleration g(*l*). We could of course use Newton's prediction for g(*l*), which experiments tell us is approximately correct near the earth. However, in general we expect this to be the correct answer only for weak gravitational fields.

On the other hand, we know that the Schwarzschild metric describes the gravitational field around a spherical object even when the field is strong. So, what we will do is to first use the Schwarzschild metric to compute the gravitational time dilation effect directly. We will then be able to use the relation between this time dilation and the gravitational acceleration to compute the corrections to Newton's law of gravity.

Gravitational Time Dilation from the Metric

Suppose we want to calculate how clocks run in this gravitational field. This has to do with proper time $d\tau$, so we should remember that $d\tau^2 = -ds^2$. For the Schwarzschild metric we have:

$$d\tau^2 = -\,ds^2 = \left(1-\frac{R_s}{r}\right)dt^2 + \frac{dr^2}{1-\frac{R_s}{r}} + r^2(d\theta^2 + \sin^2\theta\, d\phi^2)$$

The Schwarzschild metric describes any spherically symmetric gravitational field in the region outside of all the matter. So, for example, it gives the gravitational field outside of the earth. In using the Schwarzschild metric, remember that $R_s = 2MG/c^2$.

Let's think about a clock that just sits in one place above the earth. It does not move toward or away from the earth, and it does not go around the earth. It just 'hovers.'

Perhaps it sits in a tower, or is in some rocket ship whose engine is tuned in just the right way to keep it from going either up or down. Such a clock is called a static clock since, from it's point of view, the gravitational field does not change with time.

Since r, θ, and ϕ do not change, we have $dr = d\theta = d\phi = 0$. So, on our clock's worldline we have just: $d\tau^2 = \left(1 - \frac{R_s}{r}\right) dt^2$. That is,

$$d\tau = \sqrt{1 - \frac{R_s}{r}} dt.$$

Note that if the clock is at $r = \infty$ then the square root factor is equal to 1. So, we might write $d\tau_\infty = dt$. In other words, $d\tau = \sqrt{1 - \frac{R_s}{r}} d\tau_\infty$, or,

$$\frac{\Delta\tau}{\Delta\tau_\infty} = \sqrt{1 - \frac{R_s}{r}}.$$

As saw before, clocks higher up run faster. Now, however, the answer seems to take a somewhat simpler form, when we were using only the Newtonian approximation.

Corrections to Newton's Law

Note that the Schwarzschild geometry is a time independent gravitational field. The rate at which various clocks run to the acceleration of freely falling observers. In other words, we can use this to compute the corrections to Newton's law of gravity.

The relation is

$$\frac{\Delta\tau_b}{\Delta\tau_a} = \exp\left(\int_a^b \frac{\alpha(a)}{c^2} ds\right).$$

Here, $\alpha(s)$ is the acceleration of a static clock relative to a freely falling clock at s, and s measures distance. To compare this with our formula above, we want to take $a = s$ and $b = \infty$. Taking the ln of both sides gives us

$$\ln\left(\frac{\tau(s)}{\tau_\infty}\right) = \int_\infty^s \frac{\alpha(a)}{c^2} ds.$$

Now, taking a derivative with respect to s we find:

$$\frac{\alpha(s)|}{c^2} = -\frac{d}{ds} \ln\left(\frac{\tau(s)}{\tau_\infty}\right).$$

Now, it is important to know what exactly s measures in this formula. When we derived this result we were interested in the actual physical height of a tower. As a result, this s describes proper distance, say, above the surface of the earth.

On the other hand, equation is given in terms of r which, it turns out, does not describe proper distance. To see this, let's think about the proper distance ds along a radial line with $dt = d\theta = d\phi = 0$. In this case, we have

$$ds^2 = \frac{dr^2}{1 - R_s/r}, \text{ or } ds = \frac{dr}{\sqrt{1 - R_s/r}}, \text{ and}$$

$$\frac{dr}{ds} = \sqrt{1 - R_s/r}.$$

However we can deal with this by using the chain rule:

$$\alpha = c^2 \frac{d}{ds} \ln\left(\frac{\tau(s)}{\tau_\infty}\right) = c^2 \left(\frac{dr}{ds}\right) \frac{d}{dr} \ln\left(\frac{\tau(r)}{\tau_\infty}\right).$$

Going through the calculation yields:

$$\alpha = c^2 \sqrt{1 - R_s/r} \frac{d}{dr} \ln \sqrt{1 - R_s/r}$$

$$= c^2 \sqrt{1 - R_s/r} \frac{1}{2} \frac{d}{dr} \ln(1 - R_s/r)$$

$$= \frac{c^2}{2} \sqrt{1 - R_s/r} \frac{1}{1 - R_s/r} \frac{+R_s}{r^2}$$

$$= \frac{c^2}{2\sqrt{1 - R_s/r}} \frac{R_s}{r^2}.$$

Note that for $r >> R_s$, we have $\alpha \sim \frac{c^2}{2} \frac{R_s}{r^2} = \frac{MG}{r^2}$. This is exactly Newton's result.

However, for small r, α is much bigger. In particular, look at what happens when r = RS. There we have $\alpha(R_s) = \infty$! So, at $r = R_s$, it takes an infinite proper acceleration for a clock to remain static. A static person at $r = R_s$ would therefore feel infinitely heavy. This is clearly a various special value of the radius coordinate, r. This value is known as the Schwarzschild radius. Now, let's remember that the Schwarzschild metric only gives the right answer outside of all of the matter.

Suppose then that the actual physical radius of the matter is bigger than the associated Schwarzschild radius (as is the case for the earth and the Sun). In this case, you will not see the effect described above since the place where it would have occurred ($r = R_s$) in inside the earth where the matter is non-zero and the Schwarzschild metric does not apply.

But what if the matter source is very small so that its physical radius is less than Rs? Then the Schwarzschild radius R_s will lie outside the matter at a place you could actually visit. In this case, we call the object a "black hole." You will see why in a moment.

ON BLACK HOLES

Objects that are smaller than their Schwarzschild radius (i.e., black holes) are one of the most intriguing features of general relativity. We now proceed to explore them in some detail, discussing both the formation of such objects and a number of their interesting properties. Although black holes may seem very strange at first, we will soon find that many of their properties are in quite similar to features that we encountered in our development of special relativity some time ago.

Forming a Black Hole

A question that often arises when discussing black holes is whether such objects actually exist or even whether they could be formed in principle. After all, to get $R_s = 2MG/c^2$ to be bigger than the actual radius of the matter, you've got to put a lot of matter in a very small space, right? So, maybe matter just can't be compactified that much. In fact, it turns out that making black holes (at least big ones) is actually very easy.

In order to stress the importance of understanding black holes and the Schwarzschild radius in detail, we'll first talk about just why making a black hole is so easy before going on to investigate the properties of black holes. Suppose we want to make a black hole out of, say, normal rock. What would be the associated Schwarzschild radius? We know that $R_s = 2MG/c^2$. Suppose we have a big ball of rock or radius r.

How much mass in in that ball? Well, our experience is that rock does not curve spacetime so much, so let's use the flat space formula for the volume of a sphere: $V = \frac{4}{3}\pi r^3$. The mass of the ball of rock is determined by its density, ρ, which is just some number 1. The mass of the ball of rock is therefore $M = \rho V = \frac{4\pi}{3}\rho r^3$. The associated Schwarzschild

$$R_s = \frac{8\pi G}{3c^2}\rho r^3.$$

Now, for large enough r, any cubic function is bigger than r. In particular, we get $r = R_s$ at $r = \left(\frac{3c^2}{8\pi G\rho}\right)^{1/2}$ and there is a solution no matter what the value of ρ! So, a black hole can be made out of rock. without even working hard to compress it more than normal, so long as we just have enough rock. Similarly, a black hole could be made out of people, so long as we had enough of them just insert the average density of a person in the formula above.

A black hole could even be made out of very diffuse air or gas, so long so as we had enough of it. For air at normal density, we would need a ball of air 10^{13} meters across. For comparison, the Sun is 10^9 meters across, so we would need a ball of gas 10,000 times larger than the Sun (in terms of radius).

Black holes in nature seem to come come in two basic kinds. The first kind consists of small black holes whose mass is a few times the mass of the Sun. These form at the end of a stars life cycle when nuclear fusion no longer produces enough heat (and thus pressure) to hold up the star. The star then collapses and compresses to enormous densities. Such collapses are accompanied by extremely violent processes called supernovae. The second kind consists of huge black holes, whose mass is 10^6 (a million) to 10^{10} (ten billion) times the mass of the sun. Some black holes may be even larger.

Astronomers tell us that there seems to be a large black hole at the centre of every galaxy, or almost every galaxy. These large black holes are much easier to form than are small ones and do not require especially high densities. To pack the mass of 10^6 suns within the corresponding Schwarzschild radius does not require a density much higher than that of the Sun itself (which is comparable to the density of water or rock).

One can imagine such a black hole forming in the centre of a galaxy, where the stars are densely packed, just by having a few million stars wander in very close together.

The larger black holes are even easier to make: to pack a mass of ten billion suns within the corresponding Schwarzschild radius requires a density of only 10^{-5} times the density of air! It could form from just a very large cloud of very thin gas.

Matter within the Schwarzschild Radius

Since black holes exist (or at least could easily be made) we're going to have to think more about what is going on at the Schwarzschild radius. At first, the Schwarzschild radius seems like a very strange place. There, a rocket would require an infinite proper acceleration to keep from falling in. So what about the matter that first formed the black hole itself? Where is that matter and what is it doing?

Let's go through this step by step. Let us first ask if there can be matter sitting at the Schwarzschild radius (as part of a static star or ball or gas). Clearly not, since the star or ball of gas cannot produce the infinite force that would be needed to keep its atoms from falling inward. The star or ball of gas must contract. Even more than this, the star will be already be contracting when it reaches the Schwarzschild radius and, since gravitation produces accelerations, it must cause this rate of contraction to increase.

Now, what happens when the star becomes smaller than its Schwarzschild radius? The infinite acceleration of static observers at the Schwarzschild radius suggests that the Schwarzschild metric may not be valid inside Rs. As a result we cannot yet say for sure what happens to objects that have contracted within Rs. However, we would certainly find it odd if the effects of gravity became weaker when the object was compressed. Thus, since the object has no choice but to contract (faster and faster) when it is of size Rs, one would expect smaller objects also to have no choice but accelerated contraction!

It now seems that in a finite amount of time the star must shrink to an object of zero size, a mathematical point. This most 'singular' occurrence (to quote Sherlock Holmes) is called a 'singularity.' But, once it reaches zero size, what happens then? This is an excellent question, but we are getting ahead of ourselves. For the moment, let's go back out to the Schwarzschild radius and find out what is really going on there.

The Schwarzschild Radius and the Horizon

Not only does a clock require an infinite acceleration to remain static at the Schwarzschild radius, but something else interesting happens there as well. Let's look back at the formula we had for the time measured by a static clock:

$$\frac{\Delta\tau(r)}{\Delta\tau_\infty} = \sqrt{1 - R_s / r}\,.$$

Notice what happens at at the Schwarzschild radius. Since $r = R_s$, we have $\Delta\tau = 0$. Our clock stops, and no time passes at all.

Now, this is certainly very weird, but perhaps it rings a few bells? It should sound vaguely familiar.... clocks running infinitely slow at a place where the acceleration required to keep from falling becomes infinite.... You may recall that the same thing occurred for the acceleration horizons back in special relativity.

This gives us a natural guess for what is going on near the Schwarzschild radius. In fact, let us recall that any curved spacetime is locally flat. So, if our framework holds together at the Schwarzschild radius we should be able to match the region near $r = R_s$ to some part of Minkowski space. Perhaps we should match it to the part of Minkowski space near an acceleration horizon? Let us guess that this is correct and then proceed to check our answer.

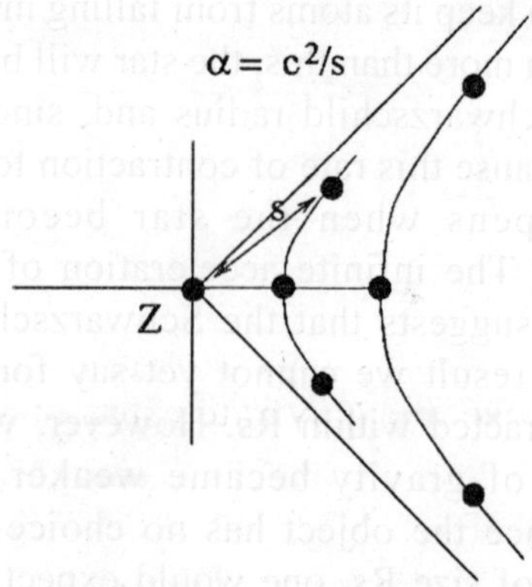

We will check our answer using the equivalence principle. The point is that an accelerating coordinate system in flat spacetime contains an apparent gravita-tional field.

There is some nontrivial proper acceleration α that is required to remain static at each position. Furthermore, this proper acceleration is not thesame at all locations, but instead becomes infinitely large as one approaches thehorizon.

What we want to do is to compare this apparent gravitational field (the proper acceleration $\alpha(s)$, where s is the proper distance from the horizon) near the acceleration horizon with the corresponding proper acceleration $\alpha(s)$ required to remain static a small proper distance s away from the Schwarzschild radius.

If the two turn out to be the same then this will mean that static observers have identical experiences in both cases. But, the experiences of static observers are related to the experiences of freely falling observers. Thus, if we then consider freely falling observers in both cases, they will also describe both situations in the same way. It will then follow that physics near the event horizon is identical to physics near an acceleration horizon - something that we understand well from special relativity.

In flat spacetime the proper acceleration required to maintain a constant proper distance s from the acceleration horizon (e.g., from event Z) is given by

$$\alpha = c^2/s.$$

Now, so far this does not look much like our result for the black hole. However, we should again recall that r and s represent different quantities. That is, r does not measure proper distance. Instead, we have

$$ds = \frac{dr}{\sqrt{1 - R_s / r}} = \sqrt{\frac{r}{\sqrt{1 - R_s}}}\,dr.$$

The behaviour when $r - R_s$ is small. To examine this, it is useful to introduce the quantity $\Delta = r - R_s$. We can then write the above formula as: $ds = \sqrt{\frac{r}{\Delta}}d\Delta$. Integrating, we get

$$s = \int_0^{\Delta}\sqrt{\frac{r}{\Delta}}d\Delta.$$

This integral is hard to perform exactly since $r = R_s + \Delta$ is a function of Δ. However, since we are only interested in small Δ (for our local comparison), r doesn't differ much from R_s. So, we can simplify our work and still maintain sufficient accuracy by replacing r in the above integral by R_s. The result is:

$$s \approx \sqrt{R_s}\int_0^{\Delta}\sqrt{\frac{r}{\Delta}}d\Delta = 2\sqrt{R_s\Delta}.$$

Let us use this to write α for the black hole (let's call this α_{BH}) in terms of the proper distance s. From above, we have

$$\alpha_{BH} = \frac{c^2}{2\sqrt{1-R_s/|r}}\frac{R_s}{r^2}$$

$$= \frac{c^2}{2}\frac{\sqrt{r}}{1-R_s}\frac{R_s}{r_2}$$

$$= \frac{c^2}{2}\frac{1}{\sqrt{\Delta}}\frac{r}{\sqrt{r}}\frac{R_s}{r^2}$$

$$= \frac{c^2}{2\sqrt{\Delta R_s}} = \frac{c^2}{s}.$$

Note that this is identical to the expression for α near an acceleration horizon It worked! Thus we can conclude:

Near the Schwarzschild radius, the black hole spacetime is just the same as flat spacetime near an acceleration horizon.

The part of the black hole spacetime at the Schwarzschild radius is known as the horizon of the black hole.

Going Beyond the Horizon

We are of course interested in what happens when we go below the horizon of a black hole. However, the connection with acceleration horizons tells us that we will need to be careful in investigating this question. In particular, so far we have made extensive use of static observers - measuring the acceleration of freely falling frames relative to them. Static observers were also of interest when discussing acceleration horizons - so long as they were outside of the acceleration horizons.

The past and future acceleration horizons divided Minkowski space into four regions: static worldlines did not enter two of these at all, and in another region static worldlines would necessarily move 'backwards in time.' The fourth region was the normal 'outside' region, and we concluded that true static observers could only exist there.

We have seen that the spacetime near the black hole horizon is just like that near an acceleration horizon. As a result, there will again be no static observers below the horizon.

We suspected this earlier based on the idea that it takes infinite acceleration to remain static at the horizon and we expected the gravitational effects to be even stronger deeper inside.

Based on our experience with acceleration horizons, we now begin to see how this may in fact be possible. It has become clear that we will need to abandon static observers in order to describe the region below the black hole horizon.

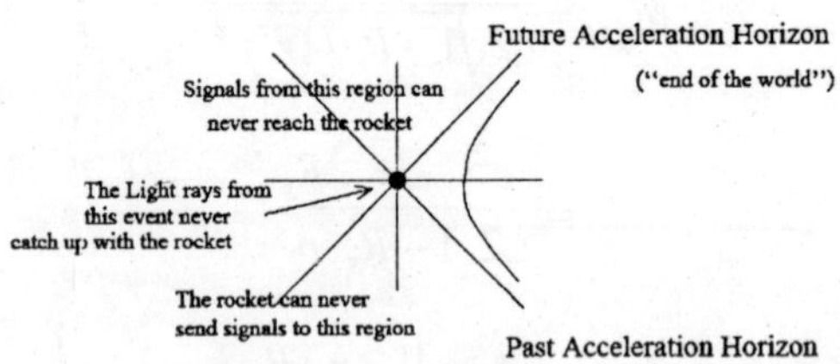

Suppose then that we think about freely falling observers instead. As we know, freely falling observers typically have the simplest description of spacetime.

Using the connection with acceleration horizons, we see immediately how to draw a (freely falling) spacetime diagram describing physics near the Schwarzschild radius. It must look just like our diagram above for flat spacetime viewed from an inertial frame near an acceleration horizon! Note that $r = R_s$ for the black hole is like $s = 0$ for the acceleration horizon since $\alpha \to \infty$ in both cases.

The important part of this is that $s = 0$ is not only the event Z, but is in fact the entire horizon! This is because events separated by a light ray are separated by zero proper distance.

It also follows from continuity since, arbitrarily close to the light rays shown below we clearly have a curve of constant r for r arbitrarily close to R_s. So, $r = R_s$ is also the path of a light ray, and forms a horizon in the black hole spacetime. In the black hole context, the horizon is often referred to as the 'event horizon of the black hole.'

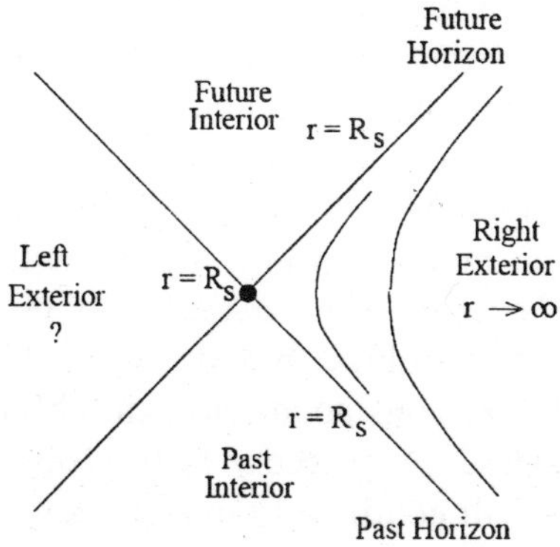

Let us review our discussion so far. We realized that, so long as we were outside the matter that is causing the gravitational field, any spherically symmetric (a.k.a. 'round') gravitational field is described by the Schwarzschild metric. This metric has a special place, at $r = R_s$, the 'Schwarzschild Radius.' Any object which is smaller than its Schwarzschild Radius will be surrounded by an event horizon, and we call such an object a black hole.

If we look far away from the black hole, at r >> R_s, then the gravitational field is much like what Newton would have predicted for an object of that mass. There is of course a little gravitational time dilation, and a little curvature4, but not much.

Indeed, the Schwarzschild metric describes the gravitational field not only of a black hole, but of the earth, the Sun, the moon, and any other round object. However, for those more familiar objects, the surface of the object is at $r >> R_s$. For example, on the surface of the Sun $r/R_s \sim 5 \times 10^5$.

So, far from a black hole, objects can orbit just like planets orbit the Sun. By the way, remember that orbiting objects are freely falling - they do not require rocket engines or other forces to keep them in orbit. However, suppose that we look closer in to the horizon. What happens then?

In one of the homework problems, you will see that something interesting happens to orbiting objects when they orbit at $r = 3R_s/2$. There, an orbiting object experiences no proper time: $\Delta\tau = 0$. This means that the orbit at this radius is a lightlike path. In other words, a ray of light will orbit the black hole in a circle at $r = 3R_s/2$. For this reason, this region is known as the 'photon sphere.' This makes for some very interesting visual effects if you would imagine traveling to the photon sphere.

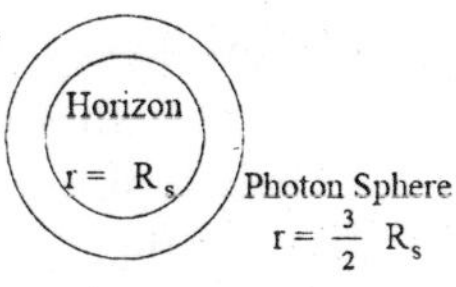

This is not to say that light cannot escape from the photon sphere. The point is that, if the light is moving straight sideways (around the black hole) then the black hole's gravity is strong enough to keep the light from moving farther away.

However, if the light were directed straight outward at the photon sphere, it would indeed move outward, and would eventually escape.

And what about closer in, at $r < 3R_s/2$? Any circular orbit closer in is spacelike, and represents an object moving faster than the speed of light. So, given our usual assumptions about physics, nothing can orbit the black hole closer than $r = 3R_s/2$. Any freely falling object that moves inward past the photon sphere will continue to move to smaller and smaller values of r. However, if it ceases to be freely falling (by colliding with something or turning on a rocket engine) then it can still return to larger values of r.

Now, suppose that we examine even smaller r, and still have not run into the surface of an object that is generating the gravitational field. If we make it all the way to $r = R_s$ without hitting the surface of the object, we find a horizon and we call the object a black hole.

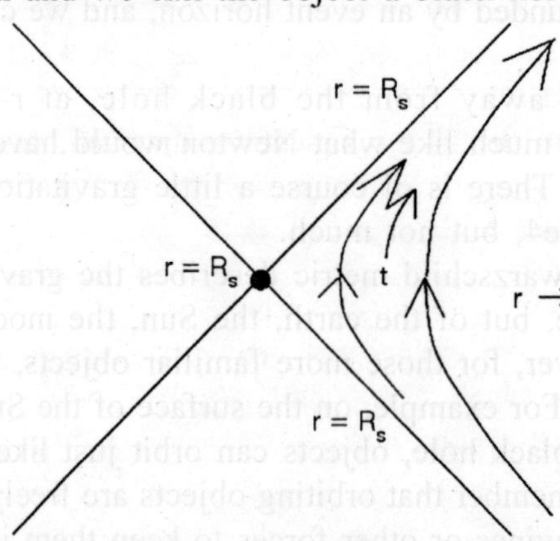

It is at a constant value of r, the horizon contains the worldlines of outward directed light rays. To see what this means, imagine an expanding sphere of light (like one of the ones produced by a firecracker) at the horizon. Although it is moving outward at the speed of light (which is infinite boost parameter.), the sphere does not get any bigger. The curvature of spacetime is such that the area of the spheres of light do not increase. A spacetime diagram looks like this:

Not only do light rays directed along the horizon remain at $r = R_s$, any light ray at the horizon which is directed a little bit sideways (and not perfectly straight outward) cannot even stay at $r = R_s$, but must move to smaller r. The diagram below illustrates this by showing the horizon as a surface made up of light rays. If we look at a light cone emitted from a point on this surface, only the light ray that is moving in the same direction as the rays on the horizon can stay in the surface. The other light rays all fall behind the surface and end up inside the black hole (at $r < R_s$).

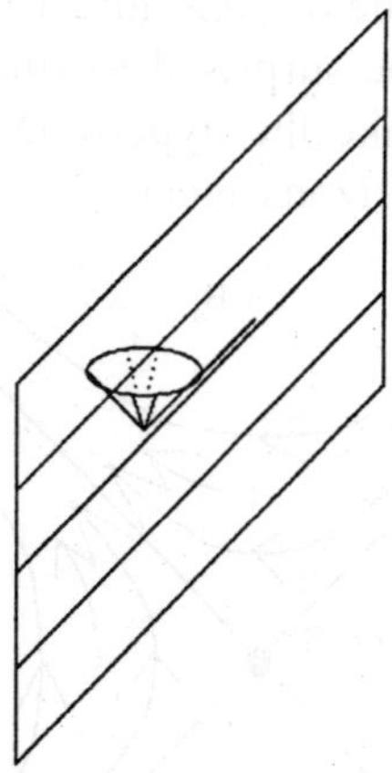

Similarly, any object of nonzero mass requires an infinite acceleration (directed straight outward) to remain at the horizon. With any finite acceleration, the object falls to smaller values of r. At any value of r less than Rs no object can ever escape from the black hole.

This is clear from the above spacetime diagram, since to move from the future interior to, say, the right exterior the object would have to cross the light ray at $r = R_s$, which is not possible.

Note that we could have started with this geometric insight at the horizon and used it to argue for the existence of the photon sphere: Light aimed sideways around the black hole escapes when started far away but falls in at the horizon.

Somewhere in the middle must be a transition point where the light neither escapes nor falls in. Instead, it simply circles the black hole forever at the same value of r.

BEYOND THE HORIZON

Of course, the question that everyone would like to answer is "What the heck is going on inside the black hole?" To understand this, we will turn again to the Schwarzschild metric.

The Interior Diagram

To make things simple, let's suppose that all motion takes place in the r, t plane. This means that $d\theta = d\phi = 0$, and we can ignore those parts of the metric. The relevant pieces are just

$$ds^2 = -(1 - R_s / r)dt^2 + \frac{dr^2}{1 - R_s / r}.$$

Let's think for a moment about a line of constant r (with $dr = 0$). For such a line, $ds^2 = -(1 - R_s/r)dt^2$. The interesting thing is that, for $r < R_s$, this is positive. Thus, for $r < R_s$, a line of constant r is spacelike.

You will therefore not be surprised to find that, near the horizon, the lines of constant r are just like the hyperbolae that are a constant proper time from where the two horizons meet.

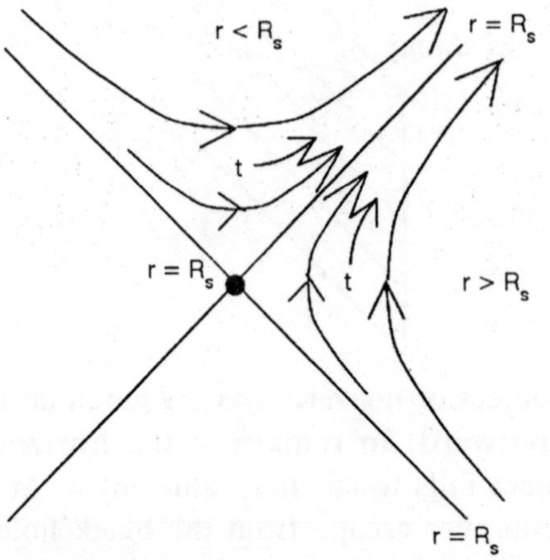

The coordinate t increases along these lines, in the direction indicated by the arrows. This means that the t-direction is actually spacelike inside the black hole. The point here is not that something screwy is going on with time inside a black hole.

Instead, it is merely that using the Schwarzschild metric in the way that we have written it we have done something 'silly' and labelled a space direction t. The problem is in our notation, not the spacetime geometry.

Let us fix this by changing notation when we are in this upper region. We introduce $t' = r$ and $r' = t$. The metric then takes the form

$$ds^2 = -(1 - R_s/t')dr'^2 + \frac{dt'^2}{1 - R_s/t'}$$

You might wonder if the Schwarzschild metric is still valid in a region where the t direction is spacelike.

It turns out that it is. Unfortunately, we were not able to discuss the Einstein equations in detail. If we had done so, however, then we could check this by directly plugging the Schwarzschild metric into equation just as we would to check that the Schwarzschild metric is a solution outside the horizon.

Finally, notice that the lines above look just the like lines we drew to describe the boost symmetry of Minkowski space associated with the change of reference frames. In the same way, these lines represent a symmetry of the black hole spacetime. After all, the lines represent the direction of increasing $t = r'$.

But, the Schwarzschild metric is completely independent of $t = r'$ – it depends only on $r = t'$! So, sliding events along these lines and increasing their value of $t = r'$ does not change the spacetime in any way. Outside of the horizon, this operation moves events in time.

As a result, the fact that it is a symmetry says that the black hole's gravitational field is not changing in time.

However, inside the horizon, the operation moves events in a spacelike direction. Roughly speaking, we can interpret the fact that this is a symmetry as saying that the black hole spacetime is the same at every place inside.

However, the metric does depend on $r = t'$, so the interior is dynamical.

We have discovered a very important point: although the black hole spacetime is independent of time on the outside, it does in fact change with time on the inside. On the inside the only symmetry is one that relates different points in space, it says nothing about the relationship between events at different times.

Now, you might ask just how the spacetime changes in time. On one of the hyperbolae drawn above there is a symmetry that relates all of the points in space. The full spacetime is 3+1 dimensional and that for every point on the diagram above the real spacetime contains an entire sphere of points.

Even inside the horizon, the spacetime is spherically symmetric Now, the fact the points on our hyperbola are related by a symmetry means that the spheres are the same size (r) at each of these points! What changes as we move from one hyperbola to another ('as time passes') is that the size of the spheres (r) decreases.

This is 'why' everything must move to smaller r inside the black hole – the whole spacetime is shrinking!

To visualize what this means, it is useful to draw a picture of the curved space of a black hole at some time.

You began this process on a recent homework assignment when you considered a surface of constant t ($dt = 0$) and looked at circumference (C) vs. radius (R) for circles in this space6. You found that the space was not flat, but that the size of the circles changed more slowly with radius than in flat space.

One can work out the details for any constant t slice in the exterior (since the symmetry means that they are all the same!). Two such slices are shown below. Note that they extend into both the 'right exterior' with which we are familiar and the 'left exterior', a region about which we have so far said little.

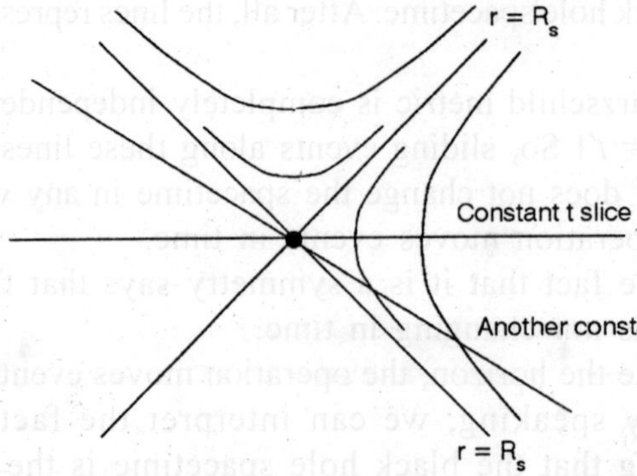

Ignoring, say, the θ direction and drawing a picture of *r* and φ (at the equator, θ = π/2), any constant *t* slice (through the two "outside" regions) looks like this:

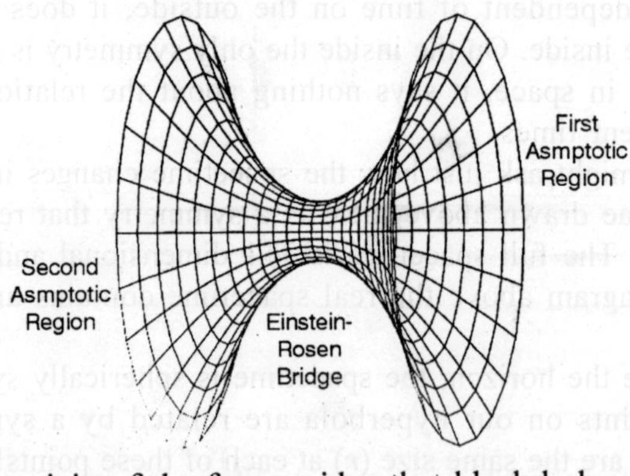

This is the origin of the famous idea that black holes can connect our universe (right exterior) to other universes (left exterior), or perhaps to some distant region of our own universe.

If this idea bothers you, don't worry too much, the other end of the tunnel is not really present for the black holes commonly found in nature. Note that the left exterior looks just like the right exterior and represents another region 'outside' the black hole, connected to the first by a tunnel. This tunnel is called a 'wormhole,' or 'Einstein-Rosen bridge.'

So, what are these spheres inside the black hole ? They are the 'throat' of the wormhole.

Gravity makes the throat shrink, and begin to pinch o. That is, if we draw the shape of space on each of the slices numbered 0,1,2 below, they would look much like the Einstein-Rosen bridge above, but with narrower and narrower necks as we move up the diagram.

Does the throat ever pinch o completely? That is, does it collapse to $r = 0$ in a finite proper time? We can find out from the metric. Let's see what happens to a freely falling observer who falls from where the horizons cross (at $r = R_s$) to $r = 0$ (where the spheres are of zero size and the throat has collapsed).

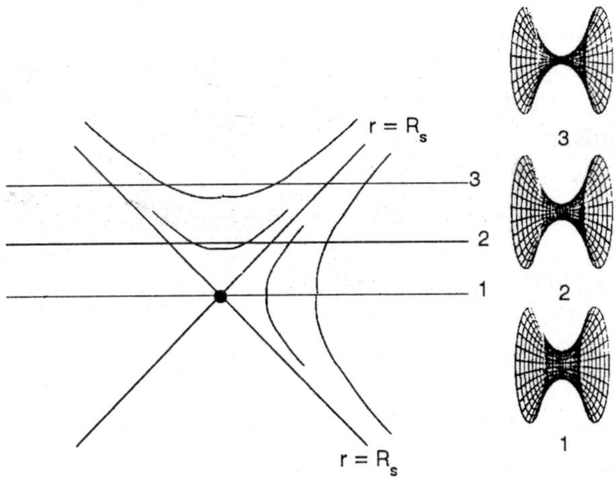

Our question is whether the proper time measured along such a worldline is finite. Consider an observer that starts moving straight up the diagram, as indicated by the dashed line in the figure below. We first need to figure out what the full worldline of the freely falling observer will be.

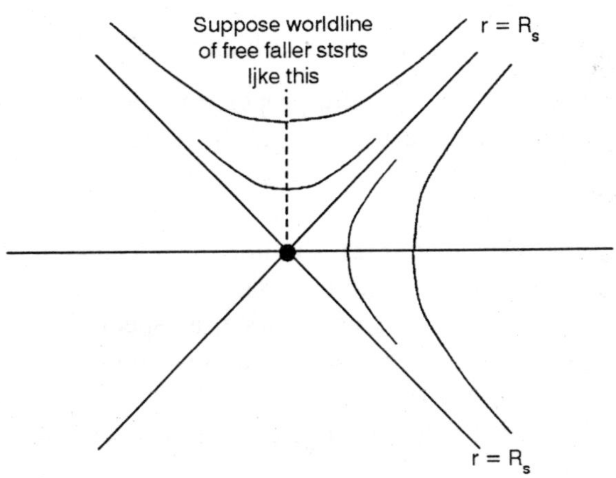

Will the freely falling worldline curve to the left or to the right? Since t is the space direction inside the black hole, this is just the question of whether it will move to larger t or smaller t. What do you think will happen?

Well, our diagram is exactly the same on the right as on the left, so there seems to be a symmetry.

In fact, you can check that the Schwarzschild metric is unchanged if we replace t by $-t$. So, both directions must behave identically. If any calculation found that the worldline bends to the left, then there would be an equally valid calculation showing that the worldline bends to the right. As a result, the freely falling worldline will not bend in either direction and will remain at a constant value of t.

Now, how long does it take to reach $r = 0$? We can compute the proper time by using the freely falling worldline with $dt = 0$. For such a worldline the metric yields:

$$d\tau^2 = \frac{dr^2}{R_s / r - 1} = \frac{r}{R_s - r} dr^2.$$

Integrating, we have:

$$\tau = \int_{R_s}^{0} dr \sqrt{\frac{r}{R_s - r}}.$$

It is not important to compute this answer exactly. What is important is to notice that the answer is finite.

We can see this from the fact that, near $r \approx R_s$ the integral is much like $\frac{dx}{\sqrt{x}}$ near $x = 0$.

This latter integral integrates to $\sqrt{x}$ and is finite at $x = 0$. Also, near $r = 0$ the integral is much like $\frac{x}{R_s} dx$, which clearly gives a finite result. Thus, our observer measures a finite proper time between $r = R_s$ and $r = 0$ and the throat does collapse to zero size in finite time.

The Singularity

This means that we should draw the line $r = 0$ as one of the hyperbolae on our digram. It is clearly going to be a 'rather singular line' (to paraphrase Sherlock Holmes again), and we will mark it as special by using a jagged line. As you can see, this line is spacelike and so represents a certain time. We call this line the singularity.

Note that this means that the singularity of a black hole is not a place at all!

The singularity is most properly thought of as being a very special time, at which the entire interior of the black hole squashes itself (and everything in it) to zero size.

Note that, since it cuts all of the way across the future light cone of any events in the interior (such as event A below), there is no way for any object in the interior to avoid the singularity.

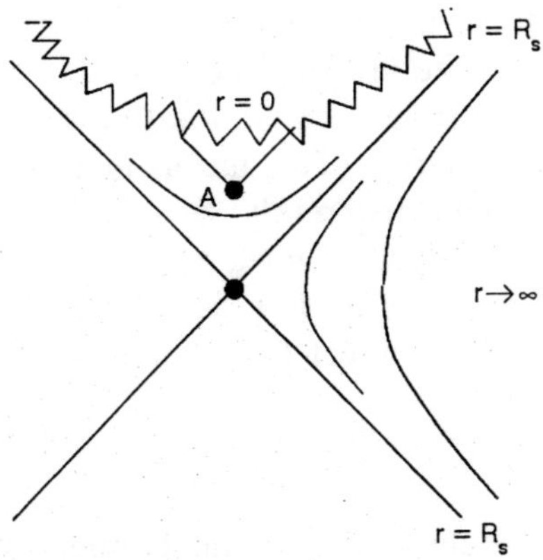

By the way, this is a good place to comment on what would happen to you if you tried to go from the right exterior to the left exterior through the wormhole. Note that, once you leave the right exterior, you are in the future interior region. From here, there is no way to get to the left exterior without moving faster than light. Instead, you will encounter the singularity. What this means is that the wormhole pinches o so quickly that even a light ray cannot pass through it from one side to the other. It turns out that this behaviour is typical of wormholes.

Let's get a little bit more information about the singularity by studying the motion of two freely falling objects. Some particularly simple geodesics inside the black hole are given by lines of constant t.

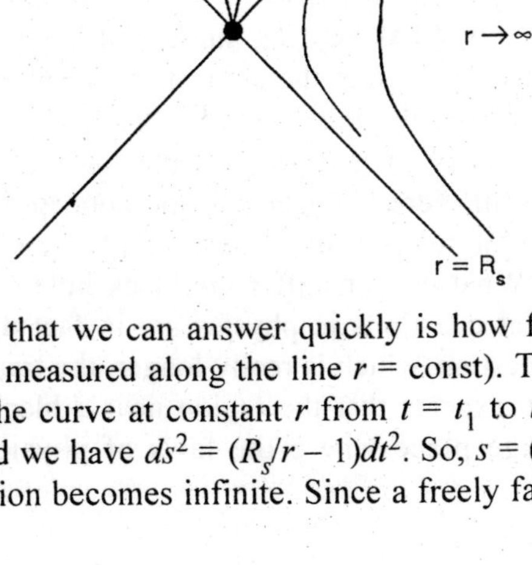

One question that we can answer quickly is how far apart these lines are at each r (say, measured along the line r = const). That is, "What is the proper length of the curve at constant r from $t = t_1$ to $t = t^2$?" Along such a curve, $dr = 0$ and we have $ds^2 = (R_s/r - 1)dt^2$. So, $s = (t_1 - t_2)\ R_s/r - 1$. As $r \to 0$, the separation becomes infinite. Since a freely falling object reaches

$r = 0$ in finite proper time, this means that any two such geodesics move infinitely far apart in a finite proper time.

It follows that the relative acceleration (a.k.a. the gravitational tidal force) diverges at the singularity. (This means that the spacetime curvature also becomes infinite.) Said differently, it would take an infinite proper acceleration acting on the objects to make them follow (nongeodesic) paths that remain a finite distance apart. Physically, this means that it requires an infinite force to keep any object from being ripped to shreds near the black hole singularity.

Beyond the Singularity?

Another favourite question is "what happens beyond (after!) the singularity?" The answer is not at all clear. The point is that just as Newtonian physics is not valid at large velocities and as special relativity is valid only for very weak spacetime curvatures, we similarly expect General Relativity to be an incomplete description of physics in the realm where curvatures become truly enormous. This means that all we can really say is that a region of spacetime forms where the theory we are using (General Relativity) can no longer be counted on to correctly predict what happens.

The main reason to expect that General Relativity is incomplete comes from another part of physics called quantum mechanics. Quantum mechanical effects should become important when the spacetime becomes very highly curved.

Roughly speaking, you can see this from the fact that when the curvature is strong local inertial frames are valid only over very tiny regions and from the fact the quantum mechanics is always important in understanding how very small things work. Unfortunately, no one yet understands just how quantum mechanics and gravity work together. We say that we are searching for a theory of "quantum gravity." It is a very active area of research that has led to a number of ideas, but as yet has no definitive answers. This is in fact the area of my own research.

Just to give an idea of the range of possible answers to what happens at a black hole singularity, it may be that the idea of spacetime simply ceases to be meaningful there. As a result, the concept of time itself may also cease to be meaningful, and there may simply be no way to properly ask a question like "What happens after the black hole singularity?" Many apparently paradoxical questions in physics are in fact disposed of in just this way (as in the question 'which is really longer, the train or the tunnel?').

In any case, one expects that the region near a black hole singularity will be a very strange place where the laws of physics act in entirely unfamiliar ways.

There still remains one region of the diagram (the 'past interior') about which we have said little. The Schwarzschild metric is time symmetric (under $t \to -t$). As a result, the diagram should have a top/bottom symmetry, and the past interior should be much like the future interior. This part of the spacetime is often called a 'white hole' as there is no way that any object can remain inside: everything must pass outward into one of the exterior regions through one of the horizons!

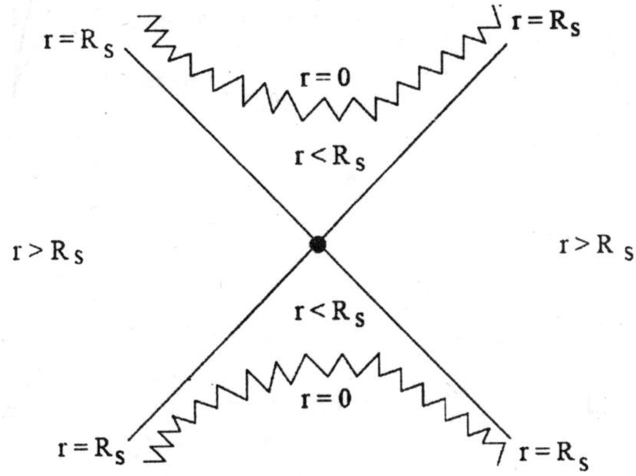

As we mentioned briefly with regard to the second exterior, the past interior does not really exist for the common black holes found in nature. Let's talk about how this works. So far, we have been studying the pure Schwarzschild solution. As we have discussed, it is only a valid solution in the region in which no matter is present. Of course, a little bit of matter will not change the picture much. However, if the matter is an important part of the story (for example, if it is matter that causes the black hole to form in the first place), then the modifications will be more important.

Let us notice that in fact the 'hole' (whether white or black) in the above spacetime diagram has existed since infinitely far in the past. If the Schwarzschild solution is to be used exactly, the hole (including the wormhole) must have been created at the beginning of the universe. We expect that most black holes were not created with the beginning of the universe, but instead formed later when too much matter came too close together. A black hole must form when, for example, too much thin gas gets clumped together.

Once the gas gets into a small enough region (smaller than its Schwarzschild radius), we have seen that a horizon forms and the gas must shrink to a smaller size. No finite force (and, in some sense, not even infinite force) can prevent the gas from shrinking.

Now, outside of the gas, the Schwarzschild solution should be valid. So, let me draw a worldline on our Schwarzschild spacetime diagram that represents the outside edge of the ball of gas. This breaks the diagram into two pieces: an outside that correctly describes physics outside the gas, and an inside that has no direct physical relevance and must be replaced by something that depends on the details of the matter:

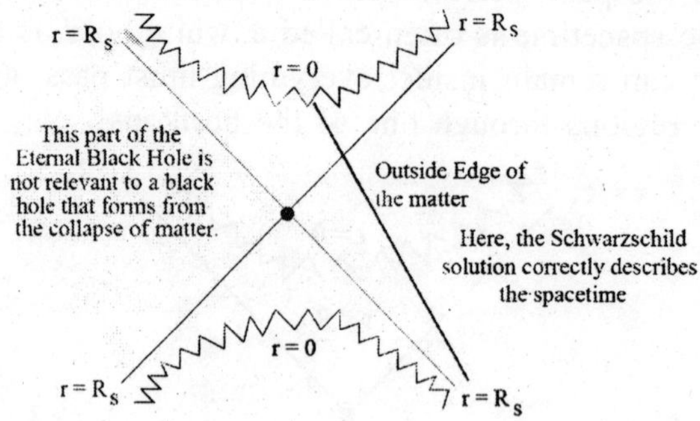

We see that the 'second exterior' and the 'past interior' are in the part of the diagram with no direct relevance to relevance to black holes that form from collapsing matter.

A careful study of the Einstein equations shows that, inside the matter, the spacetime looks pretty normal. A complete spacetime diagram including both then region inside the matter and the region outside would look like this:

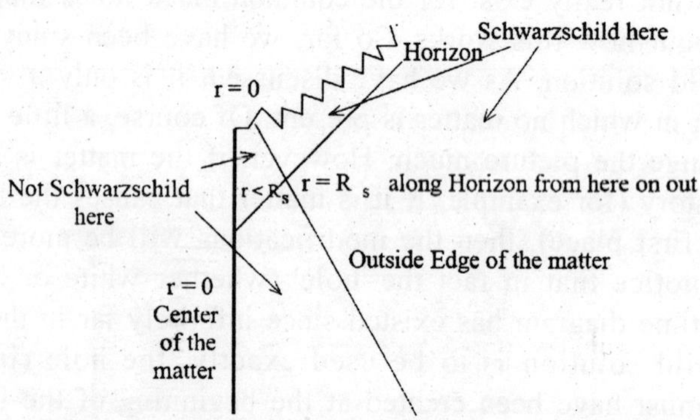

VISUALIZING BLACK HOLE SPACETIMES

We have now had a fairly thorough discussion about Schwarzschild black holes including the outside, the horizon, the inside, and the "extra regions" (second exterior and past interior). One of the things that we emphasized was that the spacetime at the horizon of a black hole is locally flat, just like everywhere else in the spacetime.

Also, the curvature at the horizon depends on the mass of the black hole. The result is that, if the black hole is large enough, the spacetime at the horizon is less curved than it is here on the surface of the earth, and a person could happily fall through the horizon without any discomfort. It is useful to provide another perspective on the various issues that we have discussed. The idea is to draw a few pictures.

The point is that the black hole horizon is an effect caused by the curvature of spacetime, and the way that our brains are most used to thinking about curved spaces is to visualize them inside of a larger flat space.

For example, we typically draw a curved (two-dimensional) sphere) as sitting inside a flat three-dimensional space.

Now, the r, *t* plane of the black hole that we have been discussing and drawing on our spacetime diagrams forms a curved two-dimensional spacetime. It turns out that this two-dimensional spacetime can also be drawn as a curved surface inside of a flat three-dimensional spacetime.

To get an idea of how this works, let me first do something very simple: A flat two-dimensional spacetime inside of a flat three-dimensional spacetime.

As usual, time runs up the diagram, and we use units such that light rays move along lines at 45o angles to the vertical. Note that any worldline of a light ray in the 3-D spacetime that happens to lie entirely in the 2-D spacetime will also be the worldline of a light ray in the 2-D spacetime, since it is clearly a curve of zero proper time. A pair of such crossed light rays are shown below where the light cone of the 3-D spacetime intersects the 2-D spacetime.

Now that we've got the idea, a picture that represents the (2-D) *r, t* plane of our black hole, drawn as a curved surface inside a 3-D flat spacetime. It looks like this:

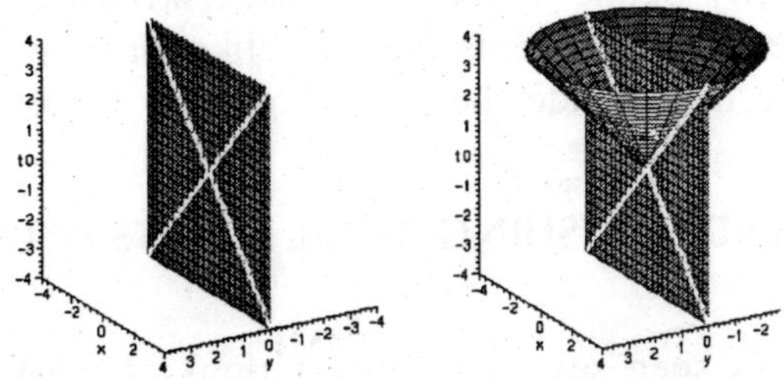

The curves of constant r so that you can visualize them more easily. Note that larger r is farther from the centre of the diagram, and in particular farther out along the 'flanges.' One flange represents the left exterior, and one represents the right exterior.

The most important thing to notice is that we can once again spot two lines that 1) are the worldlines of light rays in the 3D flat space and 2) lie entirely within the curved 2D surface.

As a result, they again represent worldlines of light rays in the black hole spacetime. They are marked with lines on the first picture *I* showed you (above) of the black hole spacetime and also on the diagrams below. Note that they do not move at all outward toward larger values of r.

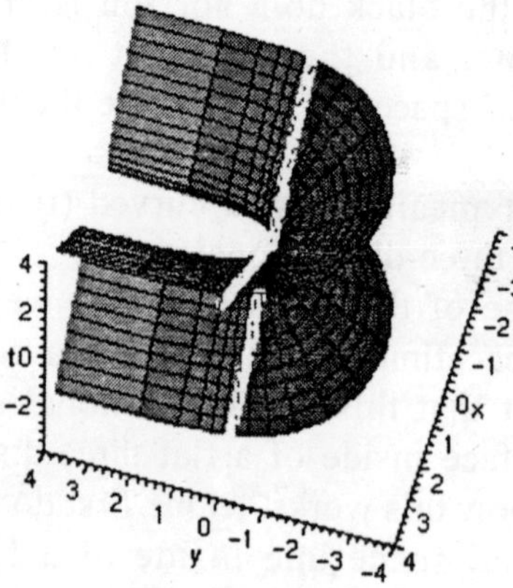

These are the horizons of the black hole.

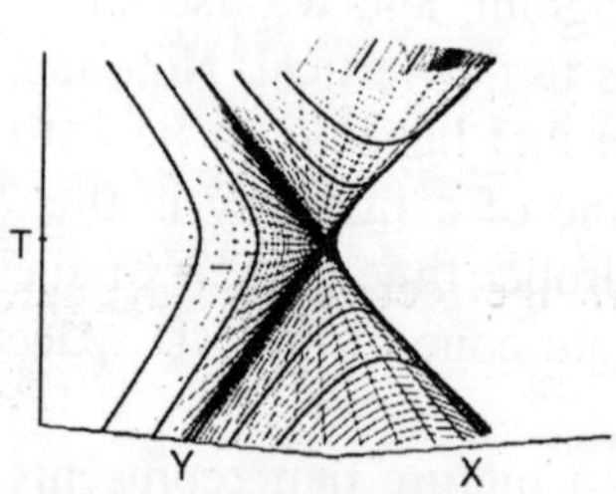

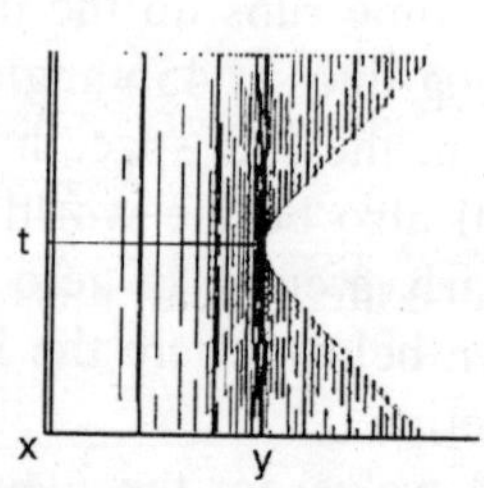

Another thing we can see from these diagrams is the symmetry we discussed. The symmetry of the 2- D black hole spacetime is the same as the boost symmetry of the larger 3D Minkowski space. Inside the black hole, this symmetry moves events in a spacelike direction. We can also see from this picture that, inside the black hole, the spacetime does change with time.

STRETCHING AND SQUISHING: TIDAL EFFECTS IN GENERAL RELATIVITY

We have now seen several manifestations of what are called 'tidal effects' in general relativity, where gravity by itself causes the stretching or squashing of an object. These a lot in homework problems 1 and 2, but even earlier our most basic observation in general relativity was that gravity causes freely falling observers to accelerate relative to each other. That is to say that, on a spacetime diagram, freely falling worldlines may bend toward each other or bend away.

This effects the ocean around the earth as the earth falls freely around the moon. The answer was that it stretches the ocean in the direction pointing toward (and away from) the moon, while it squishes (or compresses) the ocean in the perpendicular directions. This is because different parts of the ocean would like to separate from each other along the direction toward the moon, while they would like to come closer together in the other directions:

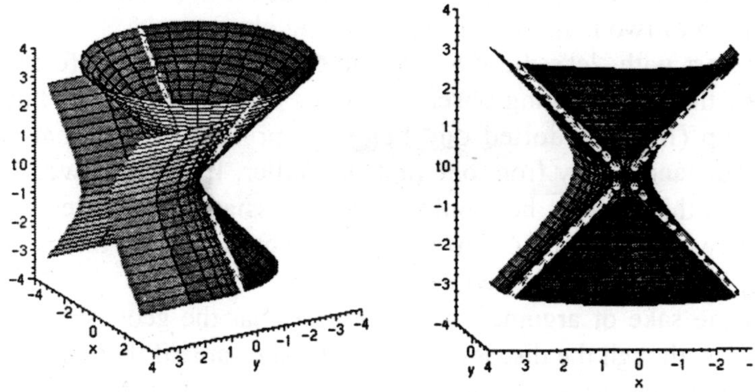

As stated in the homework solutions, this effect is responsible for the tides in the earth's oceans. (You know: if you stand at the beach for 24 hours, the ocean level rises, falls, then rises and falls again.) Whenever gravity causes freely falling observers (who start with no relative velocity) to come together or to separate, we call this a tidal effect. As we have seen, tidal effects are the fundamental signature of spacetime curvature, and in fact tidal effects are a direct measure of spacetime curvature.

Of course any other object (a person, rocket ship, star, etc.) would feel a similar stretching or squishing in a gravitational field. Depending on how you are lined up, your head might be trying to follow a geodesic which would cause it to separate from your feet, or perhaps to move closer to your feet. If this effect were large, it would be quite uncomfortable, and could even rip you into shreds (or squash you flat).

On the other hand, we argued that this tidal effect will become infinitely large at the singularity of a black hole. There the effect certainly will be strong enough to rip apart even tiny objects like humans, or cells, or atoms, or even subatomic particles.

It is therefore of interest to learn how to compute how strong this effect actually is. We know that it is small far away from a black hole and that it is large at the singularity, but how big is it at the horizon?

This last question is the key to understanding what you would feel as you fell through the horizon of a Schwarzschild black hole.

The SetUp

So, let's suppose that somebody tells us what the spacetime metric is (for example, it might be the Schwarzschild metric). For convenience, let's suppose that it is independent of time and spherically symmetric.

In this case, we discussed in class how to find the acceleration of static observers relative to freely falling observers who are at the same event in spacetime.

What we are going to do now is to use this result to compute the relative acceleration of two neighboring freely falling observers.

To start with, let's draw a spacetime diagram in the reference frame of one of the freely falling observers. What this means is that lines drawn straight up (like the dotted one below) represent curves that remain a constant distance away from our first free faller. If you followed Einstein's discussion, this is what he would call a 'Gaussian' coordinate system. We want our two free fallers to start o with the same velocity - this is analogous to using 'initially parallel geodesics'.

For the sake of argument, let's suppose that the geodesics separate as time passes, though the discussion is exactly the same if they come together. The freely falling observers are the solid lines, and the static observers are the dashed lines. To be concrete, the static observers to be accelerating toward the right, but again it doesn't really matter.

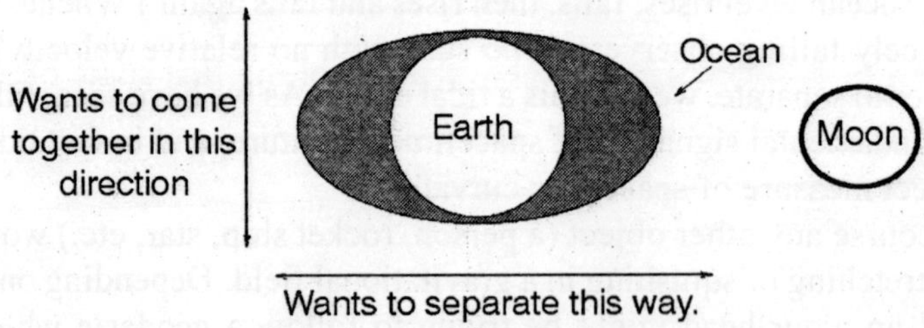

The coordinate x measures the distance from the first freely falling observer.

What we would like to know is how fast the second geodesic is accelerating away from the first. Let us call this acceleration $a_{FF\,2}$, the acceleration of the second free faller. Since we are working in the reference frame of the first free faller, the corresponding acceleration aFF 1 is identically zero.

Now, what we already know is the acceleration of the two static observers relative to the corresponding free faller. In other words, we know the acceleration as1 of the first static observer relative to the first free faller, and we know the acceleration a_{s2} of the second static observer relative to the second free faller. Note that the total acceleration of the second static observer in our coordinate system is $a_{FF2} + as_2$ - her acceleration relative to the second free faller plus the acceleration of the second free faller in our coordinate system.

This is represented pictorially on the diagram above. Actually, there is something else that we know: since the two static observers are, well, static, the proper distance between them (as measured by them) can never change. We will use this result to figure out what a_{FF2} is. The way we will proceed is to use the standard Physics/Calculus trick of looking at small changes over small regions.

Note that there are two parameters (T and L, as shown below) that tell us how big our region is. L is the distance between the two free fallers, and T is how long we need to watch the system. We will assume that both L and T are very small, so that the accelerations a_{s1} and a_{s2} are not too different, and so that the speeds involved are all much slower than the speed of light.

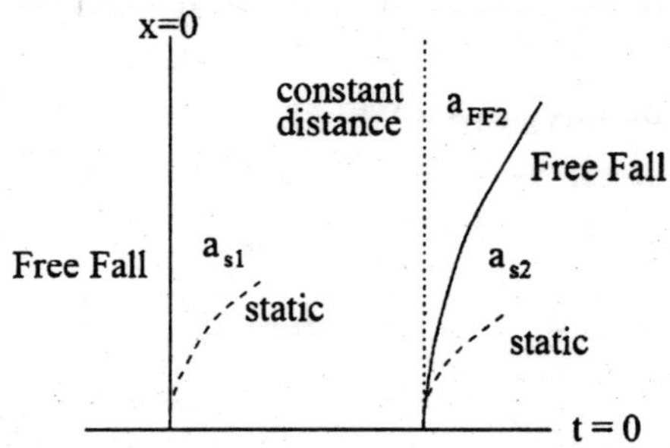

Now, pick a point (p_1) on the worldline of the first static observer. Call the coordinates of that point x_1, t_1. (We assume $t_1 < T$.) Since the velocity is still small at that point, we can ignore the difference between acceleration and proper acceleration and the Newtonian formula:

$$x_1 = \frac{1}{2}a_{s1}t_1^2 + O(T^4)$$

is a good approximation. The notation $O(T^4)$ is read "terms of order T^4." This represents the error we make by using only the Newtonian formula. It means that the errors are proportional to T^4 (or possibly even smaller), and so become much smaller than the term that we keep (t_1^2) as $T \to 0$. Note that since this is just a rough description of the errors, we can use T instead of t_1.

The two static observers will remain a constant distance apart as determined by their own measurements.

To write this down mathematically, we need to understand how these observers measure distance. Any observer will measure distance along a line of simultaneity, and called the point p_2 (where it intersects the worldline of the second static observer) x_2, t_2.

Now, since spacetime is curved, this line of simultaneity need not be perfectly straight on our diagram.

However, we also know that, in a very small region near the first Free Faller (around whom we drew our diagram), space is approximately flat. This means that the curvature of the line of simultaneity has to vanish near the line $x = 0$.

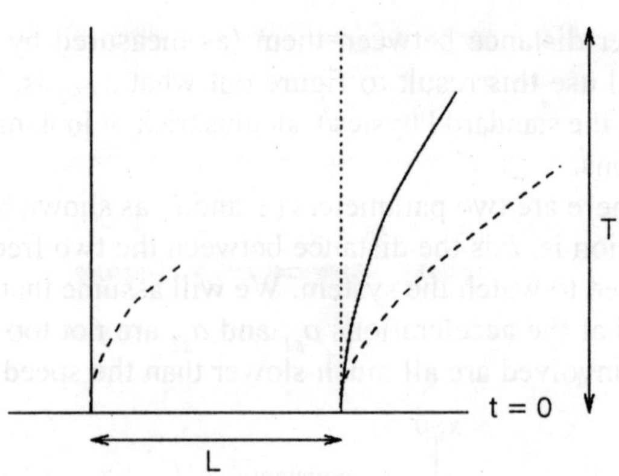

Technically, the curvature of this line (the second derivative of t with respect to x) must itself be 'of order $(x_2 - x_1)$. This means that p_1 and p_2 are related by an equation that looks like:

$$\frac{t_2 - t_1}{x_2 - x_1} = \text{slope at } p_1 + [\text{curvature at } p_1]\,(x_2 - x_1) + O([x_2 - x_1]^2)$$
$$= \text{slope at } p_1 + O([x_2 - x_1]^2) + O(T^2[x_2 - x_1]).$$

Again, we need only a rough accounting of the errors. As a result, we can just call the errors $O(L^2)$ instead of $O([x_2 - x_1]^2)$.

Remember that, in flat space, the slope of this line of simultaneity would be v_{s1}/c^2, where v_{s1} is the velocity of the first static observer. Very close to $x = 0$, the spacetime can be considered to be flat. Also, as long as t1 is small, the point p_1 is very close to $x = 0$. So, the slope at p1 is essentially v_{s1}/c^2. Also, for small t_1 we have $v_{s1} = a_{s1}t$. Substituting this into the above equation and including the error terms yields

$$t^2 = t_1 + (a_{s1}t_1/c^2)(x_2 - x_1) + O(L^3)$$
$$= t_1 + \left(1 + \frac{a_{s1}}{c^2}(x_2 - x_1)\right) + O(L^3) + O(T^2L)$$

We've already got two useful equations, and we know that a third will be the condition that the proper distance between p1 and p2 will be the same as the initial separation L between the two free fallers:

$$L^2 = (x_2 - x_1)^2 - c^2(t_2 - t_1)^2$$

In addition, there is clearly an analogue of equation for the second static observer (remembering that the second one does not start at $x = 0$, but instead starts at $x = L$):

$$x_2 = L + \frac{1}{2}(a_{s2} \mp a_{FF2})t_2^2 + O(t^4).$$

The Solution

So, let's try using these equations to solve. The way proceed is to substitute equation for t_2 in equation. That way we express both positions in terms of just t_1. The result is

$$x^2 = L+\frac{1}{2}(a_{s2}+a_{FF2})\left(1+\frac{a_{s1}}{c^2}(x_2-x_1)\right)^2 t_1^2+O(t^4)+O(L^3T^3)$$

The condition that the proper distance between the static observers does not change. This equation involves the difference $x_2 - x_1$. Subtracting equation, we get:

$$x_2 - x_1 = L+\frac{1}{2}(a_{s2}+a_{FF2}-a_{s1})t_1^2+(a_{s2}+a_{FF_2})\frac{a_{s1}}{c^2}(x_2-x_1)t_1^2$$

$$+\ L+\frac{1}{2}(a_{s2}+a_{FF2})\frac{a_{s1}^2}{c^4}(x_2-x_1)^2 t_1^2+O(T^4)+O(L^3T^2)$$

And, actually, we won't need to keep the $(x_2 - x_1)^2$ term, so we can write this as:

$$x_2 - x_1 = L+\frac{1}{2}(a_{s2}+a_{FF2}-a_{s1})t_1^2$$

$$+(a_{s2}+a_{FF_2})\frac{a_{s1}}{c^2}(x_2-x_1)^2 t_1^2+O(T^4)+O(L^3T^2)$$

Now, this equation involves $x_2 - x_1$ on both the left and right sides, so let's solve it for $x_2 - x_1$. As you can check, the result is:

$$x_2 - x_1 = \left(L+\frac{1}{2}(a_{s2}+a_{FF2}-a_{s1})t_1^2\right)$$

$$\left(1-(a_{s2}+a_{FF_2})\frac{a_{s1}}{c^2}t_1^2\right)^{-1}+O(T^4)+O(T^3L^2)$$

But there is a standard 'expansion' $(1 - x)^{-1} = 1 + x + O(x^2)$ that we can use to simplify this. We find:

$$x_2 - x_1 = L+\frac{1}{2}(a_{s2}+a_{FF2}-a_{s1})t_1^2$$

$$+L(a_{s2}+a_{FF_2})\frac{a_{s1}}{c^2}t_1^2+O(T^4)+O(T^3L^2)$$

Believe it or not, we are almost done!!!! All we have to do now is to substitute this (and also equation for the times) into the requirement that $\Delta x^2 - c^2\,\Delta t^2 = L^2$. Below, we will only keep terms up through T^2 and L^2. Note that:

$$(x_2 - x_1)^2 = L^2 + L(a_{s2}+a_{FF2}-a_{s1})t_1^2$$

$$+2L^2(a_{s2}+a_{FF_2})\frac{a_{s1}}{c^2}t_1^2+O(T^4)+O(T^2L^3)$$

while

$$(t_2 - t_1)^2 = t_1^2 a_{s1}^2 L^2 / c^2 + O(T^3 L^2).$$

So, since the proper distance between p_1 and p_2 must be L^2,

$$L^2 = \Delta x^2 - c^2 \Delta t^2$$

$$= L^2 + L(a_{s2} + a_{FF2} - a_{s1})t_1^2 + L^2(2a_{s2} + a_{FF_2} - a_{s1})\frac{a_{s1}}{c^2}t_1^2$$

$$+ O(T^4) + O(T^2 L^3)$$

Canceling the L^2 terms on both sides leaves only terms proportional to $t_1^2 L$. So, after subtracting the L^2, let's also divide by $t_1^2 L$. This will leave:

$$0 = (a_{s2} + a_{FF2} - a_{s1}) + L(2a_{s2} + a_{FF_2} - a_{s1})\frac{a_{s1}}{c^2} + O(T^2 / L) + O(T^2)$$

Reminder: What we want to do is to solve for aFF 2, the acceleration of the second free faller. In preparation for this, let's regroup the terms above to collect things with a_{FF2} in them:

$$0 = a_{FF2}(1 + 2La_{s1}/c^2) + (a_{s2} + a_{s1}) + O(T^2 / L) + O(L^2)$$

Now, before we solve for a_{FF2}. Remember that we started o by assuming that the region was very small. If it is small enough, then in fact a_{s1} and a_{s2} are not very different. In fact, we will have $a_{s1} - a_{s2} = O(L)$. This simplifies the last term a lot since $L\,(2a_{s2} - a_{s1}) = La_{s1} + O(L^2)$. Using this fact, and solving the above equation for $a_{FF\,2}$ we get:

$$a_{FF2} = \frac{a_{s2} - a_{s1} + La_{s1}^2 / c^2}{1 + La_{s1}/c^2} + O(T^2 / L) + O(L^2)$$

$$= -(a_{s2} - a_{s1}) - La_{s1}^2 / c^2 + O(T^2 / L) + O(L^2).$$

The Differential Equation

We want to convert it into a more useful form which will apply without worrying about whether our region is small. What we're going to do is to take the limit as T and L go to zero and turn this into a differential equation.

Technically, we will take T to zero faster that L so that $T^2/L^2 \rightarrow 0$. Note that we are really interested in how things change with position at $t = 0$, so that is is natural to take T to zero before taking L to zero.

Imagine not just two free fallers, but a whole set of them at every value of x. Each of these starts out with zero velocity, and each of them has an accompanying static observer.

The free faller at x will have some acceleration $a_{FF}(x)$, and the static observer at x will have some acceleration $a_s(x)$ relative to the corresponding free faller. If L is very small above, notice that $a_{s2} - a_{s1} = L\frac{La_s}{dx} + O(L^2)$ and that (since $a_{FF1} = 0$), $a_{FF2} = L\frac{d_{aFF}}{dx} + O(L^2)$. So, we can rewrite equation as:

$$L\frac{d_{aFF}}{dx} = -L\frac{da_s}{dx} - |La_s^2/c^2 + O(T^2/L) + O(L^2)$$

We can now divide by L and take the limit as T/L and L go to zero. The result is a lovely differential equation:

$$\frac{da_{FF}}{dx} = -\frac{da_s}{dx} - La_s^2/c^2$$

By the way, the important point to remember about the above expression is that the coordinate x represents proper distance.

What Does it all Mean?

One of the best ways to use this equation is to undo part of the last step. Say that you have two free falling observers close together that have no initial velocity.

Then, if their separation L is small enough, their relative acceleration is $L\frac{da_{FF}}{dx}$ or

$$\text{Relative acceleration} = -L\left(\frac{da_s}{dx} + a_s^2/c^2\right)$$

Let's take a simple example of this. Suppose that you are near a black hole and that your head and your feel are both freely falling objects. Then, this formula tells you at what acceleration your head would separate from (or, perhaps, accelerate toward) your feet.

Of course, your head and feet are not, in reality, separate freely falling objects.

The rest of your body will pull and push on them to keep your head and feet roughly the same distance apart at all times. However, your head and feet will want to separate or come together, so depending on how big the relative acceleration is, keeping your head and feet in the proper places will cause a lot of stress on your body.

For example, suppose that the relative acceleration is $10 m/s^2$ (1g) away from each other.

In that case, the experience would feel much like what you feel if you tie your legs to the ceiling and hang upside down. In that case also, your head wants to separate from the ceiling (where your feet are) at $10\ m/s^2$.

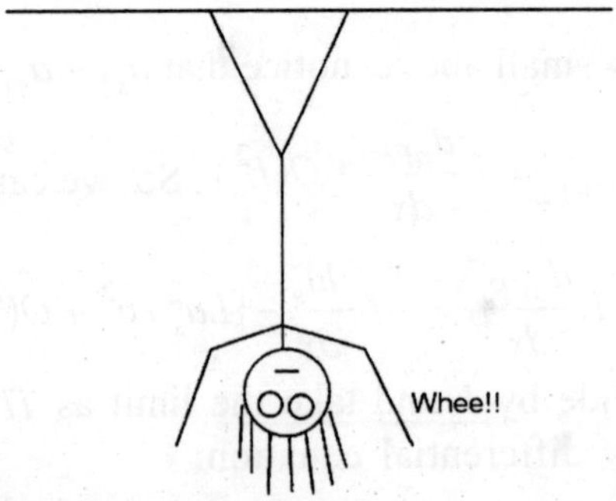

However, if the relative acceleration were a lot bigger, it would be extremely uncomfortable. In fact, a good analogy with the experience would be being on a Medieval rack - an old torture device where they pulled your arms one way and your feet in the opposite direction.

Black Holes and the Schwarzschild Metric

The acceleration of static observers (relative to freely falling observers) in the Schwarzschild metric is given by:

$$a_s = \frac{c^2}{2}\left(\frac{R_s}{r^2}\right)(1 - R_s/r)^{-1/2}.$$

We would like to take the derivative of this with respect to the proper distance S in the radial direction. That is, we will work along a line of constant t, ϕ, and θ. In this case, as we have seen before,

$$\frac{dr}{dS} = \sqrt{1 - R_s/r}.$$

So,

$$\frac{da_s}{dS} = (\sqrt{1 - R_s/r})\frac{da_s}{dr}.$$

A bit of computation yields

$$\frac{da_s}{dS} = -c^2\left(\frac{R_s}{r^3}\right) - \frac{c^2}{4}\left(\frac{R_s}{r^2}\right)^2 (1 - R_s/r)^{-1}.$$

On the other hand, we have:

$$a_s^2/c^2 = \frac{c^2}{4}\left(\frac{R_s}{r^2}\right)^2 (1 - R_s/r)^{-1}.$$

To evaluate the relative acceleration, equation tells us to add these two results together. Clearly, there is a major cancellation and all that we have left is:

$$\text{Relative acceleration} = c^2\left(\frac{R_s}{r^3}\right)L.$$

This gives the relative acceleration of two freely falling observers who, at that moment, are at rest with respect to the static observers. (The free fallers are also located at radius r and are separated by a radial distance L, which is much smaller than r.) The formula holds anywhere that the Schwarzschild metric applies. In particular, anywhere outside a black hole.

$$\text{Relative acceleration} = c^2\left(\frac{R_s}{r^3}\right)L.$$

The most important thing to notice about this formula is that the answer is finite. Despite the fact that a static observer at the horizon would need an infinite acceleration relative to the free fallers, any two free fallers have only a finite acceleration relative to each other.

The second thing to notice is that, for a big black hole (large R_s), this relative acceleration is even small. (However, for a small black hole, it can be rather large.)

BLACK HOLE ASTROPHYSICS AND OBSERVATIONS

We have now come to understand basic round (Schwarzschild) black holes fairly well. We have obtained several perspectives on black hole exteriors and interiors and we have also learned about black hole singularities. However, there are several issues associated with black holes that we have yet to discuss. Not least of these is the observational evidence that indicates that black holes actually exist.

The Observational Evidence for Black Holes

Big black holes should not be too hard to make, the question arises, are there really such things out there in the universe? If so, how do we find them? Black holes are dark after all, they themselves do not shine brightly like stars do.

Well, admittedly most of the evidence is indirect. Nevertheless, it is quite strong.

Let's begin by reviewing the evidence for a black hole at the centre of our own galaxy.

What is quite clear is that there is something massive, small, and dark at the centre of our galaxy. Modern techniques allow us to make high resolution photographs of stars orbiting near the galactic centre. One can also measure the velocities using the Doppler shift. The result is that we know a lot about the orbits of these objects, so that we can tell a lot about the mass of whatever object lies at the very centre at they are orbiting around.

The status of black hole observa-tions by Andrew Fabian, What the data shows quite clearly is that there is a mass of 2.61×10^6 solar masses

(M_O is the mass of the sun) contained in a region of size.02 parsecs (pc). Now, a parsec is around 3×10^{16}m.

So, this object has a radius of less than 6×10^{14}m. In contrast, the Schwarzschild radius for a 2.61×10^6 solar mass object is around 10^{10} *m.* So, what we get from direct observations of the orbits of stars is that this object is smaller than 10, $000R_s$.

That may not sound like a small bound (since 10,000 is a pretty big number), but an important point is that an object of that mass at r = 10, $000R_s$ could not be very dense. If we simply divide mass by volume, we would find an average density of 10?9 that of water! We know an awful lot about how matter behaves at that density and the long and short of it is that the gravitational field of this object should make such a diffuse gas of stuff contract.

You might then ask what happens when it becomes dense enough to form a solid. This brings us to another interesting observation...

It turns out that, at the very position at the centre of our galaxy where the massive object (black hole?) should be located, a strong radio signal is being emitted. The source of this signal has been named "Sagittarius A* " (Sgr A*). It therefore natural to assume that this signal is coming from the massive object that we have been discussing.

It is natural for radio signals to be emitted not from black holes themselves, but from things falling into black holes. Precision radio measurements using what is called "very-long baseline interferometry" (VLBI) tell us that the radio signal is coming from a small region. In terms of Rs for the mass we have been discussing, the region's size is about $30R_s$.

It therefore appears that the object itself is within $30R_s$. If the mass were spread uniformly over a volume of $30R_s$, it would have a density about three times greater than that of air.

However, the proper acceleration (of static observers relative to freely falling ones) would be about 100g's. Again, we know a lot about how matter behaves under such conditions. In particular, we know that matter at that density behaves like a gas. However, the 100g acceleration means that the pressure in the gas must be quite high in order to keep the gas from collapsing.

In particular, the pressure would reach one atmosphere about 1km inside the object. One hundred thousand km inside, the pressure would reach one hundred thousand atmospheres! Since we are thinking of an object of size $30R_s = 3 \times 10^{11}$m (which is 300 million km), one hundred thousand km is less than.1% of the way to the centre.

So, the vast majority of the object is under much more than one hundred thousand (10^5) atmospheres of pressure. At 10^5 atmospheres of pressure, all forms of matter will have roughly the density of a solid. The matter supports this pressure by the electrons shells of the atoms bumping up against one another.

So, using what we know about matter, the object must surely be even smaller: small enough that have at least the density of water. Such an object (for this mass) would have a size of less than $3R_s$. So, we are getting very close. At the surface of such an object, the relative accelerations of freely falling and static observers would be around 10, 000g's. At a depth of 10, 000km (again.1% of the way to the centre), the pressure would be 10^{14}N/ m^2, or roughly one billion atmospheres. At this pressure, any kind of matter will compress to more than 30 times the density of water. So, again, we should redo the calculation, but now at 30 times the density of water...

At this density, the object would be within its Schwarzschild radius. It would be a black hole. We conclude that we the experimental bounds and what we know about physics the object at the centre of our galaxy either is a black hole already or is rapidly collapsing to become one. Oh, the time such an object would it take to collapse from $30R_s$ is about 15 minutes. Astronomers have been monitoring this thing for a while.

FINDING OTHER BLACK HOLES

So, while the astronomical measurements do not directly tell us that Sagittarius A? is a black hole, when combined with what we know about (more or less) ordinary matter, the conclusion that the object is a black hole is hard to escape. Much the same story is true for other "black hole candidates" as the astronomers call them. The word candidate is added to be intentionally conservative.

Black hole candidates at the centre of other galaxies are identified in much the same way that Sagittarius A? was found. Astronomers study how stars orbit around those galactic centers to conclude that there is "massive compact object" near the centre. Typically, such objects are also associated with strong emissions of radio waves.

Similar techniques are used for finding smaller black holes as well. The small black holes that we think we have located are in so-called 'binary systems.' The way that these black holes were found was that astronomers found certain stars which seemed to be emitting a lot of high energy x - rays. This is an unusual thing for a star to do, but it is not so odd for a black hole. On closer inspection of the star, it was found that the star appeared to "wobble" back and forth.

This is just what the star would seem to do if it was in fact orbiting close to a small massive dark object that could not be seen directly. This is why they are called binary systems, since there seem to be two objects in the system. These massive dark objects have masses between 5 and 10 solar masses. Actually, there are also cases where the dark companion has a mass of less then 2 solar masses, but those are known to be neutron stars. Our knowledge of normal matter led to the conclusion that Sagittarius A* is a black hole. Well, we also have a pretty good idea of how star-like objects work in the solar mass range. In actual stars, what happens is that the objects become dense enough that nuclear fusion occurs.

This generates large amounts of heat that increases the pressure in the matter far above what it would be otherwise. It is this pressure that keeps the object from collapsing to higher density. Thus, the reason that a star has a relatively low density (the average density of the sun is a few times that of water) is that it is very hot! This of course is also the reason that stars shine.

Now, the dark companions in the binary systems do not shine. It follows that they are not hot. As a result, they must be much smaller and much more dense. Our understanding of physics tells us that massive cold objects will collapse under their own weight. In particular, a cold object greater than 1.4 times the mass of the sun will not be a star at all. It will be so dense that the electrons will be crushed into the atomic nuclei, with the result that they will be absorbed into the protons and electron + proton will turn into a neutron.

Thus the object ceases to be normal matter (with electrons, protons, and neutrons) at all, but becomes just a big bunch of neutrons. This number of 1.4 solar masses is called the Chandrasekhar limit after the physicist who discovered it. In practice, when we look at the vast numbers of stars in the universe, we have never found a cold star of more than 1.4 solar masses though we have found some that are close.

So, any cold object of more than 1.4 solar masses must be at least as strange as a big bunch of neutrons. Well, neutrons can be packed very tightly without resistance, so that in fact such 'neutron stars' naturally have the density of an atomic nucleus. What this means is that one can think of a neutron star as being essentially one incredibly massive atomic nucleus (but will all neutrons and no protons).

The density of an atomic nucleus is a huge 10^{18} kg/m^3. (This is 10^{15} times that of normal matter.) Let us ask: suppose we had a round ball of nuclear matter at this density. How massive would this ball need to be for the associated Schwarzschild radius to be larger than the ball itself? The answer is about 4 times the mass of the sun.

So, working with a very simple model in which the density is constant (and always equal to the density of normal nuclei, which are under significantly less pressure) inside the object, we find that any cold object with a mass greater than four solar masses will be a black hole! It turns out that any model where the density increases with depth and pressure yields an even stronger bound. As a result, modern calculations predict that any cold object with a mass of greater than 2.1 solar masses will be a black hole.

The dark companions in the binary systems all have masses significantly greater than 2 solar masses. By the way, it is reassuring to note that every neutron star that has been found has been in the range between 1.4 and 2.1 solar masses.

A few words on Accretion and Energy

Even with the above arguments, one might ask what direct measurements could be made of the size of the dark companions. Can we show directly that their size is comparable to the Schwarzschild radius? To do so one needs to use the energy being released from matter falling into a black hole. This leads us to a brief discussion of what are called accretion disks.

In general, matter tends to flow into black holes. This addition of matter to an object is called "accretion." Black holes (and neutron stars) are very small, so that a piece of matter from far away that becomes caught in the gravitational field is not likely to be directed straight at the black hole or neutron star, but instead is likely to go into some kind of orbit around it. The matter piles up in such orbits and then, due to various interactions between the bits of matter, some bits slowly loose angular momentum and move closer and closer to the centre.

Eventually, they either fall through the horizon of the black hole or hit the surface of the neutron star.

In cases where the compact object is in a binary system, the matter flowing in comes mostly from the shining star. This process makes the accreting matter into a disk, as shown in the picture8 below. This is why astronomers often talk about 'accretion disks' around black holes and neutron stars.

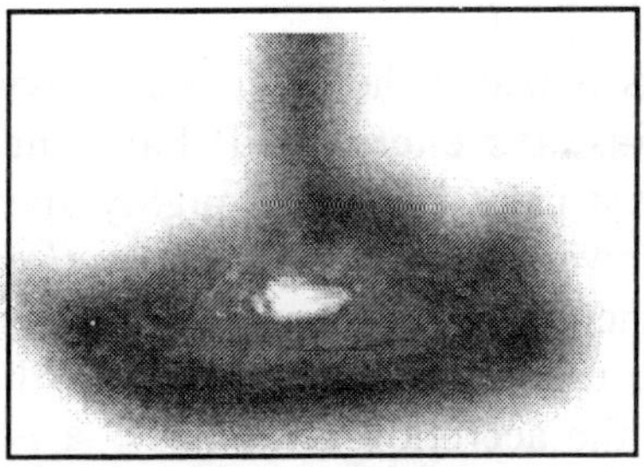

Now, an important point is that a lot of energy is released when matter falls toward a black hole. Why does this happen? Well, as an object falls, its speed relative to static observers becomes very large. When many such of matter bump into each other at high these speeds, the result is a lot of very hot matter. This is where those x -rays come from. The matter is hot enough that x -rays are emitted as thermal radiation.

By the way, it is worth talking a little bit about just how we can calculate the extra 'kinetic energy' produced when objects fall toward black holes (or neutron stars). To do so, we will run in reverse a discussion we had long ago about light falling in a gravitational field.

Do you recall how we first argued that there must be something like a gravitational time dilation effect? It was from the observation that a photon going upward through a gravitational field must loose energy and therefore decrease in frequency.

Well, let's now think about a photon that falls down into a gravitational field from far away to a radius r. Clocks at r run slower than clocks far away by a factor of $\sqrt{1-R_s/r}$. Since the lower clocks run more slowly, from the viewpoint of these clocks the electric field of the photon seems to be oscillating very quickly.

So, this must mean that the frequency of the photon (measured by a static clock at r) is higher by a factor of $1/\sqrt{1-R_s/r}$ than when the frequency is measured by a clock far away. Since the energy of a photon is proportional to its frequency, the energy of the photon has increased by $1/\sqrt{1-R_s/r}$.

Now, in our earlier discussion of the effects of gravity on light we noted that the energy in light could be turned into any other kind of energy and could then be turned back into light.

We used this to argue that the effects of gravity on light must be the same as on any other kind of energy. So, consider an object of mass m which begins at rest far away from the black hole. It contains an energy $E = mc^2$. So, by the time the object falls to a radius r, its energy (measured locally) must have increased by the same factor was would the energy of a photon; to $E = mc^2 1/\sqrt{1-R_s/r}$.

What this means is that if the object gets anywhere even close to the Schwarzschild radius, it's energy will have increased by an amount comparable to its rest mass energy. Roughly speaking, this means that objects which fall toward a black hole or neutron star and collide with each other release energy on the same scale as a star or a thermonuclear bomb. This is the source of those x -rays and the other hard radiation that we detect from the accretion disk.

Actually, there is one step left in our accounting of the energy. After all, we don't sit in close to the black hole and measure the energy of the x -rays. Instead, we are far away.

So, we also need to think about the energy that the x -rays loose as they climb back out of the black hole's gravitational field. To this end, suppose our object begins far away from the black hole and falls to r. As we said above, its energy is now $E = mc^2/\sqrt{1-R_s/r}$. Suppose that the object now comes to rest at r. The object will then have an energy $E = mc^2$ as measured at r. So, stopping this object will have released an energy of

$$\Delta E = mc^2\left(\frac{1}{\sqrt{1-R_s/r}} - 1\right).$$

as measured at r. This is how much energy can be put into x-ray photons and sent back out. But, on it's way back out, such photons will decrease in energy by a factor of $\sqrt{1-R_s/r} - 1$. So, the final energy that gets out of the gravitational field is:

$$\Delta E_\infty = mc^2\sqrt{1-R_s/r}\left(\frac{1}{\sqrt{1-R_s/r}} - 1\right)$$
$$= mc^2(1-\sqrt{1-R_s/r})$$

In other words, the total energy released to infinity is a certain fraction of the energy in the rest mass that fell toward the black hole. This fraction goes to 1 if the mass fell all the way down to the black hole horizon. Again, so long as r was within a factor of 100 or so of the Schwarzschild radius, this gives an efficiency comparable to thermonuclear reactions.

Using direct observations, how strongly can we bound the size of a black hole candidate? It turns out that one can study the detailed properties of the spectrum of radiation produced by an accretion disk, and that one can match this to what one expects from an accretion disk living in the Schwarzschild geometry. Current measurements focus on a particular (x -ray) spectral line associated with iron. In the best case, the results show that the region emitting radiation is within $25Rs$.

So, where Does all of this Energy go, Anyway?

This turns out to be a very interesting question. There is a lot of energy be- ing produced by matter falling into a black hole or a neutron star. People are working very hard with computer models to figure out just how much matter falls into black holes, and therefore just how much energy is produced. Unfor- tunately, things are sufficiently complicated that one cannot yet state results with certainty.

Nonetheless, some very nice work has been done in the last few years by Ramesh Narayan and his collaborators showing that in certain cases there appears to be much less energy coming out than there is going in. Where is this energy going?

It is not going into heating up the object or the accretion disk, as such effects would increase the energy that we see coming out (causing the object to shine more brightly). If their models are correct, one is forced to conclude that the energy is truly disappearing from the part of the spacetime that can communicate with us. In other words, the energy is falling behind the horizon of a black hole. As the models and calculations are refined over the next five years or so, it is likely that this missing energy will be the first 'direct detection' of the horizon of a black hole.

A Very few words about Hawking Radiation

Strictly speaking, Hawking Radiation is not a part of this course because it does not fall within the framework of general relativity.

Here's the story, when we discussed the black hole singularity, we said that what really happens there will not be described by general relativity? We mentioned that physicists expect a new and even more fundamental understanding of physics to be important there, and that the subject is called "quantum gravity." We also mentioned that very little is understood about quantum gravity at the present time.

Well, there is one thing that we think we do understand about quantum effects in gravity. This is something that happens outside the black hole and therefore far from the singularity. In this setting, the effects of quantum mechanics in the gravitational field itself are extremely small. So small that we believe that we can do calculations by simply splicing together our understanding of quantum mechanics (which governs the behaviour of photons, electrons, and such things) and our understanding of gravity. In effect, use quantum mechanics together with the equivalence principle to do calculations.

Stephen Hawking did such a calculation back in the early 1970's. What he found came as a real surprise. Consider a black hole by itself, without an accretion disk or any other sort of obvious matter nearby. It turns out that the region around the black hole is not completely dark! Instead, it glows like a hot object, albeit at a very low temperature. The resulting thermal radiation is called Hawking radiation.

This is an incredibly tiny effect. For a solar mass black hole the associated temperature is only 10^{-5} Kelvin, that is, 105 degrees above absolute zero. Large black holes are even colder, as the temperature is proportional to M^{-2}, where M is the black hole mass. So black holes are very, very cold. In particular, empty space has a temperature of about 3K due to what is called the 'cosmic microwave background', so a black hole is much colder than empty space.

However, if one could make or find a very tiny black hole, that black hole would be very hot.

Second, let me add that the radiation does not come directly from the black hole itself, but from the space around the black hole. This is a common misconception about Hawking radiation: the radiation does not by itself contradict our statement that nothing can escape from within the horizon.

But, you may ask, how can radiation be emitted from the space around the black hole? How can there be energy created from nothing? The answer is that, in 'quantum field theory10,' one can have negative energies as well as positive energies. However, these negative energies should always be very small and should survive only for a short time.

What happens is that the space around the black hole produces a net zero energy, but it sends a positive energy flux of Hawking radiation outward away from the black hole while sending a negative energy flux inward across the horizon of the black hole. The negative energy is visible only for a short time between when it is created and when it disappears behind the horizon of the black hole.

The net effect is that the black hole looses mass and shrinks, while positive energy is radiated to infinity. A diagram illustrating the fluxes of energy is shown below.

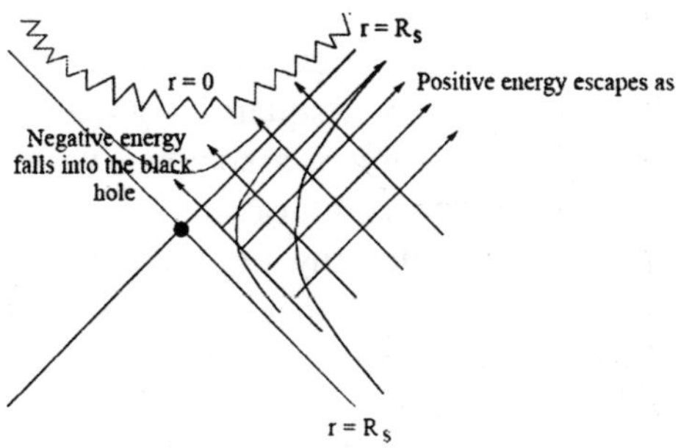

Penrose Diagrams, or "How to put Infinity in a Box"

There are a few comments left to make about black holes, and this will re- quire one further technical tool. The tool is yet another kind of spacetime diagram (called a 'Penrose diagram') and it will be useful both for discussing more complicated kinds of black holes.

The point is that, as we have seen, it is often useful to compare what an observer very far from the black hole sees to what one sees close to the black hole. We say that an observer very far from the black hole is "at infinity." Comparing infinity with finite positions is even more important for more complicated sorts of black holes that we have not yet discussed. However, it is difficult to draw infinity on our diagrams since infinity is after all infinitely far away.

How can we draw a diagram of an infinite spacetime on a finite piece of paper? Think back to the Escher picture of the Lobachevskian space. By 'squishing' the space, Escher managed to draw the infinitely large Lobachevskian space inside a finite circle. If you go back and try to count the number of fish that appear along on a geodesic crossing the entire space, it turns out to be infinite. It's just that most of the fish are drawn incredibly small. Escher achieved this trick by letting the scale vary across his map of the space.

In particular, at the edge an infinite amount of Lobachevskian space is crammed into a very tiny amount of Escher's map. In some sense this means that his picture becomes infinitely bad at the edge, but nevertheless we were able to obtain useful information from it.

We want to do much the same thing for our spacetimes. However, for our case there is one catch: As usual, we will want all light rays to travel along lines at 45 degrees to the vertical.

This idea was first put forward by (Sir) Roger Penrose[11], so that the resulting pictures are often called "Penrose Diagrams." They are also called "conformal diagrams" - conformal is a technical word related to the rescaling of size.

Let's think about how we could draw a Penrose diagram of Minkowski space. For simplicity, let's consider our favourite case of 1+1 dimensional Minkowski space. Would you like to guess what the diagram should look like? As a first guess, we might try a square or rectangle.

However, this guess has a problem associated with the picture below. To see the point, consider any light ray moving to the right in 1+1 Minkowski space, and also consider any light ray moving to the left. Any two such light rays are guaranteed to meet at some event. The same is in fact true of any pair of leftward and rightward moving objects since, in 1 space dimension, there is no room for two objects to pass each other!

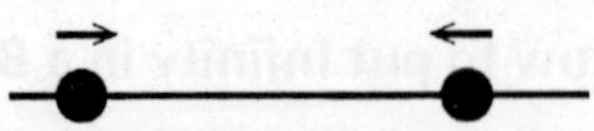

Left- and right- moving objects always collide when space has only one dimension

However, if the Penrose diagram for a spacetime is a square, then there are in fact leftward and rightward moving light rays that never meet! Some examples are shown on the diagram below.

These light rays do not meet

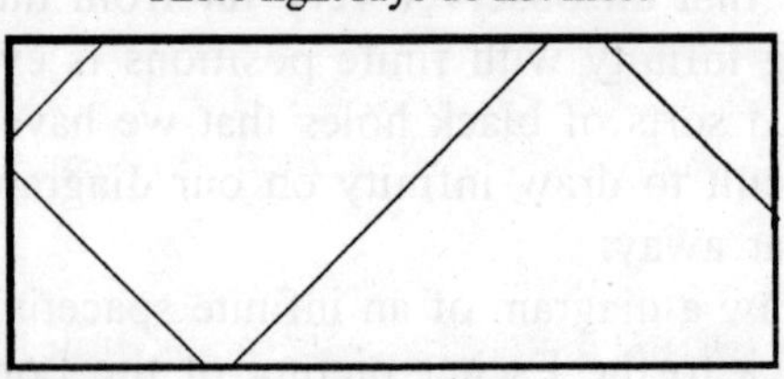

So, the rectangular Penrose diagram does not represent Minkowski space. What other choices do we have? A circle turns out to have the same problem. After a little thought, one finds that the only thing which behaves differently is a diamond:

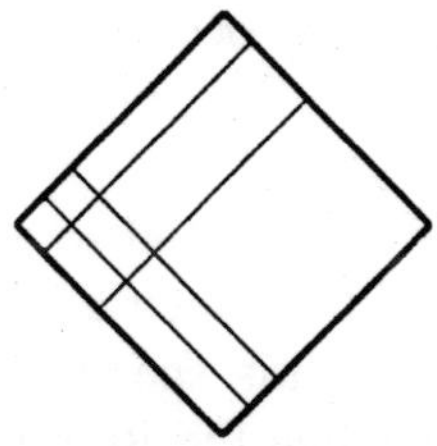

That is to say that infinity (or at least most of it) is best associated not which a place or a time, but with a set of light rays! In 3+1 dimensions, we can as usual decide to draw just the r, *t* coordinates. In this case, the Penrose diagram for 3+1 Minkowski space is drawn as a half-diamond:

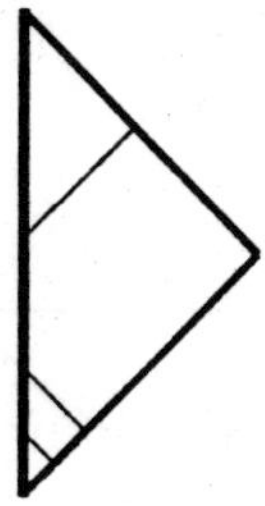

Penrose Diagrams for Black Holes

Using the same scheme, we can draw a diagram that shows the entire spacetime for the eternal Schwarzschild black hole. Remember that the distances are no longer represented accurately.

As a result, some lines that used to be straight get bent. For example, the constant r curves that we drew as hyperbolae before appear somewhat different on the Penrose diagram. However, all light rays still travel along straight 45 degree lines. The result is:

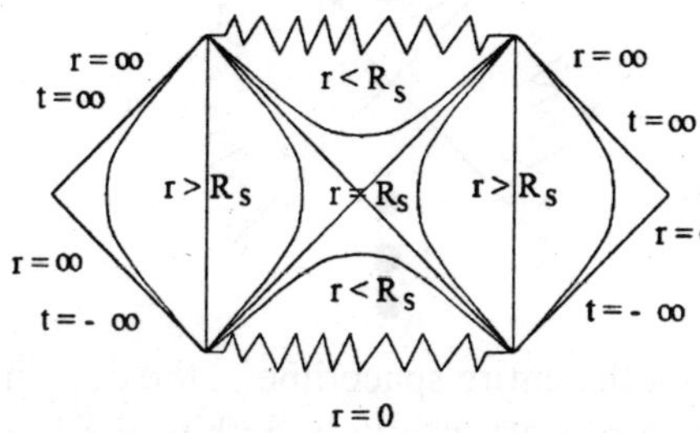

A new diagram for the Schwarzschild black hole, it turns out though that Schwarzschild black holes are not the only kind of black holes that can exist. The Schwarzschild metric was correct only outside of all of the 'matter' (which means anything other than gravitational fields) and only if the matter was spherically symmetric ('round').

Another interesting case to study occurs when we add a little bit of electric charge to a black hole. In this case, the charge creates an electric field which will fill all of space!

This electric field carries energy, and so is a form of 'matter.' Since we can never get out beyond all of this electric field, the Schwarzschild metric by itself is never quite valid in this spacetime. Instead, the spacetime is described by a related metric called the Reissner-No¨rdstr? *m* (RN) metric. The Penrose diagram for this metric is shown below:

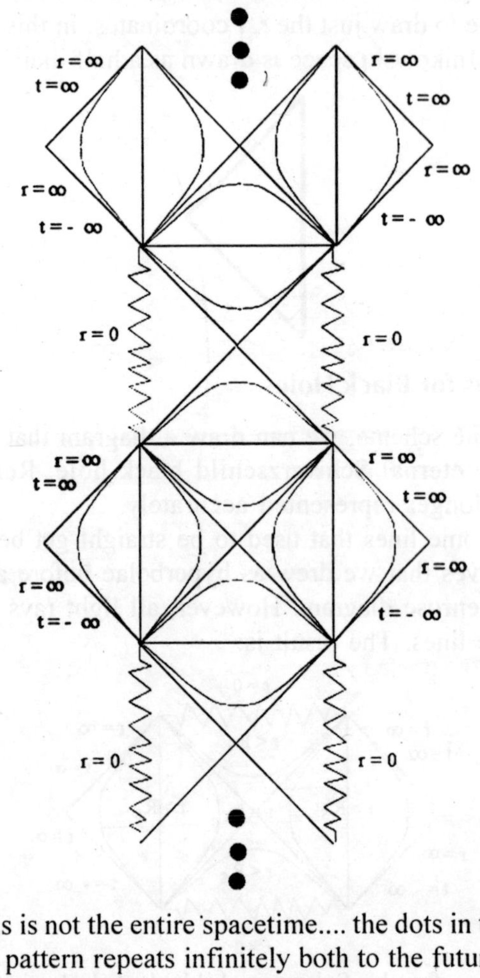

Actually, this is not the entire spacetime.... the dots in the diagram above indicate that this pattern repeats infinitely both to the future and to the past! This diagram has many interesting differences when compared to the Schwarzschild diagram. One is that the singularity in the RN metric is timelike instead of being spacelike.

Another is that instead of there being only two exterior regions, there are now infinitely many!

The most interesting thing about this diagram is that there does exist a timelike worldline (describing an observer that travels more slowly than light) that starts in one external region, falls into the black hole, and then comes back out through a 'past horizon' into another external region. Actually, is possible to consider the successive external regions as just multiple copies of the same external region.

In this case, the worldline we are discussing takes the observer back into the same universe but in such a way that they emerge to the past of when the entered the black hole!

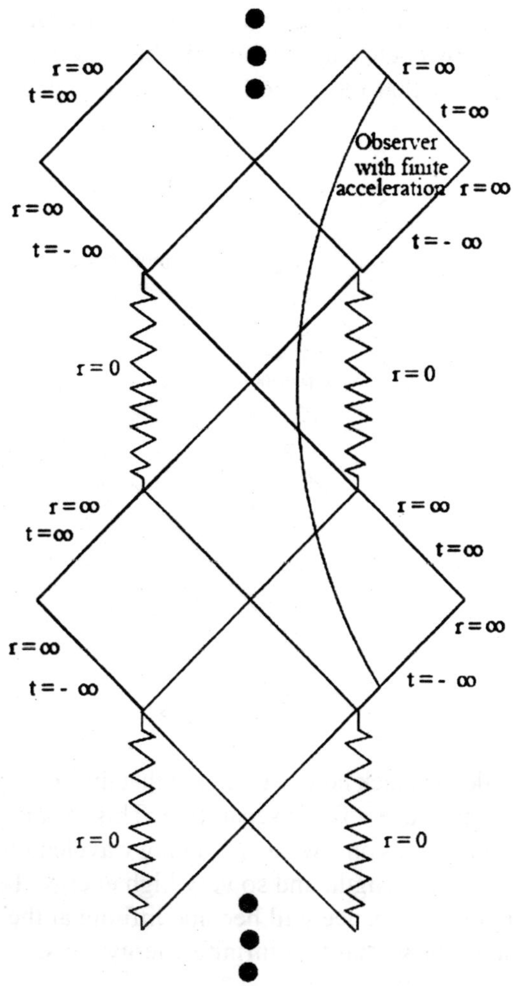

However, it turns out that there is an important difference between the Schwarzschild metric and the RN metric. The Schwarzschild metric is stable. This means that, while the Schwarzschild metric describes only an eternal black hole in a spacetime by itself (without, for example, any rocket ships near by carrying observers who study the black hole), the actual metric which would include rocket ships, falling scientists and students, and so on can be shown to be very close to the Schwarzschild metric. This is why we can use the Schwarzschild metric itself to discuss what happens to objects that fall into the black hole.

It turns out though that the RN metric does not have this property. The exterior is stable, but the interior is not. This happens because of an effect illustrated on the diagram below. Suppose that some energy (say, a light wave) falls into the black hole. From the external viewpoint this is a wave with a long wavelength and therefore represents a small amount of energy. The two light rays drawn below are in fact infinitely far apart from the outside perspective, illustrating that the wave has a long wavelength when it is far away.

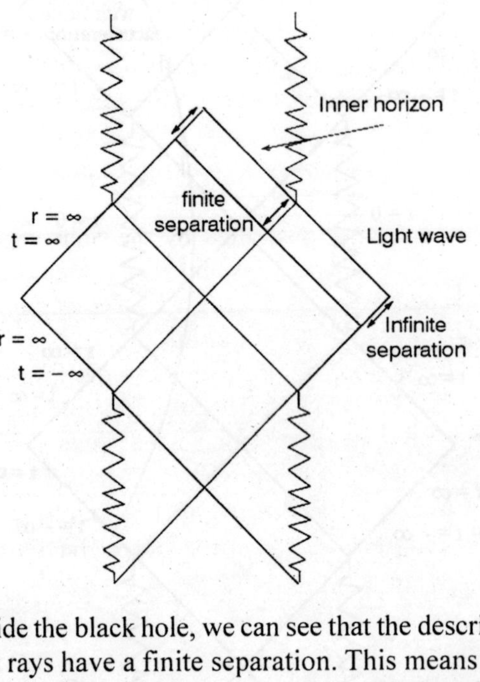

However, inside the black hole, we can see that the description is different. Now the two light rays have a finite separation. This means that that near the light ray marked "inner horizon," what was a long wavelength light ray outside is now of very short wavelength, and so very high energy! In fact, the energy created by any small disturbance will become infinite at the "inner horizon." It will come as no surprise that this infinite energy causes a large change in the spacetime.

The result is that dropping even a small pebble into an RN black hole creates a big enough e ect at the inner horizon to radically change the Penrose diagram. The Penrose diagram for the actual spacetime containing an RN black hole together with even a small disturbance looks like this:

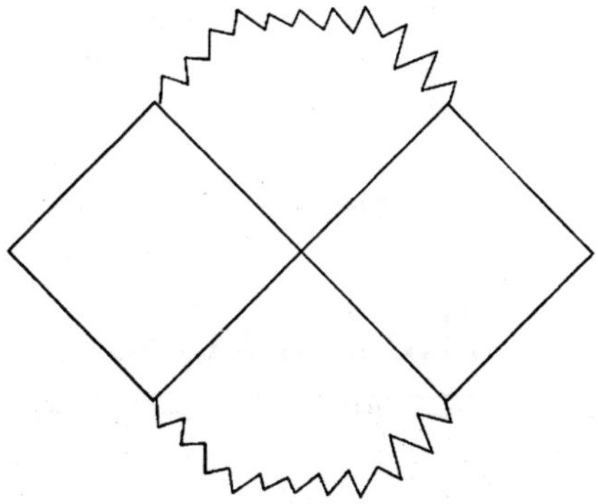

Some of the researchers who originally worked this out have put together a nice readable website that you might enjoy.

Real black holes in nature will have a significant electric charge. The point is that a black hole with a sig-nificant (say, positive charge) will attract other (negative) charges, which fall in so that the final object has zero total charge. However, real black holes do have one property that turns out to make them quite different from Schwarzschild black holes: they are typically spinning. Spinning black holes are not round, but become somewhat disk shaped.

As a result, they are not described by the Schwarzschild metric. The spacetime that describes a rotating black hole is called the Kerr metric. There is also of course a generalization that allows both spin and charge and which is called the Kerr-Newman metric.

It turns out that the Penrose diagram for a rotating black hole is much the same as that of an RN black hole, but with the technical complication that rotating black holes are not round. One finds the same story about an unstable inner horizon in that context as well, with much the same resolution. The details of the Kerr metric because of the technical complications involved, but it is good to know that things basically work just the same as for the RN metric above.

Chapter 10

The Universe

THE COPERNICAN PRINCIPLE AND RELATIVITY

Of course, in the early 1900's people did not know all that much about the universe, but they did have a few ideas on the subject. In particular, a certain philosophical tradition ran strong in astronomy, dating back to Copernicus. (Copernicus was the person who promoted the idea that the stars and planets did not go around the earth, but that instead the planets go around the sun.) This tradition held in high esteem the principle that "The earth is not at a particularly special place in the Universe".

It was this idea which had freed Copernicus from having to place the earth at the centre of the Universe. The idea was then generalized to say that, for example "The Sun is not a particularly special star," and then further to "There is no special place in the Universe." Or, said differently, the Copernican principle is that "Every place in the universe is basically the same."

So, on philosophical grounds, people believed that the stars were sprinkled more or less evenly throughout the universe. Now, one might ask, is this really true?

Well, the stars are not in fact evenly sprinkled. We now know that they are clumped together in galaxies. And even the galaxies are clumped together a bit. However, if one takes a sufficiently rough average then it is basically true that the clusters of galaxies are evenly distributed. We say that the universe is homogeneous. Homogeneous is just a technical word which means that every place in the universe is the same.

Homogeneity and Isotropy

In fact, there is another idea that goes along with every place being essentially the same. This is the idea that the universe is the same in every direction. The technical word is that the universe is isotropic. To give you an idea of what this means, a picture of a universe that is homogeneous but is not isotropic - the galaxies are farther apart in the vertical direction than in the horizontal direction:

In contrast, a universe that is both homogeneous and isotropic must look roughly like this:

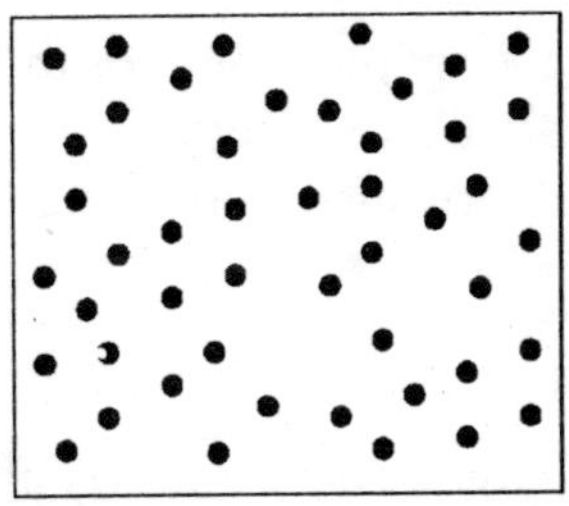

That Technical point about Newtonian Gravity in Homogeneous Space

The point is that, to compute the gravitational field at some point in space we need to add up the contributions from all of the infinitely many galaxies. This is an infinite sum. When you discussed such things in your calculus class, you learned that some infinite sums converge and some do not. Actually, this sum is one of those interesting in-between cases where the sum converges, but it does not converge absolutely. What happens in this case is that you can get different answers depending on the order in which you add up the contributions from the various objects.

To see how this works, recall that all directions in this universe are essentially the same. Thus, there is a rotational symmetry and the gravitational field must be pointing either toward or away from the centre. Now, it turns out that New-tonian gravity has a property that is much like Gauss' law in electromagnetism. In the case of spherical symmetry, the gravitational field on a given sphere de-pends only on the total charge inside the sphere. This makes it clear that on any given sphere there must be some gravitational field, since there is certainly matter inside:

But what if the sphere is very small? Then, there is essentially no matter inside, so the gravitational field will vanish. So, at the 'center' the gravitational field must vanish, but at other places it does not.

But now we recall that there is no centre! This universe is homogeneous, meaning that every place is the same. So, if the gravitational field vanishes at one point, it must also vanish at every other point. This is what physicists call a problem.

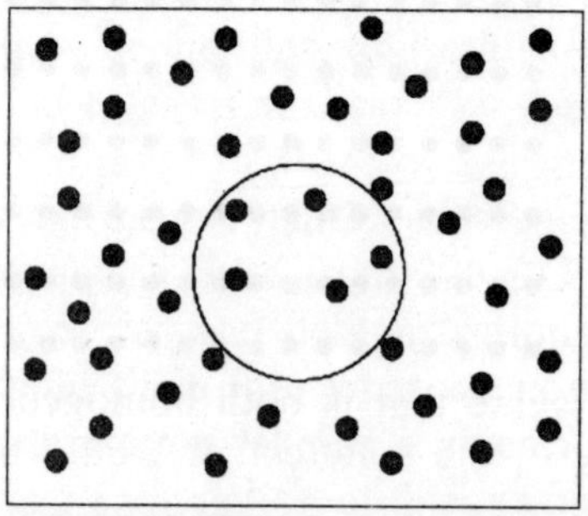

However, Einstein's theory turns out not to have this problem. In large part, this is because Einstein's conception of a gravitational field is very different from Newton's. In particular, Einstein's conception of the gravitational field is local while Newton's is not.

Homogeneous Spaces

Now, in general relativity, we have to worry about the curvature (or shape) of space. So, we might ask: "what shapes are compatible with the idea that space must be homogeneous and isotropic?" It turns out that there are exactly three answers:

- A three-dimensional sphere (what the mathematicians call S 3). This can be thought of as the set of points that satisfy $x_1^2 + x_2^2 + x_3^2 + x_4^2 = R^2$ in four-dimensional Euclidean space.
- Flat three dimensional space.
- The three dimensional version of the Lobachevskian space.

By the way, it is worth pointing out that option gives us a finite sized universe. The second and third options gives us infinite spaces. However, if we were willing to weaken the assumption of isotropy just a little bit, we could get finite sized spaces that are very much the same. To get an idea of how this works, think of taking a piece of paper (which is a good model of an infinite flat plane) and rolling it up into a cylinder. This cylinder is still flat, but it is finite in one direction. This space is homogeneous, though it is not isotropic (since one direction is finite while the other is not):

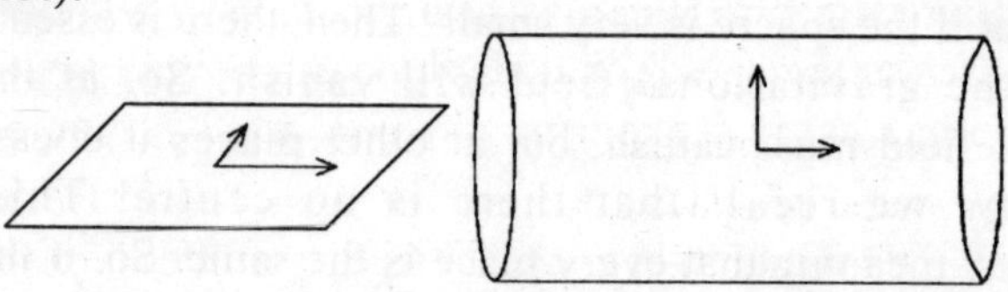

Rolling up flat three dimensional space in all three directions gives what is called a 3-torus, and is finite in all three directions. The

Lobachevskian space can also be 'rolled up' to get a finite universe. This particular detail is not mentioned in many popular discussions of cosmology.

Actually, these are not just three spaces. Instead, each possibility (sphere, flat, Lobachevskian) represents 3 sets of possibilities. To see the point, let's consider option #1, the sphere. There are small spheres, and there are big spheres. The big spheres are very flat while the tiny spheres are tightly curved. So, the sphere that would be our universe could, in principle, have had any size.

The same is true of the Lobachevskian space. Think of it this way: in Escher's picture, no one told us how big each fish actually is. Suppose that each fish is one light-year across.

Such a space can also be considered 'big,' although of course any Lobachevskian space has infinite volume (an infinite number of fish). In particular, if we consider a region much smaller than a single fish, we cannot see the funny curvature effects and the space appears to be flat. You may recall that we have to look at circles of radius 2 fish or so to see that C/R is not always 2?. So, if each fish was a light year across, we would have to look really far away to see the effects of the curvature. On the other hand, if each fish represented only a millimeter (a 'small' space), the curvature would be readily apparent just within our class room. The point is again that there is really a family of spaces here labelled by a length - roughly speaking, this length is the size of each fish.

What about for the flat space? After all, flat is flat..... Here, making the universe bigger does not change the geometry of space at all - it simply remains flat. However, it will spread out the galaxies, stars, and such. (The same is, of course, also true in the spherical and Lobachevskian contexts.) So, for the flat space case, one easy effect to visualize is the change in the density of matter. However, there is more to it than this: the spacetime is curved, and the curvature depends on the rate of expansion.

We can see this because observers at different places in 'space' who begin with no relative velocity nevertheless accelerate apart when the universe 'expands' !

Dynamics (a.k.a. Time Evolution)

So, homogeneity and isotropy restrict the shape of space to be in one of a few simple classes. That is to say, at any time (to the extent that this means anything) the shape of space takes one of these forms. But what happens as time passes? Does it maintain the same shape, or does it change? The answer must somehow lie inside Einstein's equations (the complicated ones that we have said rather little about), since they are what control the behaviour of the spacetime metric.

Luckily, the assumptions of homogeneity and isotropy simplify these equations a lot. Let's think about what the metric will look like. It will certainly have a dt^2 part. If we decide to use a time coordinate which measures proper time directly then the coefficient of dt^2 will just be 1. We can always decide to make such a choice.

The rest of the metric controls the metric for space1, which must be the metric for one of the three spaces described above. Now, the universe cannot suddenly change from, say, a sphere to a Lobachevskian space. So, as time passes the metric for space can only change by changing the the overall size (a.k.a. 'scale') of the space. In other words, the space can only get bigger or smaller.

What this means mathematically is that the metric must take the general form:

$$ds^2 = -d\tau^2 + a^2(t)(\text{metric for unit} - \text{sized space}).$$

The factor $a(t)$ is called the 'scale factor' or 'size of the universe.' When a is big, all of the spatial distances are very big. When a is small, all of the spatial distances are very small. So, a space with small a will have a highly curved space and very dense matter. Technically, the curvature of space is proportional to $1/a^2$, while the density of matter is proportional to $1/a^3$.

Note that the only freedom we have left in the metric is the single function $a(t)$. Einstein's equations must therefore simplify to just a single equation that tells us how $a(t)$ evolves in time.

Expanding and Contracting Universes

Before diving into Einstein's equations themselves, let's first take a moment to understand better what it means if a changes with time. To do so, let's consider a case where a starts o 'large' but then quickly decreases to zero:

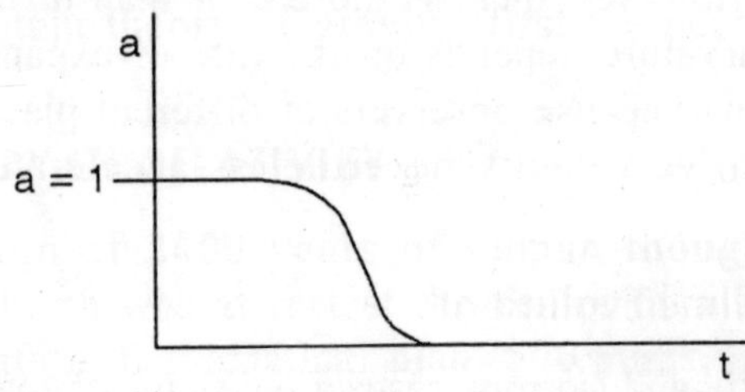

This represents any reasonable solution of Einstein's equations. Neverthe less, let's think about what happens to a freely falling object in this universe that begins 'at rest', meaning that it has zero initial velocity in the reference frame used in equation. If it has no initial velocity, then we can draw a spacetime diagram showing the first part of its worldline as a straight vertical line:

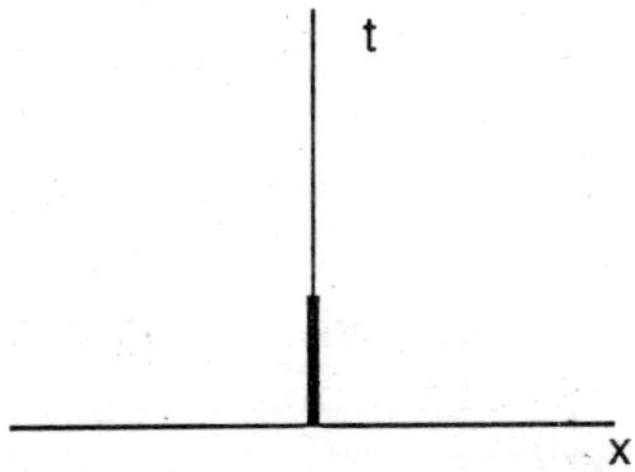

Now, when a shrinks to zero, what happens to the worldline? Will it bend to the right or to the left? Well, we assumed that the Universe is isotropic, right? So, the universe is the same in all directions. This means that there is a symmetry between right and left, and there is nothing to make it prefer one over the other. So, it does not bend at all but just runs straight up the diagram. In other words, an object that begins at $x = 0$ with zero initial velocity will always remain at $x = 0$.

Of course, since the space is homogeneous, all places in the space are the same and any object that begins at any $x = x_0$ with zero initial velocity will always remain at $x = x_0$. From this perspective it does not look like much is happening.

However, consider two such objects: one at x_1 and one at x_2. The metric ds^2 contains a factor of the scale a. So, the actual proper distance between these two points is proportional to a. Suppose that the distance between x_1 and x_2 is L when $a = 1$ (at $t = 0$). Then, later, when the scale has shrunk to $a < 1$, the new distance between this points is only aL. In other words, the two objects have come closer together.

Clearly, what each object sees is another object that moves toward it. The reason that things at first appeared not to move is that we chose a funny sort of coordinate system (if you like, you can think of this as a funny reference frame, though it is nothing like an inertial reference frame in special relativity). The funny coordinate system simply moves along with the freely falling objects cosmologists call it the 'co-moving' coordinate system. It is also worth pointing out what happens if we have lots of such freely falling objects, each remaining at a different value of x. In this case, each object sees all of the other objects rushing toward it as a decreases. Furthermore, an object which is initially a distance L away (when $a = 1$) becomes only a distance aL away.

So, the object has 'moved' a distance (1 – a) L. Similarly, an object which is initially a distance $2L$ away becomes $2aL$ away and 'moves' a distance 2(1 – a) L – twice as far.

This reasoning leads to what is known as the 'Hubble Law.' This law states that in a homogeneous universe the relative velocity between any two objects is proportional to their distance:

$$v = H(t) \cdot d,$$

where v is the relative 'velocity', d is the distance, and $H(t)$ is the 'Hubble constant' - a number that does depend on time but does not depend on the distance to the object being considered.

It is important to stress again that the Hubble constant is constant only in the sense of being independent of d. There is no particular reason that this 'constant' should be independent of time and, indeed, we will see that it is natural for H to change with time. The above relation using H (t) to emphasize this point. The Hubble constant is determined by the rate of change of a: $H(t) = \frac{1}{a}\frac{da}{dt}$.

There is no special object that is the 'center' of our collapsing universe. Instead, every object sees itself as the centre of the process. As usual, none of these objects is any more 'right' about being the centre than any other. The difference is just a change of reference frames.

A Flat Spacetime Model

In case this is hard to grasp, it is worth mentioning that you have seen something similar happen even in flat spacetime.

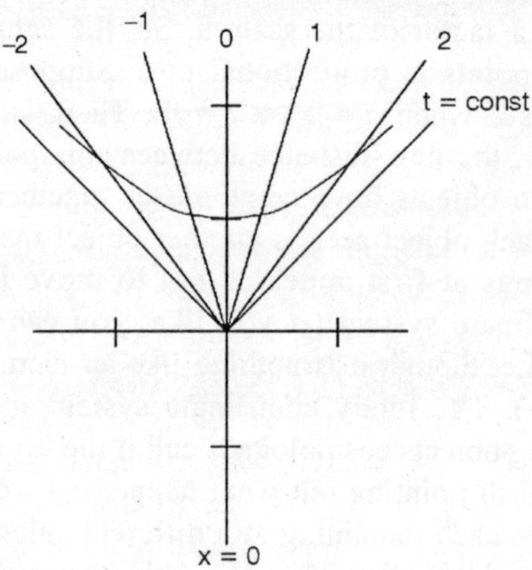

Suppose an infinite collection of inertial observers all of whom pass through some special event. Let me suppose that observer #1 differs from observer 0 by the same boost parameter as any other observer n + 1 differs from observer n. We could draw a spacetime diagram showing these observers as below:

Note that this is not the $k = 0$ Universe which has flat space. Instead, the entire spacetime is flat here when viewed as a whole, but the slice representing space on the above diagram is a hyperboloid, which is most definitely not flat.

Instead, this hyperboloid is a constant negative curvature space ($k = -1$). Since the spacetime here is flat, we have drawn the limit of the $k = -1$ case as we take the matter density to zero. It is not physically realistic as a cosmology, but

The co-moving coordiante system used in cosmology. In addition, for $k = -1$ the matter density does become vanishingly small in the distant future (if the cosmological constant vanishes; see below). Thus, for such a case this diagram does become accurate in the limit $t \to \infty$.

Shown here in the reference frame of observer 0, that observer appears to be the centre of the expansion. However, we know that if we change reference frames, the result will be:

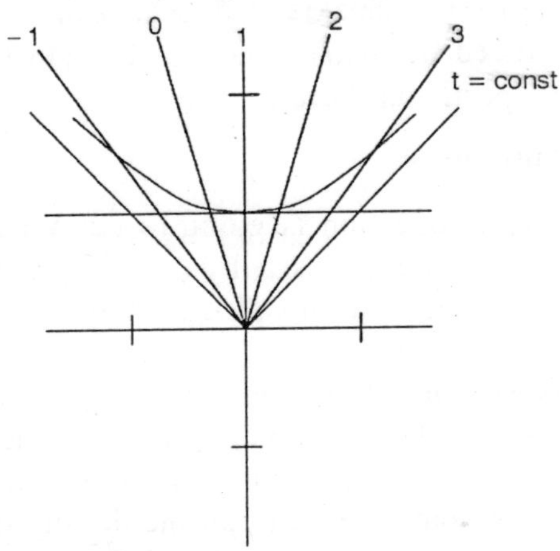

In this new reference frame, now another observer appears to be the 'center.' These discussions in flat spacetime illustrate three important points: The first is that although the universe is isotropic (spherically symmetric), there is no special 'center.' Note that the above diagrams even have a sort of 'big bang' where everything comes together, but that it does not occur any more where one observer is than where any other observer is. The second important point that the above diagram illustrates is that the surface that is constant t in our co-moving cosmological coordinates does not represent the natural notion of simultaneity for any of the co-moving observers. The 'homogeneity' of the universe is a result of using a special frame of reference in which the t = const surfaces are hyperbolae. As a result, the universe is not in fact homogeneous in any inertial reference frame (or any similar reference frame in a curved spacetime).

This is related to the third point: When discussing the Hubble law, a natural question is, "What happens when d is large enough that $H(t)$. d is greater than the speed of light?" In general relativity measurements that are not local are a subtle thing. For example, in the flat spacetime example above, in the coordinates that we have chosen for our homogeneous metric, the t = const surfaces are hyperbolae. They are not in fact the surfaces of simultaneity for any of the co-moving observers.

Now, the distance between co-moving observers that we have been discussing is the distance measured along the hyperbola (i.e., along the homogeneous slice), which is a very different notion of distance than we are used to using in Minkowski space.

This means that the 'velocity' in the Hubble law is not what we had previously called the relative velocity of two objects in Minkowski space. Instead, in our flat spacetime example, the velocity in the Hubble law turns out to correspond directly to the boost parameter θ.

However, for the nearby galaxies (for which the relative velocity is much less than the speed of light), this subtlety can be safely ignored (since v and θ are proportional there).

On to the Einstein Equations

So, the all important question is going to be: What is the function a(t)? What do the Einstein equations tell us about how the Universe will actually evolve? Surely what Newton called the attractive 'force' of gravity must cause something to happen!

As you might expect, the answer turns out to depend on what sort of stu you put in the universe. For example, a universe filled only with light behaves somewhat differently from a universe filled only with dirt.

In particular, it turns out to depend on the density of energy (ρ) and on the pressure (P). [You may recall that we briefly mentioned earlier that, in general relativity, pressure is directly a source of gravity.]

For our homogeneous isotropic metrics, it turns out that the Einstein equations can be reduced to the following two equations:

$$\frac{3}{a^2}\left(\frac{da}{dt}\right)^2 = \frac{8\pi G}{c^2}\rho - 3\frac{kc^2}{a^2},$$

$$\frac{3}{a}\frac{d^2a}{dt^2} = -\frac{8\pi G}{c^2}(\rho + 3P).$$

In the first equation, the constant k is equal to +1 for the spherical (positively curved) universe, $k = 0$ for the flat universe, and $k = -1$ for the Lobachevskian (negatively curved) universe.

We're not going to derive these equations, but let's talk about them a bit. The second one is of a more familiar form. It looks kind of like Newton's second law combined with Newton's law of Universal Gravitation - on the left we have the acceleration $d^2 a/dt^2$ while the right provides a force that depends on the amount of matter present (ρ).

Interestingly though, the pressure P also contributes. The reason that Newton never noticed the pressure term is that ρ is the density of energy and, for an object like a planet, the energy is mc^2 which is huge due to the factor of c^2. In comparison, the pressure inside the earth is quite small. Nevertheless, this pressure contribution can be important in cosmology.

A changes it tells us whether the (co-moving) bits of matter are coming closer together or spreading farther apart. This means that, in the present context, the Einstein equations tell us what the matter is doing as well as what the spacetime is doing. Thought of this way, the second equation should make a lot of sense.

The left hand side is an acceleration term, while the right hand side is related to the sources of gravity. Under familiar conditions where the particles are slowly moving, the energy density is roughly c^2 times the mass density. This factor of c^2 nicely cancels the c^2 in the denominator, leaving the first term on the right hand side as G times the density of mass.

The pressure has no hidden factors of c^2 and so P/c^2 is typically small. Under such conditions, this equation says that gravity causes the bits of matter to accelerate toward one another (this is the meaning of the minus sign) at a rate proportional to the amount of mass around. That sounds just like Newton's law of gravity, doesn't it?

In fact, we see that gravity is attractive in this sense whenever energy density ρ and pressure (P) are positive. In particular, for positive energy and pressure, a must change with time in such a way that things accelerate toward each other. Under such conditions it is impossible for the universe to remain static.

Now, back in the early 1900's people in fact believed (based on no particular evidence) that the universe had been around forever and had been essentially the same for all time. So, the idea that the universe had to be changing really bothered Einstein. In fact, it bothered him so much that he found a way out.

Negative Pressure, Vacuum Energy, and the Cosmological Constant

Physicists do expect that (barring small exceptions in quantum field theory) the energy density ρ will be always be positive. However, the is no reason in principle why the pressure P must be positive. Let's think about what a negative pressure would mean. A positive pressure is an effect that resists something being squeezed.

So, a negative pressure is an effect that resists something being stretched. This is also known as a 'tension.' Imagine, for example, a rubber band that has been stretched. We say that it is under tension, meaning that it tries to pull itself back together. A sophisticated relativistic physicist calls such an effect a 'negative pressure.'

We see that the universe can in fact 'sit still' and remain static if $\rho + 3P = 0$. If $\rho + 3P$ is negative, then gravity will in fact be repulsive (as opposed to attractive) the various bits of matter will accelerate apart.

Now, because ρ is typically very large (since it is the density of energy and $E = mc^2$) this requires a truly huge negative pressure. The kinds of matter that we are most familiar with will never have such a large negative pressure. However, physicists can imagine that their might possibly be such a kind of matter.

The favourite idea along these lines is called "vacuum energy." The idea is that empty space itself might somehow have energy. At first, this is a rather shocking notion. If it is empty, how can it have energy? But, some reflection will tell us that this may simply be a matter of semantics: given the space that we think is empty (because we have cleared it of everything that we know how to remove), how empty is it really? In the end, like everything else in physics, this question must be answered experimentally. We need to find a way to go out and to measure the energy of empty space.

Now, what is clear is that the energy of empty space must be rather small. Otherwise, it's gravitational effects would screw up our predictions of, for example, the orbits of the planets. However, there is an awful lot of 'empty' space out there. So, taken together it might still have some nontrivial effect on the universe as a whole.

Why should vacuum energy (the energy density of empty space) have negative pressure? Well, an important fact here is that energy density and pressure are not completely independent. Pressure, after all is related to the force required to change the size of a system: to smash it or to stretch it out. On the other hand, force is related to energy: for example, we must add energy to a rubber band in order fight the tension forces and stretch it out. The fact that we must add energy to a spring in order to stretch it is what causes the spring to want to contract; i.e., to have a negative pressure when stretched.

Now, if the vacuum itself has some energy density ρ and we stretch the space (which is just what we will do when the universe expands) then the new (stretched) space has more vacuum and therefore more energy. So, we again have to add energy to stretch the space, so there is a negative pressure. It turns out that pres-sure is (minus) the derivative of energy with respect to volume $P = -dE/dV$. Here, $E = \rho V$, so $P = -\rho$.

Clearly then for pure vacuum energy we have $\rho + 3P < 0$ and gravity is repulsive. On the other hand, combining this with the appropriate amount of normal matter could make the two effects cancel out and could result in a static universe.

Since $P = -\rho$ for vacuum energy, we see that vacuum energy is in fact characterized by a single number. It is traditional to call this number Λ, and to define Λ so that we have

$$\rho = \frac{\Lambda}{8\pi G}$$

$$P = -\frac{\Lambda}{8\pi G}.$$

Such a Λ is called the 'cosmological constant.' We have, in fact seen it before. You may recall that, during our very brief discussion of the Einstein equations, we mentioned that Einstein's assumptions and the mathematics in fact allowed two free parameters. One of these we identified as Newton's Universal Gravitational Constant G. The other was the cosmological constant *L*. This is the same cosmological constant: as we discussed back then, the cosmological constant term in the Einstein equations could be called a funny sort of 'matter.' In this form, it is none other than the vacuum energy that we have been discussing.

We mentioned that Λ must be small to be consistent with the observations of the motion of planets. However, clearly matter is somewhat more clumped together in our solar system than outside. Einstein hoped that this local clumping of normal matter (but not of the cosmological constant) would allow the gravity of normal matter to completely dominate the situation inside the solar system while still allowing the two effects to balance out for the universe overall.

Anyway, Einstein thought that this cosmological constant had to be there otherwise the universe could not remain static.

However, in the early 1920's, something shocking happened: Edwin Hubble made detailed measurements of the galaxies and found that the universe is in fact not static. He used the Doppler effect to measure the motion of the other galaxies and he found that they are almost all moving away from us. Moreover, they are moving away from us at a rate proportional to their distance! This is why the rule $v = H(t).\ d$ is known as the 'Hubble Law.'

The universe appeared to be expanding..... The result was that Einstein immediately dropped the idea of a cosmological constant and declared it to be the biggest mistake of his life.

OUR UNIVERSE: PAST, PRESENT, AND FUTURE

The other galaxies are running away from ours at a rate proportional to their distance from us. The implication is that the universe is expanding, and that it has been expanding for some time. In fact, since gravity is generally attractive, we would expect that the universe was expanding even faster in the past.

To find out more of the details we will have to look again to the Einstein equations. We will also need to decide how to encode the current matter in the universe in terms of a density ρ and a pressure *P*. Let's first think about the pressure.

Most matter today is clumped into galaxies, and the galaxies are quite well separated from each other. How much pressure does one galaxy apply to another? Essentially none. So, we can model the normal matter by setting $P = 0$.

When the pressure vanishes, one can use the Einstein equations to show that the quantity: $\varepsilon = 8\pi G\rho a^3/3$ is independent of time. Roughly speaking, this is just conservation of energy (since ρ is the density of energy and a^3 is proportional to the volume). As a result, assuming that $\Lambda = 0$ the Einstein equations can be written:

$$\frac{1}{c^2}\left(\frac{da}{dt}\right)^2 - \frac{\varepsilon}{c^2 a} + k = 0.$$

k is a constant that depends on the overall shape of space: $k = +1$ for the spherical space, $k = 0$ for the flat space, and $k = -1$ for the Lobachevskian space.

In the above form, this equation can be readily solved to determine the behaviour of the universe for the three cases $k = -1, 0, +1$. We don't need to go into the details here, but let me draw a graph that gives the idea of how a changes with *t* in each case:

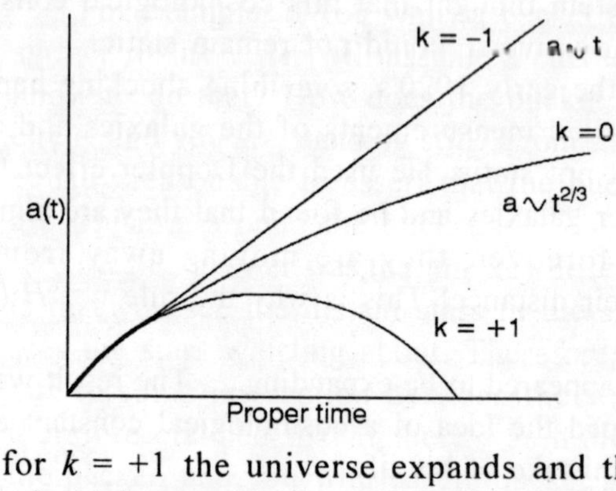

Note that for $k = +1$ the universe expands and then recontracts, whereas for $k = 0, -1$ it expands forever. In the case $k = 0$ the Hubble constant goes to zero at very late times, but for $k = -1$ the Hubble constant asymptotes to a constant positive value at late times.

Note that at early times the three curves all look much the same. Roughly speaking, our universe is just now at the stage where the three curves are beginning to separate. This means that, the past history of the universe is more or less independent of the value of *k*.

OBSERVATIONS AND MEASUREMENTS

So, which is the case for our universe? How can we tell? Well, one way to figure this out is to try to measure how fast the universe was expanding at various times in the distant past. This is actually not as hard as you might think: you see, it is very easy to look far backward in time. All we have to do is to look at things that are very far away. Since the light from such objects takes such a very long time to reach us, this is effectively looking far back in time.

Runaway Universe?

The natural thing to do is to try to enlarge on what Hubble did. If we could figure out how fast the really distant galaxies are moving away from us, this will tell us what the Hubble constant was like long ago, when the light now reaching us from those galaxies was emitted. The redshift of a distant galaxy is a sort of average of the Hubble constant over the time during which the signal was in transit, but with enough care this can be decoded to tell us about the Hubble constant long ago.

By measuring the rate of decrease of the Hubble constant, we can learn what kind of universe we live in.

However, it turns out that accurately measuring the distance to the distant galaxies is quite difficult. (In contrast, measuring the redshift is easy.) Until recently, no one had seriously tried to measure such distances with the accuracy that we need. However, a few years ago it was realized that there may be a good way to do it using supernovae.

The particular sort of supernova of interest here is called 'Type Ia.' Astrophysicists believe that type Ia supernovae occur when we have a binary star system containing one normal star and one white dwarf. We can have matter flowing from the normal star to the white dwarf in an accretion disk, much as matter would flow to a neutron star or black hole in that binary star system. But remember that a white dwarf can only exist if the mass is less than 1.4 solar masses.

When extra matter is added, bringing the mass above this threshold, the electrons in the core of the star get squeezed so tightly by the high pressure that they bond with protons and become neutrons. This releases vast amount of energy in the form of neutrinos (another kind of tiny particle) and heat which results in a massive explosion: a (type Ia) supernova.

Anyway, it appears that this particular kind of supernova is pretty much always the same. It is the result of a relatively slow process where matter is gradually added to the white dwarf, and it always explodes when the total mass hits 1.4 solar masses. In particular, all of these supernovae are roughly the same brightness (up to one parameter that astrophysicists think they know how to correct for). As a result, supernovae are a useful tool for measuring the distance to far away galaxies. All we have to do is to watch a galaxy until one of these supernovae happens, and then see how bright the supernova appears to be. Since it's actual brightness is known, we can then figure out how far away it is. Supernovae farther away appear to be much dimmer while those closer in appear brighter.

About two years ago, the teams working on this project released their data. The result came as quite a surprise.

Their data shows that the universe is not slowing down at all. Instead, it appears to be accelerating!

As you might guess, this announcement ushered in the return of the cosmological constant. By the way, the cosmological constant has very little effect when the universe is small (since vacuum energy is the same density whether the universe is large or small while the density of normal matter was huge when the universe was small).

However, with a cosmological constant, the effects of the negative pressure get larger and larger as time passes (because there is more and more space, and thus more and more vacuum energy). As a result, a cosmological constant makes the universe expand forever at an ever increasing rate. Adding this case to our graph, we get:

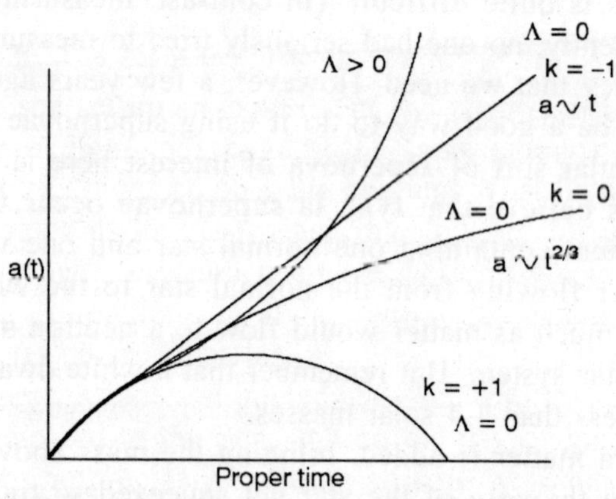

The line for $\Lambda > 0$ is more or less independent of the constant k.

So, should we believe this? The data in s upport of an accelerating universe has held up well for three years now. However, there is a long history of problems with observations of this sort.

There are often subtleties in understanding the data that are not apparent at first sight, as the various effects can be much more complicated than one might naively expect. Physicists say that there could be significant 'systematic errors' in the technique.

All this is to say that, when you measure something new, it is always best to have at least two independent ways to find the answer. Then, if they agree, this is a good confirmation that both methods are accurate.

Once Upon a time in a Universe Long Long Ago

It turns out that one way to get an independent measurement of the cosmological constant is tied up with the story of the very early history of the universe. This is of course an interesting story in and of itself.

Let's read the story backwards. Here we are in the present day with the galaxies spread wide apart and speeding away from each other. Clearly, the galaxies used to be closer together. As indicated by the curves in our graphs, the early history of the universe is basically independent of the value of L or k.

So, imagine the universe as a movie that we now play backwards. The galaxies now appear to move toward each other. They collide and get tangled up with each other. At some point, there is no space left between the galaxies, and they all get scrambled up together - the universe is just a mess of stars.

Then the universe shrinks some more, so that the stars all begin to collide. There is no space left between the stars and the universe is filled with hot matter, squeezing tighter and tighter. The story here is much like it is near the singularity of a black hole: even though squeezing the matter increases the pressure, this does not stop the spacetime from collapsing. In fact, as we have seen, pressure only adds to the gravitational attraction and accelerates the collapse.

As the universe squeezes tighter, the matter becomes very hot. At a certain point, the matter becomes so hot that all of the atoms ionize: the electrons come o and separate from the nuclei. Something interesting happens here. Because ionized matter interacts strongly with light, light can no longer travel freely through the universe. Instead, photons bounce around between nuclei like ping pong balls! It it the cosmic equivalent of trying to look through a very dense fog, and it becomes impossible to see anything in the universe.

This event is particularly important because, as we discussed earlier, the fact that it takes light a long time to travel across the universe means that when we look out into the universe now, looking very far away is effectively looking back in time.

So, this ionization sets a limit on how far away and how far back in time we can possibly see. On the other hand, ever since the electrons and nuclei got together into atoms (deionization) the universe has been more or less transparent. For this reason, this time is also called 'decoupling.' [Meaning that light 'decouples' or 'disconnects' from matter.] As a result, we might expect to be able to see all the way back to this time.

What would we see if we could see that far back? Well, the universe was hot, right? And it was all sort of mushed together. So, we might expect to see a uniform glow that is kind of like looking into a hot fire. In fact, it was quite hot: several thousand degrees.

Another way to discuss this glow is to remember that the universe is homo- geneous.

This means that, not only was stu "way over there" glowing way back when, but so was the stuff where we are. What we are saying is that the whole universe (or, if you like, the whole electromagnetic field) was very hot back then.

A hot electromagnetic field contains a lot of light.... Anyway, the point about light barely interacting with matter since decoupling means that, since that time, the electromagnetic field (i.e., light) should just have gone on and done its thing independent of the matter. In other words, it cannot receive energy (heat) from matter or loose energy (heat) by dumping it into matter. It should have pretty much the same heat energy that it had way back then.

So, why then is the entire universe today not just one big cosmic oven filled with radiation at a temperature of several thousand degrees? The answer is that the expansion of the universe induces a redshift not only in the light from the distant galaxies, but in the thermal radiation as well. The effect is similar to the fact that a gas cools when it expands. Here, however, the gas is a gas of photons and the expansion is due to the expansion of the universe. The redshift since decoupling is about a factor of 2000, with the result that the radiation today has a temperature of a little over 3 degrees Kelvin (i.e. 3 degrees above absolute zero).

At 3 degrees Kelvin, electromagnetic radiation is in the form of microwaves (in this case, think of them as short wavelength radio waves). This radiation can be detected with what are basically big radio telescopes or radar dishes. Back in the 60's some folks at Bell Labs built a high quality radio dish to track satellites. Two of them (Penzias and Wilson) were working on making it really sensitive, when the discovered that they kept getting a lot of noise coming in, and coming in more or less uniformly from all directions. It appeared that radio noise was being produced uniformly in deep space!

This radio 'noise' turned out to be thermal radiation at a temperature of 2.7 Kelvin. Physicists call it the 'Cosmic Microwave Background (CMB).' It's discovery is one of the greatest triumphs of the 'big bang' idea. After all, that is what we have been discussing. Long ago, before decoupling, the universe was very hot, dense, and energetic.

It was also in the process of expanding, so that the whole process bears a certain resemblance (except for the homogeneity of space) to a huge cosmic explosion: a big bang. The discovery of the CMB verifies this back to an early stage in the explosion, when the universe was so hot and dense that it was like one big star.

By the way, do you remember our assumption that the universe is homogeneous? We said that it is of course not exactly the same everywhere (since, for example, the earth is not like the inside of the sun) but that, when you measure things on a sufficiently large scale, the universe does appear to be homogeneous.

Well, the cosmic microwave background is our best chance to test the homogeneity on the largest possible scales since, as we argued above, it will not be possible to directly 'see' anything coming from farther away. The microwaves in the CMB have essentially traveled in a straight line since decoupling. We will never see anything from farther away since, for the light to be reaching us now, it would have had to have been emitted from an distant object before decoupling - back when the universe was filled with thick 'fog.'

When we measure the cosmic microwave background, it turns out to be incredi-bly homogeneous. The departures from homogeneity in the CMB are only about 1 part in one hundred thousand.

An important point about the early universe. It was not like what we would get if we simply took the universe now and made all of the galaxies come together instead of rushing apart. If we pushed all of the galaxies together we would, for example, end up with a lot of big clumps (some related to galactic black holes, for example). While there would be a lot of general mushing about, we would not expect the result to be anywhere near as homogeneous as one part in one hundred thousand.

It appears then that the universe started in a very special, very uniform state with only very tiny fluctuations in its density. So then, why are there such large clumps of stuff today? Today, the universe is far from homogeneous on the small scale. The reason for this is that gravity tends to cause matter to clump over time.

Places with a little higher density pull together gravitationally and become even more dense, pulling in material from neighboring under-dense regions so that they become less dense. It turns out that tiny variations of one part in one hundred thousand back at decoupling are just the right size to grow into roughly galaxy-style clumps today.

This is an interesting fact by itself: Galaxies do not require special 'seeds' to start up. They are the natural consequence of gravity amplifying teeny tiny variations in density in an expanding universe.

Well, that's the rough story anyway. Making all of this work in detail is a little more complicated, and the details do depend on the values of *L,* k, and so on. As a result, if one can measure the CMB with precision, this becomes an independent measurement of the various cosmological parameters. The data from COBE confirmed the whole general picture and put some constraints on

L. The results were consistent with the supernova observations, but by itself COBE was not enough to measure *L* accurately. A number of recent balloonbased CMB experiments have improved the situation somewhat, and in the next few years two more satellite experiments (MAP and PLANCK) will measure the CMB in great detail. Astrophysicists are eagerly awaiting the results.

A Cosmological 'Problem'

Actually, the extreme homogeneity of the CMB raises another issue: how could the universe have ever been so homogeneous? For example, when we point our radio dish at one direction in the sky, we measure a microwave signal at 2.7 Kelvin coming to us from ten billion light-years away. Now, when we point our radio dish in the opposite direction, we measure a microwave signal at the same temperature (to within one part in one hundred thousand) coming at us from ten billion light-years away in the opposite direction! Now, how did those two points so far apart know that they should be at exactly the same temperature?

Ah! You might say, "Didn't the universe used to be a lot smaller, so that those two points were a lot closer together?" This is true, but it turns out not to help. The point is that all of the models we have been discussing have a singularity where the universe shrinks to zero size at very early times. An important fact is that this singularity is spacelike (as in the black hole). The associated Penrose diagram looks something like this:

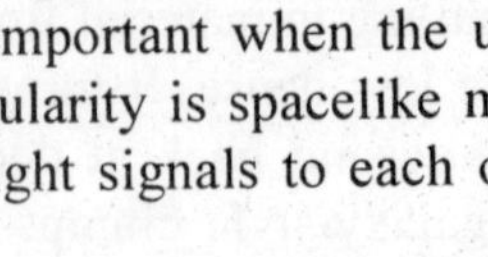

The Penrose diagram including a cosmological constant, but the part describing the big bang singularity is the same in any case (since, as we have discussed, Λ is not important when the universe is small).

The fact that the singularity is spacelike means that no two points on the singularity can send light signals to each other (even though they are zero distance apart).

Thus, it takes a finite time for any two 'places' to be able to signal each other and tell each other at what temperature they should be2. In fact, we can see that if the two points begin far enough apart then they will never be able to communicate with each other, though they might both send a light (or microwave) signal to a third observer in the middle.

The light rays that tell us what part of the singularity a given event has access to form what is called the 'particle horizon' of that event and the issue we have been discussing (of which places could possibly have been in thermal equilibrium with which other places) is called the 'horizon problem.'

There are two basic ways out of this, but it would be disingenuous to claim that either is understood at more than the most vague of levels. One is to simply suppose that there is something about the big bang itself that makes things incredibly homogeneous, even outside of the particle horizons. The other is to suppose that for some reason the earliest evolution of the universe happened in a different way than we drew on our graph above and which somehow removes the particle horizons.

The favourite idea of this second sort is called "inflation." Basically, the idea is that for some reason there was in fact a truly huge cosmological constant in the very earliest universe - sufficiently large to affect the dynamics. Let us again think of running a movie of the universe in reverse. In the forward direction, the cosmological constant makes the universe accelerate. So, running it backward it acts as a cosmic brake and slows things down.

The result is that the universe would then be older than we would otherwise have thought, giving the particle horizons a chance to grow sufficiently large to solve the horizon problem. The resulting Penrose diagram looks something like this:

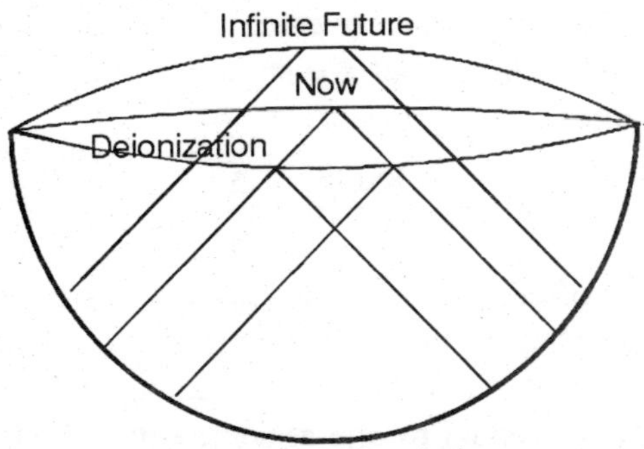

The regions we see at decoupling now have past light cones that overlap quite a bit. So, they have access to much of the same information from the singularity. In this picture, it is easier to understand how these entire universe could be at close to the same temperature at decoupling.

Oh, to be consistent with what we know, this huge cosmological 'constant' has to shut itself o long before decoupling. This is the hard part about making inflation work.

Making the cosmological constant turn o requires an amount of fine tuning that many people feel is comparable to the one part in one-hundred thousand level of inhomogeneities that inflation was designed to explain.

Luckily, inflation makes certain predictions about the detailed form of the cosmic microwave background. The modern balloon experiments are beginning to probe the interesting regime of accuracy, and it is hoped that MAP and PLANCK will have some definiti ve commentary on whether inflation is or is not the correct explanation.

Looking for Mass in all the Wrong Places

The cosmological constant, turns out that the supernovae results and the CMB do not really measure *L* directly, but instead link the cosmological constant to the overall density of matter in the universe.

So, to get a real handle on things, one has to know the density of more or less regular matter in the universe as well. Before we get into how much matter there actually is (and how we find out), The Hubble expansion rate $H = \frac{1}{a}\frac{da}{dt}$. $H^2 - \frac{8\pi G\rho}{3} - \frac{\Lambda}{3} + ka^{-2}c^2 = 0.$

The cosmologists like to reorganize this equation by dividing by H^2. This gives $H^2 - \frac{8\pi G\rho}{3} - \frac{\Lambda}{3} + ka^{-2}c^2 = 0.$

Now, the three interesting cases are $k = -1, 0, +1$. The middle case is $k = 0$. overall density of stu(matter or cosmological constant) in 'Hubble units' must be one! So, this is a convenient reference point. If we want to measure *k*, it is this quantity that we should compute. So, cosmologists give it a special name:

$$\Omega \equiv \frac{8\pi G\rho}{3H^2} + \frac{\Lambda}{3H^2}.$$

This quantity is often called the 'density parameter,' but we see that it is slightly more complicated than that name would suggest. In particular, (like the Hubble 'constant') Ω will in general change with time. If, however, Ω happens to be exactly equal to one at some time, it will remain equal to one. So, to tell if the universe is positively curved ($k = +1$), negatively curved ($k = -1$), or [spatially] flat ($k = 0$), what we need to do is to measure W and to see whether it is bigger than, smaller than, or equal to one.

By the way, cosmologists in fact break this Ω up into two parts corresponding to the matter and the cosmological constant.

$$\Omega_{\text{matter}} \equiv \frac{8\pi G\rho}{3H^2}$$

$$\Omega_{\Lambda} \equiv \frac{\Lambda}{3H^2}$$

Not only do these two parts change with time, but their ratio changes as well. The natural tendency is for Ω_Λ to grow with time at the expense of Ω_{matter} as the universe gets larger and the vacuum energy becomes more important Anyway, when cosmologists discuss the density of matter and the size of the cosmological constant, they typically discuss these things in terms of Ω_{matter} and Ω_Λ.

So, just how does one start looking for matter in the universe? Well, the place to start is by counting up all of the matter that we can see - say, counting up the number of stars and galaxies. Using the things we can see gives about $\Omega = 0.05$.

But, there are more direct ways to measure the amount of mass around - for example, we can see how much gravity it generates! Remember our discussion of how astronomers find black holes at the centers of galaxies? They use the stars orbiting the black hole to tell them about the mass of the black hole. Similarly, we can use stars orbiting at the edge of a galaxy to tell us about the total amount of mass in a galaxy.

It turns out to be much more than what we can see in the 'visible' matter. Also, recall that the galaxies are a little bit clumped together. If we look at how fast the galaxies in a given clump orbit each other, we again find a bit more mass than we expected.

It turns out that something like 90% of the matter out there is stuff that we can't see. For this reason, it is called 'Dark Matter.' Interestingly, although it is attached to the galaxies, it is spread a bit more thinly than is the visible matter. This means that a galaxy is surrounded by a cloud of dark matter than is a good bit larger than the part of the galaxy that we can see. All of these measurements of gravitational effects bring the matter count up to about $\Omega_{matter} = .4$.

Now, there is of course a natural question: Just what is this Dark Matter stu anyway? Well, there are lots of things that it is not. For example, it is not a bunch of small black holes or a bunch of little planet-like objects running around. At least, the vast majority is not of that sort.

That possibility has been ruled out by studies of gravitational lensing. Briefly, recall that general relativity predicts that light 'falls' in a gravitational field and, as a result, light rays are bent toward massive objects.

This means that massive objects actually act like lenses, and focus the light from objects shining behind them. When such a 'gravitational lens' passes in front of a star, the star appears to get brighter. When the lens moves away, the star returns to its original brightness. By looking at a large number of stars and seeing how often they happen to brighten in this way, astronomers can 'count' the number of gravitational lenses out there. To make a long story short, there are too few such events for all of the dark matter to be clumped together in black holes or small planets. Instead, most of it must be spread out more evenly. Even more interestingly, it cannot be just thin gas..... That is, there are strong arguments why the dark matter, whatever it is, cannot be made up of protons and neutrons like normal matter! To understand this, we need to continue the story of the early universe as a movie that we run backward in time. We discussed earlier how there was a very early time (just before decoupling) when the Universe was so hot and dense that the electrons were detached from the protons. Well, continuing to watch the movie backwards the universe becomes even more hot and dense. Eventually, it becomes so hot and dense that the nuclei fall apart.

Now there are just a bunch of free neutrons and protons running around, very evenly spread throughout the universe. It turns out that we can calculate what should happen in such a system as the universe expands and cools. As a result, one can calculate how many of these neutrons and protons should stick together and form Helium vs. how many extra protons should remain as Hydrogen.

This process is called 'nucleosynthesis.' One can also work out the proportions of other light elements like Lithium... (The heavy elements were not made in the big bang itself, but were manufactured in stars and supernovae.) To cut short another long story, the more dense the stuff was, the more things stick together and the more Helium and Lithium should be around. Astronomers are pretty good at measuring the relative abundance of Hydrogen and Helium, and the answers favour roughly $\Omega_{\text{normal matter}}$ =.1, – the stuff we can see plus a little bit more.

As a result, this means that the dark matter is not made up of normal things like protons and neutrons. By the way, physicists call such matter 'baryonic3 matter' so that this fact is often quoted as Ω_{baryon} =.1. A lot of this may be in the form of small not-quite stars and such, but the important point is that at least 75% of the matter in the universe really has to be stuff that is not made up of protons and neutrons.

So, what is the dark matter then? That is an excellent question and a subject of much debate. It may well be the case that all of this unknown dark matter is some strange new kind of tiny particle which simply happens not to interact with regular matter except by way of gravity.

A number of ideas have been proposed, but it is way too early to say how likely they are to be right.

Putting it all Together

The last part of our discussion is to put all of this data together to see what the implications are for Ω_Λ and Ω_{matter}. Many of these graphs (and some other stu) come from from a talk given by Sean Carroll.

These graphs show that that each of the three measurements put some kind of constraint on the relationship between Ω_{matter} and Ω_Λ, corresponding to a (wide) line in the Ω_{matter}? Ω_{Λ} plane. You can see that, taken together, the data strongly favors a value near Ω_{matter} =.4, Ω_{Λ} =.6. That is, 60% of the energy in the universe appears to be vacuum energy!

Now, what is really impressive here is that any two of the measurements would predict this same value. The third measurement can then be thought of as a double-check. As the physicists say, any two lines in a plane intersect somewhere, but to get three lines to intersect at the same point you have to do something right. This means that the evidence for a cosmological constant is fairly strong we have not just one experiment that finds it, but in fact we have another independent measurement that confirms this result. However, the individual measurements are not all that accurate and may have unforeseen systematic errors. So, we look forward to getting more and better data in the future to see whether these results continue to hold up.

We are in fact expecting to get a lot more data over the next few years. Two major satellite experiments (called 'MAP' and 'PLANCK') are going to make very detailed measurements of the Cosmic Microwave Background which should really tighten up the CMB constraints on Wmatter and WL. It is also hoped that these experiments will either confirm or deny the predictions of inflation.

By the way, it is a rather strange picture of the universe with which we are left. There are several confusing issues. One of them is "where does this vacuum energy come from?" It turns out that there are some reasonable ideas on this subject coming from quantum field theory... However, while they are all reasonable ideas for creating a vacuum energy, they all predict a value that is 10^{120} times too large.

A moment to state the obvious: 10^{120} is an incredibly huge number. A billion is ten to the ninth power, so 10^{120} is one billion raised to the thirteenth power. Physicists are always asking, "Why is the cosmological constant so small?"

Another issue is that, as we mentioned, WL and Wmatter do not stay constant in time. They change, and in fact they change in different ways. There is a nice diagram (also from Sean Carroll) showing how they change with time. What you can see is that, more or less independently of where you start, the universe naturally evolves toward Ω_{Λ} = 1. On the other hand, back at the big bang Ω_{Λ} was almost certainly near zero.

So, an interesting question is: "why is Ω_{Λ} only now in the middle ground (Ω_{Λ} =.6), making it's move between zero and one?"

For example, does this argue that the cosmological constant is not really constant, and that there is some new physical principle that keeps it in this middle ground? Otherwise, why should the value of the cosmological constant be such that Ω_{Λ} is just now making it's debut? It is not clear why Λ

should not have a value such that it would have taken over long ago, or such that it would still be way too tiny to notice.

THE BEGINNING AND THE END

Well, we are nearly finished with our story but we are not yet at the end. We traced the universe back to a time when it was so hot and dense that the nuclei of atoms were just forming. We have seen that there is experimental evidence (in the abundances of Hydrogen and Helium) that the universe actually was this hot and dense in its distant past. Well, if our understanding of physics is right, it must have been even hotter and more dense before. So, what was this like? How hot and dense was it? From the perspective of general relativity, the most natural idea is that the farther back we go, the hotter and denser it was.

Looking back in time, we expect that there was a time when it was so hot that protons and neutrons themselves fell apart, and that the universe was full of things called quarks. Farther back still, the universe so hot that our current knowledge of physics is not sufficient to describe it. All kinds of weird things might have happened, like maybe the universe had more than four dimensions back then. Maybe the universe was filled with truly exotic particles. Maybe the universe underwent various periods of inflation followed by relative quiet.

Anyway, looking very far back we expect that one would find conditions very similar to those near the singularity of a black hole. This is called the 'big bang singularity.' Just as at a black hole, general relativity would break down there and would not accurately describe what was happening.

Roughly speaking, we would be in a domain of quantum gravity where, as with a Schwarzschild black hole, our now familiar notions of space and time may completely fall apart. It may or may not make sense to even ask what came 'before.' Isn't that a good place to end our story?

Index

A

Accelerated objects 117, 146
Accelerating 8, 10, 11, 12, 33, 43, 61, 68, 86, 87, 151, 158, 159, 162, 164, 172, 180, 185, 200, 201, 202, 205, 207, 208, 210, 216, 232, 249, 280, 288, 291
Astrophysics 257

B

Starlight 224, 225
Black hole 229, 230, 232, 233, 234, 235, 236, 237, 238, 239, 240, 241, 242, 243, 244, 245, 246, 247, 248, 249, 255, 257, 258, 259, 260, 261, 262, 263, 264, 265, 267, 268, 269, 270, 271, 285, 287, 290, 293, 296
Boost parameters 130, 134, 145, 188
Building intuition 211

C

Conceptual premises 59
Coordinate systems 64, 70, 72, 114, 115, 218
Copernican 7, 272
Cosmological 219, 279, 281, 283, 286, 287, 290, 291, 292, 293, 295
Covariant derivative 68, 69
Curved surfaces 203, 204

D

Derivatives 26, 64, 84, 85
Differential equation 254, 255
Dimensions 25, 26, 37, 67, 135, 138, 139, 141, 171, 188, 189, 199, 214, 267, 296
Doppler effect 170, 283
Dynamic implications 41
Dynamics 55, 142, 156, 219, 275, 291

E

Einstein 1, 9, 11, 12, 13, 14, 17, 28, 29, 30, 38, 39, 44, 70, 78, 90, 91, 100, 113, 137, 142, 156, 157, 160, 161, 162, 163, 164, 177, 178, 180, 183, 184, 188, 196, 199, 200, 201, 205
Electromagnetic 29, 30, 40, 49, 50, 53, 54, 62, 63, 82, 83, 84, 86, 88, 97, 111, 160, 288
Electromagnetism 81, 83, 86, 157, 160, 273
Elusive ether 86
Energy 41, 42, 43, 44, 45, 46, 47, 48, 49, 51, 52, 53, 56, 62, 68, 139, 156, 157, 158, 159, 160, 161, 162, 163, 164, 165, 166, 167, 168, 169, 170, 171, 176, 177, 181, 182, 183, 185, 218, 219, 220, 221, 224, 259, 261, 262, 263, 264, 265, 268, 270, 280, 281, 282, 283, 284, 285, 286, 288, 293
Equilibrium 85, 291
Evidence 6, 78, 222, 257, 281, 295, 296

F

Forces 5, 9, 11, 46, 60, 64, 78, 79, 80, 81,

82, 92, 111, 143, 156, 159, 162, 177, 178, 180, 181, 197, 235, 282
Frequency shifts 55

G

Geometry 1, 8, 9, 113, 114, 119, 125, 126, 128, 129, 131, 133, 134, 145, 149, 166, 177, 199, 204, 205, 211, 217, 218, 219, 222, 227, 238, 263
Global frames 196
Gravitational 9, 13, 27, 60, 61, 62, 64, 67, 80, 156, 177, 178, 180, 181, 182, 183, 185, 186, 187, 188, 189, 190, 191, 193, 194, 195, 196, 197, 198, 199, 200, 203, 204, 205, 206, 209, 210, 211, 218, 219, 220, 221, 222, 224, 225, 226, 227, 232, 234, 235, 236, 239, 244, 249, 258, 261, 262, 263, 264, 268, 273, 274, 282, 283
Gravity and locality 185

H

Hawking radiation 264, 265
History of relativity 2
Homogeneity 272, 275, 276, 288, 289, 290
Horizon 14, 149, 152, 153, 231, 232, 233, 234, 235, 236, 237, 238, 239, 245, 246, 247, 249, 257, 261, 263, 264, 265, 270, 271, 291
Hyperbolae 127, 128, 129, 131, 238, 239, 242, 267, 279, 280
Hyperbolic trigonometry 130, 216

I

Indistinguishable 61, 196
Inertial frame 13, 30, 31, 33, 35, 36, 37, 38, 52, 53, 57, 78, 79, 80, 82, 91, 92, 94, 97, 98, 99, 100, 109, 116, 117, 120, 121, 122, 127, 128, 142, 143, 144, 145, 146, 147, 150, 152, 154, 155, 171, 175, 185, 186, 187, 188, 189, 190, 192, 193, 196, 197, 204
Ingredient 181
Interior diagram 237
Investigating 137, 193, 226, 233
Isotropy 272, 274, 275, 276

L

Length contraction 32, 33, 35, 45, 1074, 113, 117, 120, 121
Light clocks 103, 169
Light speed 24
Lorentz coordinate transformation 29, 31, 32, 33, 38, 54

M

Mass energy 49, 262
Minkowski spacetime 35
Momentum 41, 42, 43, 44, 45, 46, 47, 48, 49, 56, 62, 156, 157, 158, 159, 160, 161, 162, 165, 166, 167, 168, 169, 170, 171, 172, 173, 174, 175, 176, 177, 180, 219, 221, 261
Multiple boosts 139

N

Negative pressure 281, 282, 286
Newtonian 12, 30, 53, 60, 61, 70, 74, 77, 78, 79, 81, 91, 113, 114, 137, 147, 156, 158, 159, 160, 164, 165, 166, 171, 172, 176, 178, 181, 187, 193, 219, 220, 227, 244, 251, 273
Newton's law 143, 180, 193, 220, 223, 226, 227, 280, 281

O

Observations 5, 6, 78, 81, 137, 180, 257, 258, 263, 283, 285, 286, 290
Our universe 2, 6, 9, 10, 11, 12, 17, 27, 67, 240, 275, 283, 284, 285

P

Paradoxes 33, 35, 50
Penrose diagrams 265, 266, 267
Planet mercury 222

Principle and relativity 272
Proper distance 109, 117, 119, 126, 127, 128, 129, 132, 147, 148, 149, 150, 152, 153, 191, 228, 232, 233, 234, 250, 252, 253, 254, 255, 256, 277
Proper time 105, 118, 126, 145

R

Reference frames 72, 76, 78, 79, 86, 91, 92, 94, 98, 103, 106, 107, 113, 116, 117, 125, 126, 127, 128, 133, 134, 135, 137, 142, 143, 144, 156, 157, 196, 217, 238, 278, 279
Relativity 1, 2, 5, 8, 9, 11, 12, 13, 14, 23, 24, 26, 28, 29, 30, 31, 33, 35, 38, 40, 41, 42, 43, 50, 52, 53, 55, 59, 60, 61, 62, 63, 64, 67, 68, 70, 72, 81, 90, 100, 105, 114, 125, 126, 130, 135, 137, 141, 142, 143, 147, 156, 157, 160, 164, 166, 171, 176, 177, 178, 179, 180, 182, 190, 195, 196, 197, 199, 206, 208, 210, 211, 217, 218, 219, 220, 222, 225, 229, 231, 232, 244, 248, 264, 272, 274, 277, 279, 280
Runaway universe 285

S

Schwarzschild metric 220, 222, 226, 228, 231, 235, 237, 238, 239, 241, 245, 250, 256, 257, 267, 268, 269, 270, 271
Schwarzschild radius 228, 229, 230, 231, 232, 233, 234, 245, 258, 259, 260, 261, 262, 263
Singularity 242, 243, 244, 249, 264, 268, 287, 290, 291, 296
Spacetime 35, 65, 66, 67, 71, 72, 76, 92, 94, 96, 97, 98, 108, 109, 110, 114, 116, 119, 120, 122, 125, 126, 129, 130, 134, 135, 136, 142, 145, 147, 149, 151, 157, 167, 168, 172, 175, 176, 180, 184, 186, 187, 188, 189, 190, 196, 199, 200, 201, 202, 205, 206, 209, 210, 211, 214, 217, 218, 219, 221, 222, 225, 229, 231, 232, 233, 234, 236, 237, 238, 239, 244
Stellar aberration 88, 90, 91, 107, 113, 135

T

Tall tower 190, 192
Time dilation 23, 30, 32, 33, 34, 43, 44, 53, 58, 59, 60, 106, 107, 108, 113, 115, 120, 154, 155, 171, 193, 194, 195, 222, 226, 235, 262
Tiny tower 190, 191
Twin paradox 34, 109, 119, 125, 142, 151

U

Uniform acceleration 143, 144, 145, 146, 147, 149
Uniformly 142, 147, 148, 149, 150, 152, 153, 156, 164, 169, 185, 196, 258, 288

V

Vacuum energy 281, 282, 283, 286, 293, 295
Velocities 9, 55, 77, 78, 86, 113, 130, 131, 142, 143, 157, 160, 166, 171, 172, 187, 244, 257
Verification 185, 222
Visualizing 246

W

Worldline 72, 76, 94, 95, 96, 99, 100, 105, 106, 115, 117, 118, 119, 122, 123, 124, 125, 127, 129, 132, 142, 144, 145, 147, 148, 149, 150, 152, 153, 155, 156, 191, 201, 205, 222, 227, 241, 242, 245, 247, 251, 269, 276, 277